T0399594

Sundarbans Mangrove Systems

Sundarbans Mangrove Systems
A Geo-Informatics Approach

Edited by
Anirban Mukhopadhyay, Debashis Mitra, and
Sugata Hazra

CRC Press
Taylor & Francis Group
Boca Raton London New York

CRC Press is an imprint of the
Taylor & Francis Group, an **informa** business

First edition published 2022
by CRC Press
6000 Broken Sound Parkway NW, Suite 300, Boca Raton, FL 33487-2742

and by CRC Press
2 Park Square, Milton Park, Abingdon, Oxon, OX14 4RN

ISBN: 978-0-367-53881-1 (hbk)
ISBN: 978-0-367-53883-5 (pbk)
ISBN: 978-1-003-08357-3 (ebk)

Typeset in Times
by codeMantra

Contents

SECTION 1   Remote Sensing of Mangroves

SECTION 2   Ecology of Sundarbans a Geo-Informatics Approach

SECTION 3   Ecosystem Services Analysis using GIS

SECTION 4   Vulnerability of Sundarbans through Geospatial Analysis

Preface

Among the several natural terrestrial ecosystems that thrive throughout the globe, forest ecosystems are perhaps one of the most complex ecosystems owing to their variability in composition, structure and functioning. Among the various types of forest ecosystems, mangroves are one of the most productive and bio-diverse ecosystems that thrive in and around the sea coasts, river mouths and estuaries situated in the tropical and subtropical intertidal zones The Sundarbans is one of the most dynamic coastal wetlands, which comprises more than 200 islands and world's largest single block of mangrove forest. The region is estimated to have come into existence 4,000 years ago and is shared between India ($\approx$4,000 km^2) and Bangladesh ($\approx$6,000 km^2). Mangroves are the third most productive and bio-diverse ecosystem in the world after tropical rain forest and coral reefs. Sundarbans Mangroves provide a wide range of ecosystem services ranging from mitigation of global climate change by carbon capture to the sustenance of local communities, whose livelihood depends upon the mangrove forest products. They also act as biological filters, maintaining water quality in coastal regions and providing nursing grounds for a number of diverse flora and fauna species particularly fish. Being a natural barrier to hazards such as tropical cyclones and tsunamis, Sundarbans help in the protection of shoreline areas, prevent coastal erosion and act as shelter for small coastal hamlets. However, due to anthropogenic impacts and imminent consequence of climate change, the ecology of Sundarbans is under huge threat. Around one-third of the world's mangroves have been lost in the last 50 years and some scientists have suggested that the entire mangrove community might be lost by the end of the twenty-first century. The need for assessment of Sundarbans mangrove forest trends in the context of global change is therefore having paramount importance. Geo-informatics is the most useful tool for analyzing, assessment and future scenario generation of the most complex dynamic ecosystem like Sundarbans. Its usefulness in the large domain of earth and environmental science is already well established. Interpretation of various satellite images and geospatial analysis and modelling have been used to assess the past, present and future scenarios of Sundarbans. The primary aim of the proposed book is to create the basic awareness of the readers about the Sundarbans and its future in the wake of climate change. Scientifically, it is very important to understand Sundarbans as a whole to predict future course of action for better preservation of its ecological components. The proposed book will impart an overall understanding about the Sundarbans deltaic system to the researchers. This book will also help the decisions makers/nature conservators to take conscious decision for protecting the fragile ecosystem of Sundarbans.

The opinion of the individual chapters is the sole responsibility of the authors of the respective chapters. Any consequences arising from the use of the information contained in the chapters; the views and opinions expressed do not necessarily reflect those of the Publisher and Editors.

Anirban Mukhopadhyay
Debashis Mitra
Sugata Hazra

Acknowledgments

This book would not have seen the light without the assistance of the Indian Society of Remote Sensing and Indian Institute of Remote Sensing. The active participation of School of Oceanographic Studies Jadavpur University, West Bengal, is also humbly acknowledged. We are indebted to Centre for Earth Observation Science (CEOS), University of Manitoba. We also acknowledge Department of Forest, Government of West Bengal for their support. The editors put on record their gratitude to all the authors who have contributed for this book by submitting chapters within time. Last but not the least the editors extend their sincere gratitude to the reviewers of this book.

Editors

Dr. Anirban Mukhopadhyay is a Geospatial researcher. After completing two masters in Marine Science (Calcutta University) and Remote Sensing and GIS (Indian Institute of Remote Sensing), he has achieved his doctoral degree from Jadavpur University. After working on different multinational environmental projects with Jadavpur University, International Growth Centre, The World Bank, etc., he has joined as a researcher at the Centre for Earth Observation Science (CEOS), University of Manitoba, Canada. Dr. Mukhopadhyay has numerous publications on remote sensing and GIS. He has vast experience in geospatial analysis of Mangroves and the environment of Sundarbans. Dr. Mukhopadhyay also has an interest in understanding the land-ocean dynamics and their effect on the mangrove ecosystem. He has a particular interest in the study of the impact of climate change on this fragile ecosystem. He has worked on both Indian and Bangladesh Sundarbans.

Dr. Debashis Mitra is a geologist by profession. He passed his Master's in Applied Geology and M.Tech in Engineering Geology from IIT (ISM), Dhanbad, and completed his doctorate from ITC, The Netherlands, in sandwich mode. Dr. Mitra is working with the Indian Space Research Organization (ISRO) for the last 25 years. He is presently heading the Department of Marine and Atmospheric Science at the Indian Institute of Remote Sensing (IIRS), Dehradun. He is responsible for capacity building activities on remote sensing and GIS applications on Ocean and atmospheric sciences at IIRS. Dr. Mitra has vast experience in Remote Sensing and GIS application for Coastal Zone Management, Coastal hazards, and Coastal Geology and Geomorphology. He has more than 80 publications in peer-reviewed journals, along with several chapters in international books. He has also edited a book on Environmental Remote Sensing from Springer Publication. Dr. Mitra leads several projects sponsored by ISRO and other organizations on geospatial applications for the coastal zone. Dr. Mitra has guided numerous students in Ph.D., M.Tech, and Postgraduate Diploma level. He is a reviewer of several national and international journals and an editorial board member of GI Science and Remote Sensing. Dr. Mitra is working on geospatial analysis of Mangroves and Sundarbans for the last two decades. He is a lead scientist in this field with a vast experience in Sundarbans.

Dr. Sugata Hazra, a Geologist by profession, is a Professor of Coastal Zone Management and former Director of the School of Oceanographic Studies at Jadavpur University, India. He has done pioneering research on Sundarbans, Climate Change and Biophysical impact. He has led several international and national level research projects on Sundarbans and coastal oceans and has more than 100 international publications, edited volumes, monographs to his credit. He has mentored over 30 Ph.D. students.

Dr. Hazra was a member of the Indian Antarctic Expedition in the year 1996–1997. He is a recipient of the Gold Medal from the Asiatic Society and is a member of the West Bengal State Coastal Zone Management Authority, nominated by the Government of India.

Some of his ongoing projects include 'Opportunities and trade-offs between the SDGs for food, welfare and the environment in deltas(UKIERI-DBT)', Tidal energy for village electricity supply in Indian Sundarbans biosphere (UKIERI-DST), Vulnerability Assessment of Mangroves and Corals of West Bengal, Odisha and Andaman Islands (DST), Study of carbon dynamics in estuaries and nearshore waters of Hugli Estuary (NRSC), Geospatial assessment of mangrove's species discrimination in Indian Sundarbans, their health and its effect on the environment and climate using airborne hyperspectral (AVIRIS NG) and RISAT-I remote sensing data (SAC), Enhancing adaptive capacity and resilience to climate change of small and marginal farmers in Purulia and Bankura district West Bengal (NABARD, India through Development Research Communication and Services Centre).

Contributors

Rituparna Acharyya
Department of Geography, School of
Earth Science
Central University of Karnataka
Karnataka, India

Azmery Iqbal Afnan
Department of Disaster Science and
Management
University of Dhaka
Dhaka, Bangladesh

Md. Ashik-Ur-Rahman
Khulna University
Khulna, Bangladesh

Arunima Sarkar Basu
School of Architecture, Planning and
Environmental Policy
University College Dublin
Dublin, Ireland

Atreya Basu
Centre for Earth Observation Science
(CEOS)
University of Manitoba
Winnipeg, Canada

Bidroha Basu
School of Architecture, Planning and
Environmental Policy
University College Dublin
Dublin, Ireland

Apratim Biswas
Department of mining Engineering
Indian Institute of Engineering Sciences
and Technology, Shibpur
Howrah, India

Abhra Chanda
School of Oceanographic Studies
Jadavpur University
Kolkata, India

Sourav Das
School of Oceanographic Studies
Jadavpur University
Kolkata, India

Premanondo Debnath
Remote sensing Division
Center for Environmental and
Geographic Information Services
(CEGIS)
Dhaka, Bangladesh

Manoj Kumer Ghosh
Department of Geography and
Environmental Studies
University of Rajshahi
Rajshahi, Bangladesh

Tuhin Ghosh
School of Oceanographic Studies
Jadavpur University
Kolkata, India

Kaushik Gupta
School of Oceanographic Studies
Jadavpur University
Kolkata, India
and
Centre for Earth Observation Science
(CEOS)
University of Manitoba
Winnipeg, Canada

Rituparna Hajra
Department of Geography
Polba Mahavidyalaya
Polba, India

Dewan Mohammad Enamul Haque
Department of Disaster Science and
Management
University of Dhaka
Dhaka, Bangladesh

Md Ashraful Islam
Department of Geology
Dhaka University
Dhaka, Bangladesh

Sheikh Tawhidul Islam
Institute of Remote Sensing and GIS
Jahangirnagar University
Dhaka, Bangladesh

P.K. Joshi
Spatial Analysis and Informatics Lab
(SAIL), School of Environmental
Sciences (SES)
Jawaharlal Nehru University (JNU)
New Delhi, India

Lalit Kumar
Department of Ecosystem Management
School of Environmental and Rural
Science, University of New England
Armidale, Australia

S.P.S. Kushwaha
Indian Institute of Remote Sensing
Indian Space Research Organisation
Dehradun, India

Tonoy Mahmud
Department of Disaster Science and
Management
University of Dhaka
Dhaka, Bangladesh

Sayani Datta Majumdar
School of Oceanographic Studies
Jadavpur University
Kolkata, India

Shreyashi S. Mitra
Department of Civil Engineering
Haldia Institute of Technology
Haldia, India

Subrata Mitra
School of Oceanographic Studies
Jadavpur University
Kolkata, India

Krishna Prosad Mondal
Institute of Remote Sensing and GIS
Jahangirnagar University
Dhaka, Bangladesh

Subrata Nandy
Indian Institute of Remote Sensing
Indian Space Research Organisation
Dehradun, India

Indrajit Pal
Mitigation and Management
Asian Institute of Technology
Khlong Nueng, Thailand

Afshana Parven
Asian Institute of Technology
Khlong Nueng, Thailand

Francesco Pilla
School of Architecture, Planning and
Environmental Policy
University College Dublin
Dublin, Ireland

Niloy Pramanick
School of Oceanographic Studies
Jadavpur University
Kolkata, India

Khan Ferdousour Rahman
Asian Institute of Technology
Khlong Nueng
Thailand

P.S. Roy
Sustainable Landscapes and
   Restoration
World Recourses Institute India
New Delhi, India

Subrota Kumar Saha
Department of Geology
University of Dhaka
Dhaka, Bangladesh

Tumpa Saha
Beach Sand Minerals Exploitation
   Center
Bangladesh Atomic Energy
   Commission
Cox's Bazar, Bangladesh

Chalantika Laha Salui
Department of mining Engineering
Indian Institute of Engineering Sciences
   and Technology, Shibpur
Howrah, India

Srikanta Sannigrahi
School of Architecture, Planning and
   Environmental Policy
University College Dublin
Dublin, Ireland

Abhisek Santra
Dept. of Civil Engineering
Haldia Institute of Technology
Haldia, India

Shamima Ferdousi Sifa
Department of Disaster Science and
   Management
University of Dhaka
Dhaka, Bangladesh

Muna Tamang
Indian Institute of Remote Sensing
Indian Space Research Organisation
Dehradun, India

Md. Kamruzzaman Tusar
Department of Disaster Science and
   Management
University of Dhaka
Dhaka, Bangladesh

Mahmud Al Noor Tushar
Department of Geology
University of Dhaka
Dhaka, Bangladesh

Mohammad Sofi Ullah
Department of Geography and
   Environment
University of Dhaka
Dhaka, Bangladesh

Introduction

Human beings have been on Earth for millions of years. However, the definition of development underwent a radical change in the last three centuries. Ever since the Industrial Revolution, we have indiscriminately exploited the Earth's finite resources to fulfil several demands and necessities. A suite of anthropogenic activities has changed this planet's landforms, its environment and the quality of the biosphere, hydrosphere and ecosphere. Such human intervention has taken a toll on the health of this planet, and it has started to hit us back in various forms, like global warming, climate change, rising sea level and many more phenomena that have become trepidations of the present day. While seeking avenues to minimize the ill effects of such human-induced alterations, we realized that it is time we looked into the natural ecosystems of this planet. To reduce the excessive burden of atmospheric greenhouse gases and reverse the climatic change in favour of human beings, we must conserve, restore and manage the various natural ecosystems in a sustainable manner. We have realized in the past decades that these measures can ultimately enable us, in the end, to maintain a harmonic and ecological balance between man and the environment, disrupting which can lead to unimaginable catastrophes.

The present-day environmentalists have listed a wide variety of ecosystems, both in the terrestrial as well as the marine sectors, which need human attention. Mangroves, in this regard, deserve a special mention owing to their plenty of qualities and benefits that human beings can reap. The term mangroves denote a group of taxonomic flora that are halophytic and thrive in the tidal terrains of the estuaries and nearshore coasts in the tropics and subtropics. However, mangrove ecosystem, overall, encompasses a wide range of flora and fauna. This ecosystem is one of the most diverse, fragile and highly productive ecosystems of the world. The mangroves, along with some other marine ecosystems, like seagrasses and saltmarshes, store and sequester substantial quantities of carbon, which we collectively refer to as the Blue Carbon. These ecosystems are capable of offsetting the anthropogenic CO_2 emission, given proper conservation and restoration efforts take place. Mangroves also provide a multitude of ecosystem services. Coastal peripheries blessed with mangroves act as windbreakers and shelters human beings as well as prevents the loss of properties from deadly storms, like tropical cyclones. It also minimizes the effect of deadly tsunamis. The estuarine water bodies adjoining the mangroves act as an abode to a wide variety of fishes, crabs, prawns and shrimps as these waters are usually rich in nutrients. Mangroves can accumulate and stabilize sediments in the deltaic regions, owing to their broad and deepened root structure. Besides, these ecosystems act as a filter to various pollutants, which generate upstream and remain trapped in the mangrove sediments before entering the oceans. Mangroves also act as a source of litter, wood and honey to the local dwellers, who are extensively dependent upon these forests for maintaining their daily livelihood. Lastly, their recreational and aesthetic value has earned the ecosystems special significance, as tourism around mangroves has flourished worldwide in the last century.

Sundarbans is the world's largest mangrove forest shared by the two neighbouring countries, India and Bangladesh. The counterparts of both the countries' forested area encompass almost $10,000\,km^2$, and it is a UNESCO World Heritage Site. Recently, it has become a Ramsar Wetland site as well. The location of this vast forest marks the end of the world's largest delta, the Ganges–Brahmaputra–Meghna (GBM) Delta. Interspersed with many large rivers and innumerable creeks of varying width and depth, this region shelters a wide variety of mangrove floral species. Sundarbans host around 36 of the 70 mangrove floral species reported from the world. It is perhaps the only mangrove forest in the entire globe where a terrestrial predator species like that of the majestic Royal Bengal Tigers roam freely. The faunal abundance of this region includes estuarine crocodiles, spotted dears, wild boars and what more. The avian biodiversity of the Sundarbans is also remarkable, as many nature lovers have acknowledged this place as a birdwatcher's heaven. Adjacent to such wilderness, there reside millions of people both in the Indian as well the Bangladesh counterparts of Sundarbans. The population adjoining the Sundarbans has been increasing at an alarming rate in the past decades. This population-outburst is threatening the ecological equilibrium of this forest, and this region has become vulnerable to various anthropogenic activities. The Sundarbans have witnessed large-scale deforestation, forestland conversion to agriculture and aquaculture, increase in pollutants load from upstream, and tourism-related intervention. The man-animal conflict has also become a point of severe concern. Altogether, the mesh formed between the wild and the human beings in this part of the world makes this place unique and at the same time worthy of investigation.

The Sundarbans has been the point of interest for scholars belonging to diversified disciplines that include environmentalists, geologists, oceanographers, botanists, pharmacists, meteorologists, geographers and social scientists. A substantial number of scholarly articles and research outcomes are present on the internet, and still, several aspects of this crucial ecosystem require more detailed study and analysis. Irrespective of the domain under consideration, the scholar community has ubiquitously agreed upon the fact that geo-informatics has become a fruitful tool to generate a holistic concept about this kind of ecosystem. To this date, geographical information systems and satellite remote sensing find their use in almost all sorts of nature-related studies dealing with a particular study area. For research works on a regional scale and in terrains that are difficult to access physically, implementing a geo-informatics approach has become indispensable. In the mid of twentieth century, humankind embarked upon a journey whereby they started capturing images of mother earth using high-resolution cameras and sensors on-board flights and satellites, respectively. These ventures marked the beginning of a new era, where remote sensing became an essential aspect in nature-based studies. Today, we can acquire data of multiple parameters from every nook and corner of the Earth's surface and the processes that are taking place in the top layer of the crust and aquatic columns without any physical contact. The development of satellite technology and advanced sensors has enabled us to achieve such art. So far, many academicians and scientists have implemented this technology to study various aspects of Sundarbans. This book has tried to ensemble the different scholarly pieces of research conducted so far on the Sundarbans. A few chapters of this book have reviewed the previous works done so far in some of the specialized aspects that dealt with geo-informatics. Some chapters have presented original work on the Sundarbans. We classified the entire book into four sub-themes: *Remote Sensing of Mangroves; Ecology of Sundarbans A Geoinformatics Approach; Ecosystem Services Analysis using GIS* and *Vulnerability of Sundarbans through Geospatial Modelling*. In this Introduction, we have given a brief overview of all the chapters in this book.

Remote sensing of mangroves is perhaps the most vital aspect surrounding which all the other subcategories revolve. Mapping of forest cover and demarcation of species assemblage has always been a tedious endeavour. The geographically challenging terrain of mangroves makes this work even more difficult. Since the last centuries, several studies have focused on mapping the mangrove cover. Mangrove all over the world witnessed the use of both aerial photography and remotely sensed satellite images for accurate delineation of their spatial extent. Remote sensing and GIS also enable to monitor the temporal changes in the canopy cover and health of the mangrove forests. The first section of this book covered the aspects of remote sensing in characterizing mangroves and their properties. Haque et al. have devoted the very first chapter to the application of remote sensing in quantifying the mangrove forest biomass and monitoring its health with the help of geo-informatics. They have discussed the recent trend and scope of remote sensing technology concerning the different techniques of forest mapping. They also emphasized the LiDAR and SAR techniques in the estimation of mangrove floral biomass. Besides, they discussed the use of allometric features like canopy density and tree height in estimating the aboveground and belowground tree biomass. They also discussed the application of geo-informatics in characterizing the stress of the mangrove forests. Ghosh and Kumar, in the second chapter, have discussed the applicability of freely available low-resolution images in accurately mapping the mangrove cover. They employed a maximum likelihood algorithm to identify and map out mangrove species composition using open source mid-resolution Landsat data on the Bangladesh Sundarbans. Their results showed that such a mid-resolution image could yield acceptable accuracy levels required for the image analysis at the species level. They could successfully differentiate five abundant species of the Bangladesh Sundarbans, namely, *Heritiera fomes, Ceriops decandra, Excoecaria agallocha, Sonneratia apetala* and *Xylocarpus mekongensis*. Using mid-resolution satellite data, they obtained an overall accuracy of 85% deemed suitable to use in most of the natural resource mapping applications. Remotely sensed images derived from all the sensors in use today require rigorous image correction before using such images for further analysis. Various types of error originate due to different conditions of solar insolation, shadow effect and atmospheric column. Santra et al., in the third chapter of this book, discussed the role of *relative radiometric normalization* in minimizing the impact of solar illumination conditions and atmospheric attenuation on remotely sensed images. They applied six statistical relative radiometric normalizations (SRRN) techniques and compared the outputs to normalize the bi-temporal Sentinel-2A multi-spectral data. They advocated the use of the Iteratively Re-weighted Multivariate Alteration Detection (IR-MAD) method as it identifies the set of time-invariant pixels between the images based on the Iterative statistical Canonical Component Analysis (ICCA). They interpreted from their study that the use of such radiometric correction could enable us to enhance the accuracy in monitoring and estimating several crucial aspects of the Sundarbans mangrove forest, given the heterogeneous character of this forest cover. Finally, in the fourth and the last chapter of this sub-theme, Gupta et al. reviewed the application of different remote sensing techniques for mangrove-based studies, throwing light on the various Remote sensing tools, sensors, indices, and technologies used for monitoring of the Sundarbans mangroves over the past few years. Their main intention behind reviewing these pieces of research was to establish the applicability of Remote Sensing as a tool for Mangrove ecosystem mapping and monitoring.

The second sub-theme of this book focused on the ecological aspects of Sundarbans. The functioning of the Sundarbans is central to the ecology of the mangroves. The ecology of the mangrove systems principally depends on the floral composition. Portraying the species assemblage of the mangrove ecosystems has become an utmost necessity for taking proper steps of management and conservation. Nandy et al. in the fifth chapter have discussed the effectiveness of remote sensing in assessing the plant biodiversity of mangroves in the Sundarbans. They also discussed the important findings on this aspect, recorded so far, from the Sundarbans. Different abiotic factors such as coastal geomorphology, temperature, salinity, tidal amplitude and duration, dissolved oxygen, and nutrients govern the ecological setting of the mangrove ecosystem. In the modern marine environment, foraminifera constitutes an abundant diverse group of shelled microorganisms. These foraminifera are the most ancient organisms. The changes in the distribution of these microorganisms in the paleo sediments reflect the changes the mangrove environment has witnessed in the past. Saha et al. in the sixth chapter have discussed the spatial variability of the foraminifera. They inferred from their analysis the degree of environmental stress that this mangrove system has suffered in the past centuries. Besides gaining information about the past, characterizing the present health of an ecosystem is also essential. The scientific community has studied and established many such parameters, which gives us an indication of the state of health of a mangrove ecosystem. Biswas and Salui in the seventh chapter have discussed two such parameters, namely, Leaf Area Index (LAI) and Percentage Canopy Cover (PCC). They have also discussed the role of geo-informatics in delineating these two parameters in the Sundarbans. They mainly emphasized two indices, namely, the Normalized difference vegetation index and the Modified simple ratio vegetation index in characterizing these biophysical parameters, which in turn tells us about the ecological health of the mangroves.

Mangrove ecosystems provide a multitude of services to nature as well as human beings. However, quantifying the ecosystem services is a challenging endeavour. Very few studies exist at present, which implemented geo-informatics in quantifying these services provided by a forest ecosystem. The third sub-theme of this book comprised the characterization of ecosystem services. Sannigrahi et al., in the eighth chapter, demonstrated the application of satellite remote sensing data in measuring the economic importance of the Indian Sundarbans Natural Reserve Region. They implemented ten machine learning (ML) supervised classification models for land use land cover (LULC) classification and subsequently interpreted the results in terms of ecosystem services. The authors argued that the valuation approaches and methods adopted in this study could be a reference for evaluating the economic importance of natural capitals and could replicate easily for similar research interests across the ecosystems. In the last three to four decades, the tourism sector in the Sundarbans has witnessed a boom. The revenue generated from tourism serves as an alternate livelihood to the thousands, who reside in this region. Tourism has become one of the most crucial ecosystem services that the Sundarbans can provide owing to its pristine beauty. Hajra and Ghosh in the ninth chapter of this book have discussed the role of tourism in the Sundarbans, taking Sagar Island, as a case study. They showed that amalgamation of data acquired from primary surveys and secondary literature with the geo-informatics platform can fetch us valuable information on this typical ecosystem service. The authors tried to suggest an implementable and comprehensive strategy to formulate sustainable coastal tourism according to the guidelines of Global Sustainable Tourism Criteria (GSTC) through this chapter. Besides tourism, another crucial ecosystem service that the Sundarbans furnish is the fisheries. The livelihood of hundreds of thousands is dependent on the small-scale fisheries of this region. However, the state of health of the small-scale fisheries sector has substantially degraded in the recent past. Pal et al., in the tenth chapter, have taken up this issue and discussed the importance of disaster management practices and coastal governance in developing the integrated management program that is economically viable and socially acceptable for the fishers.

The fourth sub-theme of this book dealt with the most concerning aspect of the Sundarbans, the vulnerability. Worldwide mangroves are under potential threat due to the impacts of climate change and a suite of anthropogenic interventions. The Sundarbans is no exception. A rise in sea level followed by increased salinization and changing pattern of storms and rainfall have either threatened some species, degraded the health of trees, or resulted in a loss of mangrove forest cover throughout the Sundarbans. In the eleventh chapter of this book, Datta Majumdar et al. have assessed the biophysical vulnerability of the mangroves of Indian Sundarban in response to climatic change and associated sea-level rise in the northern part of the Bay of Bengal. They made use of remotely sensed images and a GIS platform to analyze the degree of vulnerability of these mangrove stands. By implementing the definition of vulnerability laid down in the IPCC AR IV framework, they developed a vulnerability index (VI) based on the indicators selected from the aspects of exposure, sensitivity and adaptive capacity. Using the GIS platform, they illustrated the variability of the vulnerability index within the Sundarbans. The authors argued that this study would provide a database and an apt protocol, which would be beneficial in framing long-term management strategies. Among the various natural factors, tropical cyclones are one of the deadly disasters that bring with them an incalculable loss of life and property. Sundarbans frequently experience the wrath of such cyclones, and these storms leave a disastrous signature on this region. On 20 May 2020, the cyclone Amphan ravaged the Sundarbans and destroyed mangrove stands in various areas of Sundarbans. Acharyya et al. in the twelfth chapter of this book analyzed the impacts of Amphan on the mangrove stands of Sundarbans. They implemented the supervised classification using the algorithm of Random Forest (RF) Classifier over the Sentinel-1 C-band SAR imageries acquired before and after the cyclone. They examined the changes in the images and assessed the impact of this cyclonic storm. Islam et al. also followed a similar approach in characterizing the ill effects of cyclonic storms on the Bangladesh Sundarbans, in the thirteenth chapter of this book. Geomorphology is such a feature that can potentially regulate the dynamics of a mangrove forest, and at the same time, the mangroves can govern the geomorphological dynamics of the terrain. The salinization of groundwater resources has made thousands of people residing on the fringes of the Sundarbans utterly vulnerable. Islam and Mitra have discussed the applicability of the GALDIT method in analyzing the saltwater intrusion scenario in three select aquifers of the Sundarbans in the fourteenth chapter. They inferred that the GALDIT-based outcomes could contribute to policies for conservation and management of groundwater resources, identification of sensitive recharge zones, and regional management planning, thereby setting a platform to attain sustainable use of groundwater resources. Among the various biogeochemical factors, salinity is perhaps the most crucial parameter that is capable of governing the overall well-being and functioning of a mangrove forest. Mitra et al., in this regard, have collated the entire historical database on salinity observations from the Hooghly-Matla estuarine complex, covering the Indian part of Sundarbans, in the fifteenth chapter. They demarcated the Indian Sundarbans into seven hypothetical compartments and analyzed the long-term temporal trends of salinity in each one of them. They fitted the collated data in a GIS platform and prepared interpolation maps for the complete study area showing the variability in salinity in three different seasons, the pre-monsoon, monsoon and post-monsoon. They also critically reviewed the physiological manifestations that the mangroves of Sundarban have shown towards salt-stress. They further discussed the plausible change in species assemblage that this ecosystem has undergone and likely to face in the future.

Thus, summarizing the statements mentioned above, we state that this book has discussed and tried to cover all the aspects relevant to Sundarbans through the lens of geo-informatics. We hope this book will enrich the knowledge gained on Sundarbans by various researchers using various techniques. In the present date, researchers from varied domains converge with an inter-disciplinary aptitude to learn and execute fruitful pieces of research with the help of geo-informatics. All such future researchers and present students would find interest in reading this book. This book, though based on Sundarbans, can provide an idea and scope of work for similar study areas all-round the globe. To make this book easy to read, all the authors have strived to write their respective chapters following a lucid style. We sincerely believe that readers from varied disciplines would find this book interesting and at the same time, a ready reference for many scholarly works.

Anirban Mukhopadhyay
Debashis Mitra
Sugata Hazra

Section 1

Remote Sensing of Mangroves

1 Remote Sensing of Mangroves

Dewan Mohammad Enamul Haque,
Tonoy Mahmud, Azmery Iqbal Afnan,
Shamima Ferdousi Sifa, and
Md. Kamruzzaman Tusar
University of Dhaka

CONTENTS

1.1   APPLICATION OF REMOTE SENSING ON MANGROVES

Mangrove forests at the union of sea and land in the world's tropical and subtropical region are unique in the context of ecological attributes and biodiversity. Mangroves are known as shrubs or trees that grow best at low wave energy and secure further deposition of fine particles initiating these woody plants to form roots and grow. Mangroves are valuable ecological and economic resource and are crucial nursery grounds and breeding sites for fish, crustaceans, reptiles, birds and mammals. It also acts as the major source of wood, aggregation sites for sediment, nutrients, contaminants, and carbon; and also offers protection against coastal erosion and other hazards (Alongi, 2002). Ecosystems of mangrove forests are estimated to be amongst the most carbon rich ecosystems within the tropical region (Donato et al., 2011) and very importantly, mangrove forests provide numerous ecosystem services that sustain the livelihood of millions of people (Barbier EB et al., 2011). Hence, mangroves play a substantial role in maintaining the carbon cycle and consequently are becoming economically viable to protect communities (Pendleton et al., 2012; Jerath, 2012). Figure 1.1 is showing the current distribution of mangrove forests worldwide.

In spite of their importance for biodiversity, supporting indigenous local communities and carbon storage, mangroves are generally threatened across their entire range (Thomas et al., 2017). Mangroves extent monitoring is therefore important to identify spatio-temporal distribution of mangrove forests to measure the impact caused by numerous threats like overexploitation, urbanization, sea level rise and natural disturbance by cyclone and increase in precipitation intensity. Geoinformatics could play an important role in quantifying the mangrove change, monitoring the mangrove health and identifying the responsible factors for mangrove degradation. Fortunately, the prospect of geoinformatics in the context of mangrove mapping and monitoring is becoming more promising with increasing availability of required satellite data and user-friendly tools. Nowadays, Landsat, ASTER and ESA's Sentinel missions are

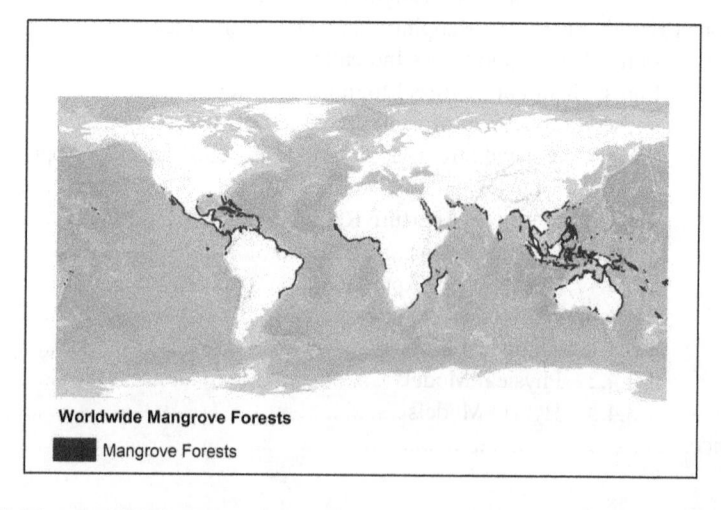

Worldwide Mangrove Forests
■ Mangrove Forests

FIGURE 1.1   Worldwide mangrove forest distributions. (From Global Mangrove Watch.)

providing satellite imagery for free. ALOS imagery is also made available by JAXA for the principal investigator who submit proposal to the earth observation collaborative project. As optical and Synthetic Aperture RADAR (SAR) imagery are mostly utilized for mangrove mapping and stress monitoring, the recent trend of milestone satellite missions is shown in Figure 1.2 in a simplified manner.

The application horizon will broaden with increasing data availability in future through the launching of important satellite missions like the NISAR (NASA-ISRO SAR) and the European Space Agency's (ESA's) Biomass mission. Moreover, researchers and practitioners are moving away from downloading and processing satellite

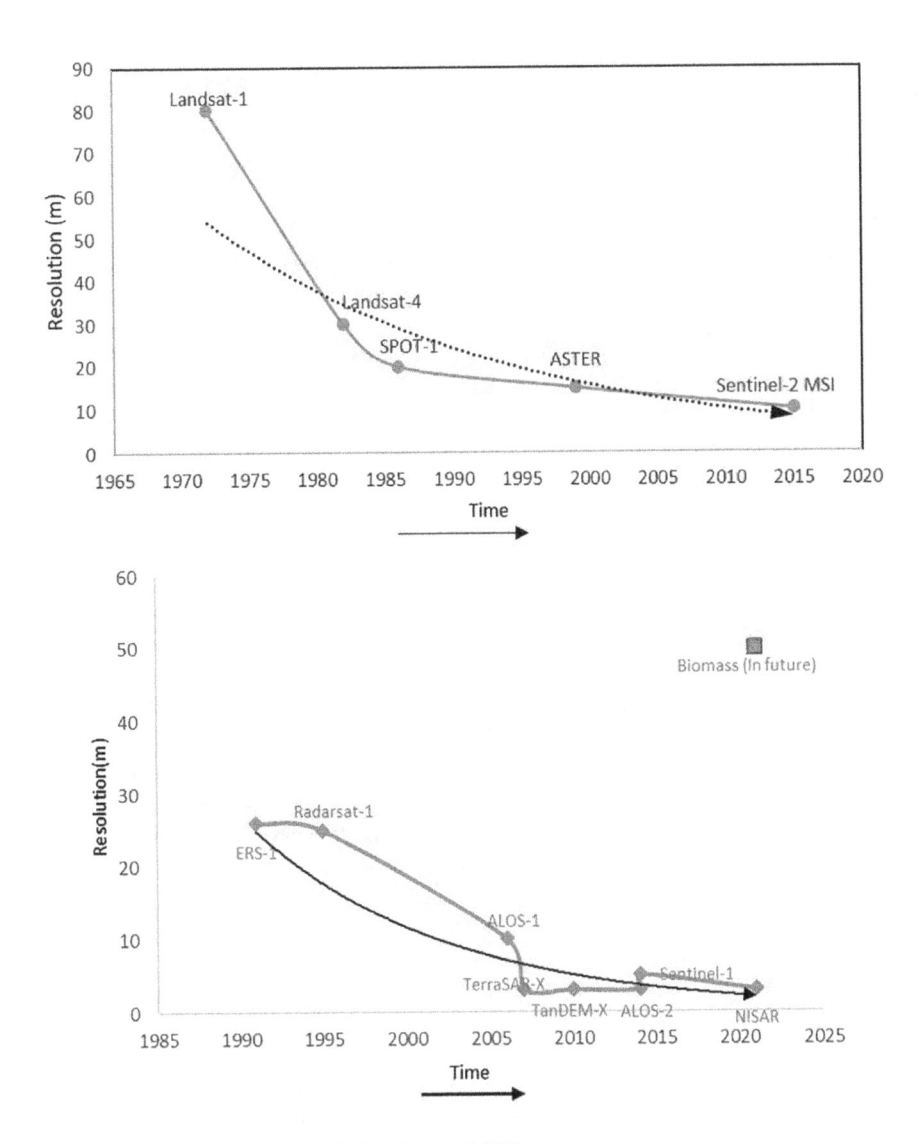

FIGURE 1.2 The current trend of optical and SAR sensors.

imagery on desktop to the user friendly cloud computing environment like Alaska Satellite Facility's (ASF's) Hybrid Pluggable Processing Pipeline (HyP3): http://hyp3. asf.alaska.edu, ESA's Thematic Exploitation Platforms: https://tep.eo.esa.int, Google Earth Engine (GEE): https://earthengine.google.com/. The following sections outline the remote sensing application on forest mapping and current approach of mangrove mapping using remote sensing techniques.

1.1.1 REMOTE SENSING APPLICATION ON FOREST MAPPING

The aspects that need to be considered while performing forest mapping are the delineation of forest from other cover types, measuring the forest vigor temporally and identifying the abrupt and gradual forest changes. Remote Sensing (RS) techniques could be categorized into three classes to address these aspects, that is, feature delineation, vegetation indices and change detection including time series analysis.

1.1.1.1 Feature Delineation

The vegetation area delineation has been the primary application of remote sensing techniques in the past. Image classification is a common practice for feature delineation. Pixel-based supervised, unsupervised classification and a combination of both classifications (hybrid classification) on optical images have been used to delineate the forests worldwide (Giri et al., 2011). The pixel-based optical image classification has been coupled with object-based classification to obtain similar end (Wang, Sousa and Gong, 2004). Not only the extent mapping, similar techniques such as fuzzy classification have been used to determine the different tree species in a mangrove forest (Neukermans et al., 2008). Furthermore, species identification is performed through the subpixel classification method such as neural network. Different approaches such as convolutional neural network and cluster-based neural network have been used for mangrove species delineation (Wang, Silván-Cárdenas and Sousa, 2008; Wan et al., 2019).

The optical image classification suffers mainly from the limitations such as cloud coverage and noisy pixels along with the disadvantages associated with individual methods. The pixel-based classification presupposes that each pixel is associated with a single land cover type or vegetation type which is invalid for relatively coarse resolution optical images (Giri et al., 2011). Although subpixel and object-based classification can overcome such barriers, other optical image classification suffers from the above-mentioned limitations. Moreover, object-based classification requires even higher resolution image to be able to separate objects, otherwise the object separation becomes erroneous, especially in the boundary of two different objects or land cover types (Wang, Sousa and Gong, 2004).

The use of SAR techniques has the edge over optical images due to the all-day acquisition and cloud penetration capability. The SAR backscatter values are generally determined by two main groups of characteristics: sensor and target characteristics. The sensor-related characteristics include the frequency/wavelength of the SAR, polarization of the transmitted and received SAR signal, incidence angle of the radar beam interacting with the ground and look direction of the sensor. The target characteristics deal with the SAR backscatter of forests and from other natural and manmade targets. The combination of these characteristics needs to be considered while interpreting and analyzing SAR imagery (Ulaby and Dobson, 1989).

1.1.1.2 Vegetation Indices

Vegetation indices are the ratios of two or more bands of remotely sensed satellite imagery. The basic indices are Normalized Difference Vegetation Index (NDVI) (Rouse et al., 1974) and Radar Vegetation Index (RVI) (Jordan, 1969). They are normal ratio between the Red and Near Infrared (NIR) band. Indices such as soil-adjusted vegetation index (SAVI) are adjusted for soil background and let better delineation for vegetation area (Huete, 1988). Similarly, atmospheric effects are also addressed in some indices such as atmospherically resistant vegetation index where the difference of Red and Blue band is used instead of more atmospheric effect sensitive to Red band (Kaufman and Tanre, 1992). Both soil and atmospheric effects are considered in Enhanced Vegetation Index (EVI), where a feedback loop is introduced in the equation that limits the soil and atmospheric effects (Liu and Huete, 1995). Moreover, leaf area index (LAI) and tasseled cap transformation are also used for forest vigor assessment. However, these indices could not differentiate mangrove forest from other forest type. Therefore, Section 1.1.2 is brought to explain the most effective mangrove indices.

1.1.1.3 Change Detection

Forest may change abruptly by wildfire, cyclone hit, etc., or gradually due to sea level rise, suffering overexploitation, expansion of urbanization and the like. Most common change detections are performed through automated algorithms; however, purely image interpretation-based change detection has been performed too (Vogelmann, Tolk and Zhu, 2009). Automated algorithm used for change detection is mainly bi-temporal. To produce the best possible results, the pre-processing of satellite images is a prerequisite. Furthermore, the spatial registration is also very important, especially in multi-date change detection. The success of digital change detection depends on the algorithms applied. Most common method is classification. This is widely used in bi-temporal change detection. The classification methods have explained in the previous section; however, from change detection perspective, there are some other advantages. The classification is performed on the pre- and post-images separately and thus pre-processing step such as radiometric correction is not a requirement. It is worthy to note that pinpointing the classes is difficult and miss-registration between images can often lead to errors. On the other hand, composite analysis-based change detection algorithm registers two images and inspects the statistics per pixel between the two images. However, prior knowledge of the study area is required for successful analysis (Jensen, 1981).

Univariate image differencing is the most used change detection algorithm. Here, two similarly pre-processed images are registered and subtracted from one another. This analysis can be formed using raw images or transformed product such as vegetation indices (Lyon et al., 1998). In some cases, the transformed images produce better result than the original ones (Healey et al., 2006). Performing matching processes other than registration such as histogram matching and band-to-band normalization enhances the result. A similar concept for change detection is Image rationing. Here, instead of differencing, the registered images are divided. However, extensive radiometric correction is required. There are statistical model-based change detection algorithms where the image values are fitted into a statistical model. Bi-temporal linear data transformation and image regression is example of these methods. Statistical methods such as principle component analysis and support vector machine often provide good results (Hayes and Sader, 2001; Shimu et al., 2019). To learn more about these algorithms, the readers are directed to Coppin et al. (2004) and Singh (1989).

The main challenge of implementing change detection and time series analysis is to perform radiometric correction to address the temporality and atmospheric issue during image acquisition. However, SAR imagery is less influenced by atmospheric interruption compared to its optical counterpart. As typically the SAR backscatter tends to decrease with deforestation resulted biomass loss or forest degradation, the application of cumulative sum analysis to SAR time series data seems potentially simple, yet powerful (Manogaran and Lopez, 2018). Image change detection methods for bi-temporal image comparison can be applied to well-calibrated and radiometrically corrected SAR data, but L- and C-band data often present a challenge, as surface roughness and moisture components can lead to significant SAR signal ambiguities (Nielsen, 2007).

1.1.2 Remote Sensing Application on Mangrove Mapping

Mangrove indices are proven to be efficient in current years to accurately and rapidly map mangrove extent utilizing satellite imagery (Baloloy et al., 2020; Gupta et al., 2018; Jia et al., 2019). Given the fact that mangroves are wetland forests, the remote sensing indices should be sensitive to the greenness and wetness. The |NIR-Green| part of the mangrove indices is distinctly sensitive to mangrove vegetation and, therefore, could efficiently delineate mangrove from other terrestrial vegetation type. The |SWIR-Green| part of index expresses the distinct moisture of mangrove and thus does not necessitate additional intertidal and water indices. Baloloy et al. (2020) has shown that mangrove indices have successfully separated mangrove from other terrestrial vegetation with an overall index accuracy of 92%.

Section 1.1.1 has explained other vegetation indices like NDVI, SAVI, LAI and tasselled cap transformation. However, these indices are not unique for mangrove discrimination. Fortunately, numerous mangrove forest indices are now available to rapidly and accurately map mangrove forest. Interested learners are requested to go through Baloloy et al. (2020) and Jia et al. (2019) for further details.

Mangrove index (MI) has been used to identify degraded and nondegraded forest cover in Indonesia. High MI value represents healthy mangrove forest. Mangrove Recognition Index (MRI) is developed using multi-temporal Landsat TM imagery which accounts for intertidal effects and is sensitive to wetness, greenness and changes in greenness (Zhang and Tian, 2013). Higher value of MRI corresponds to pixels bracketing mangrove forest and lower value corresponds to other typical terrestrial landcover like water, vegetation and bare soil. However, the performance of MRI is constrained by tidal variation. To overcome this tidal fluctuations, a combined mangrove recognition index (CMRI) is used which requires to implement a classification approach utilizing NDVI, Normalized Difference Water Index (NDWI) and differences of these two indices. Aside from Landsat, a new index namely mangrove forest index (MFI) is developed to map submerged mangrove forest from water background. In addition, hyperspectral imagery has also been used for mangrove forest mapping. Mangrove probability vegetation index (MPVI) identifies the pixels corresponding to mangroves by calculating their correlation coefficient with a candidate

TABLE 1.1
Numerous Mangrove Forest Indices

Index Name	Author	Formula
Mangrove Index (MI)	Winarso et al. (2014)	MI = (NIR − SWIR/NIR × SWIR) × 10000
Mangrove Recognition Index (MRI)	Zhang and Tian (2013)	MRI = |GVIL − GVIH| × GVIL × (WIL + WIH) where GVI − Green vegetation index; WI − Wetness index; Subscript L − at low tide; Subscript H − at high tide
Combine Mangrove Recognition Index (CMRI)	Gupta et al. (2018)	NDVI − NDWI where NDVI is the Normalized Difference Vegetation Index and NDWI is the Normalized Difference Water Index
Mangrove Probability Vegetation Index (MPVI)	Kumar et al. (2019)	$$MPVI = \frac{n\sum_{n=1}^{n} R_i r_i - \sum_{n=1}^{n} R_i \sum_{n=1}^{n} r_i}{\sqrt{n\sum_{n=1}^{n} R_i^2 - \left(\sum_{n=1}^{n} R\right)^2} \sqrt{n\sum_{n=1}^{n} r_i^2 - \left(\sum_{n=1}^{n} r_i^2\right)^2}}$$ where n = total number of bands in the image, R_i is the reflectance value at band if or a pixel of the reflectance image, and r_i is the reflectance value at band if or candidate spectrum of mangrove forest.
Normalized Difference Wetland Vegetation Index (NDWVI)	Kumar et al. (2019)	NDWVI = $(R_{2203} − R_{559})/(R_{2203} + R_{559})$
Mangrove Forest Index (MFI)	Jia et al. (2019)	MFI = $[(\rho\lambda_1 − \rho B\lambda_1) + (\rho\lambda_2 − \rho B\lambda_2) + (\rho\lambda_3 − \rho B\lambda_3) + (\rho\lambda_4 − \rho B\lambda_4)]/4$ where $\rho\lambda$ is the reflectance of the band center of λ, and i ranged from 1 to 4; $\lambda_1, \lambda_2, \lambda_3, \lambda_4$ are the center wave lengths at 705, 740, 783 and 865 nm, respectively.

mangrove spectrum (Table 1.1). In addition to MPVI, the same authors used Hyperion bands to discriminate mangrove and non-mangrove classes.

It is appreciable that MI-based methods are quite straight-forward. However, a few indices require tidal and hyperspectral imagery which might be challenging to arrange. In connection, both SAR and optical data can provide complementary information for forest monitoring based on SAR backscatter principles and optical multispectral reflectance measurements. SAR measures changes in vegetation and soil moisture content as well as the structural composition of the vegetation (lifeforms). Remote sensing of optical data measures changes in the chemical composition of leaves and their reflectance when illuminated by sunlight, also including measurements of shadow fractions within the canopies. Studies on SAR backscatter and NDVI can be used to compare time series of optical and SAR data. Fusion of SAR and optical data can also be used in analyses of deforestation and forest degradation (Reiche et al., 2015).

1.2 BIOMASS ESTIMATION OF MANGROVE FOREST

Above-ground biomass (AGB) generally includes stems and branches whereas belowground contains the roots which are known as belowground biomass. AGB is responsible for about 70%–90% of the forest biomass (Kumar and Mutanga, 2017). Figure 1.3 shows the global forest AGB distribution.

Many methods have been proposed and conducted in estimating ABG in various areas using remote sensing techniques with different data types such as optical remote sensing data, hyperspectral/multispectral data, SAR data and LIDAR data. Lidar and radar remote sensing techniques are currently recognized as the best approaches for quantifying and monitoring AGB. Therefore, the following sections will deal with forest biomass estimation utilizing SAR and Lidar data. Section 1.2.2 will particularly emphasize on mangrove AGB estimation.

1.2.1 Forest Biomass Estimation

1.2.1.1 Forest Biomass Estimation Using Lidar Technique

Airborne Lidar measurement is a common national forest inventory technique because of its accuracy and high-resolution characteristics. Lidar data are used to measure tree height, vertical structure and horizontal distribution of tree crowns (Ferraz et al., 2016). Various allometric models have been developed for diverse

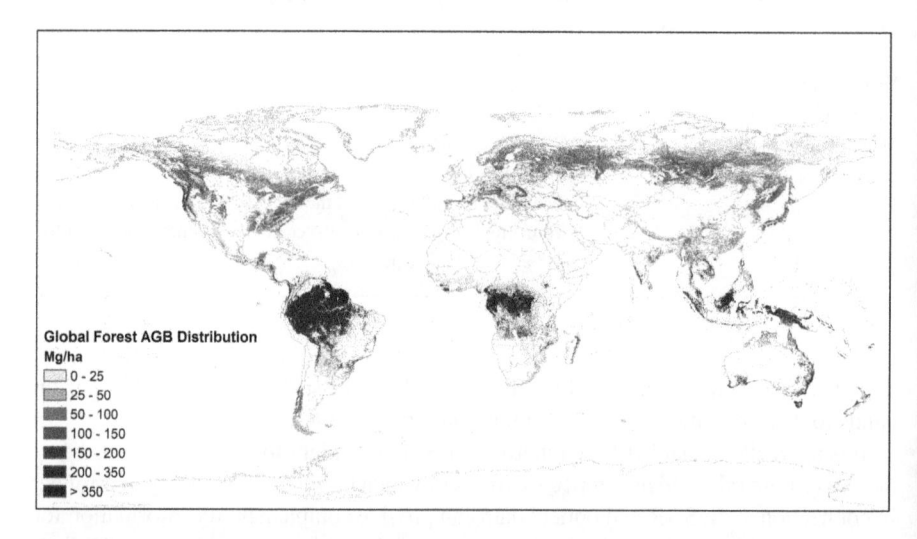

FIGURE 1.3 Global Forest AGB distributions. Tropical rainforest regions exhibit high biomass and comparatively lower biomass in extra tropics extending to temperate and boreal regions. Avitabile et al. (2016) produced this map at 1-km spatial resolution by combining two existing AGB data sets, namely (Saatchi et al., 2011) and (Baccini et al., 2012), both of which are originally produced by integrating satellite observation and LiDAR data. This is also available online at https://www.wur.nl/en/Research-Results/Chair-groups/Environmental-Sciences/Laboratory-of-Geo-information-Science-and-Remote-Sensing/Research/Integrated-landmonitoring/Forest_Biomass.htm?fbclid=IwAR0fo25ZdE27_pQvY6NeQJfzrR6pkGobJ3aAblV6BufI2iq_tN3yORkHoxY.

range of forest types globally in order to convert Lidar measurements of forest height or vertical structure into AGB (Asner & Mascaro, 2014; Nelson et al., 2009; Næsset et al., 2011). One of the common approaches of Lidar measurement is utilizing the mean top canopy height (MCH) from small footprint Lidar or utilizing percentile height from large footprint Lidar either from space-borne or air-borne sensors (Drake et al., 2002; Asner and Mascaro, 2014). Despite the development of innovative models, various Lidar-based models are established based on the conversion of ground plot level estimates of biomass and Lidar height metrics.

Nevertheless, there is no accurate universal model that can be used to convert the Lidar height measurements into AGB at all levels. For Lidar-based biomass estimation model development, the following six steps are a must to take into account:

1. Large plot size selection is important to include a large number of trees (50–100) as ground estimation of biomass depends strongly on the plot size.
2. The varying plot size depends on the forest types and size of the trees. For boreal forests which are dominated by conifers, plots of >0.1 ha may hold enough trees for accurate biomass estimation. However, for tropical forests, the plot sizes must be >0.25 ha to assure enough trees for biomass ground estimates with low uncertainty and Lidar metrics that characterizes the forest structure.
3. Plot shape could also impact the accuracy of the Lidar-biomass models. Generally, square plots are recommended as square plots are easier to create and have smaller edge lengths compared to rectangular plots. In contrast, large circular plots are difficult to develop for forests.
4. Due to different edge effects of the plots and large variations of biomass at small scales, models that are established from the plots may allow large distortion in ground biomass estimates. This is common in unmanaged forests in temperate and tropical regions specifically (Chave et al., 2004; Meyer et al., 2013).
5. In order to develop a Lidar-biomass mode, inclusion of proper height metrics is very important. This is because forest height metrics is a crucial parameter in obtaining the AGB across the landscape of the forest. For a given example, the MCH is known to be a strong metric for attaining the biomass variations as the MCH carries information about gaps and spatial extent of tree cover (Lefsky, 2010; Meyer et al., 2013; Asner and Mascaro, 2014).
6. The form of the model is also important to consider as this can take part in assessing the errors and also estimate the biomass. The use of a power-law between the AGB and the height metrics delivers the most consistent model for converting forest structure to biomass in almost all applications (Brown, 2002; Chave et al., 2005).

1.2.1.2 Forest Biomass Estimation Using SAR Technique

Forest structure is sensitive to different radar parameters like wavelength, SAR backscatter, polarization and look angle. Space-borne SAR observation can therefore be used for mapping AGB globally. However, AGB change monitoring is also influenced by other aspects, for example, surface topography, vegetation structure and density, and environmental factors like soil moisture and vegetation phenology.

SAR backscattering from forest is primarily controlled by wavelength and frequency. In case of forest, L/S-band backscatter arises from trunk and branches of trees, whereas C/X-band backscatter arises from identical sized leaves and needles. Longer wavelength SAR acquisition is characterized by deeper penetration while shorter wavelength gets reflection from the canopy level. It is also very important to consider the polarization of the radar waves because copolarized backscatter (VV, HH) (i.e. same transmit and receive components) is naturally stronger for surface scattering components; however, cross-polarized backscatter (VH or HV) (i.e. energy measured at 90° return offset to transmitting wave) is stronger for volume scattering measurements. Hence, cross-polarized observations with SAR imagery are important for biomass applications, forest degradation tracking, and identifying land changes of different volumes to surfaces. Moreover, SAR backscatter is also highly influenced by incidence angle, as it regulates scattering in the tree trunks, crown layers and on the ground surface. The SAR backscatter will be stronger if the slopes are tilted towards the sensor and weaker backscatter is expected if the slope is tilted away from the sensor. The look direction denotes the flight direction of the sensor's radar antenna. Applications of forest monitoring found it useful from combining different look directions, as corresponding backscatter information can be obtained, and different regions will be mapped for various applications. The role of moisture in SAR data is crucial in determining backscatter signals as it is highly sensitive and shows high backscatter values in increased soil and vegetation moisture. Standing open water can show dark image characteristics when wind, currents, or boat engines rough up water surfaces allowing strong backscatter to originate. Other factors like orientation of trees, branches, leaves, trunks and other structure with respect to SAR observation also influence the backscatter energy.

After accounting for the above factors, SAR indices and SAR algorithm could be implemented to estimate forest biomass. Few popular methods are described here in detail.

The two polarization indices are as follows: The RVI and the Radar Forest Degradation Index (RFDI) (Saatchi, 2019) are used to measure forest degradation level. They are both proposed below for monitoring different forest structures and can also be readily evolved from SAR satellite systems:

$$\text{RVI} = \frac{8\gamma_{\text{HV}}^0}{\gamma_{\text{HH}}^0 + \gamma_{\text{VV}}^0 + 2\gamma_{\text{HV}}^0}.$$

where γ^0 signifies the radiometrically and geometrically corrected SAR backscattering coefficient for all polarization combination in linear units (m^2/ m^2). RVI is a measure of randomness of scattering and is the ratio of cross-polarization which is used to approximate the total power from all polarization channels, which generally ranges between 0 and 1. For smooth bare surfaces, the RVI stands near 0 and it increases with vegetation growth as it has enhanced sensitivity to vegetation cover and biomass. As RVI is a ratio, it has less sensitivity to radar measurement geometry and topography and remains insensitive to absolute calibration errors in radar data.

The other index RFDI is calculated as

$$\text{RFDI} = \frac{\gamma_{HH}^0 - \gamma_{HV}^0}{\gamma_{HH}^0 + \gamma_{HV}^0}.$$

Even though the ratio is considered for all radiometrically corrected imagery, it can also be used before correcting SAR imagery radiometrically or geometrically. The value of RFDI generally ranges between 0 and 1: the values <0.3 resembles to dense forest, 0.4 or more indicates degraded forest and more than 0.6 denotes to deforested landscape.

Besides, numerous physically based and non-parametric models are being used for estimating forest biomass. A detail explanation of these models is outlined in Chapter 5 of SAR handbook (Flores et al., 2019). These models are Radar-Biomass Physically Based Models, Radar-Biomass Nonparametric Models, MaxEnt Model, Random Forest Model. Although computationally intensive, these models account for a number of forest characteristics/attributes for estimating biomass in an improved manner.

1.2.2 MANGROVE BIOMASS ESTIMATION

1.2.2.1 Mapping Mangrove Canopy Structure

Forest structures can be analyzed in terms of its tree cover, spatial extent, canopy height profile, spatial heterogeneity, and AGB. Interested learners are suggested to go through the SAR handbook (Flores et al., 2019) for further detail about mangrove biomass estimation.

Different parameters are used to measure forest canopy height and to estimate AGB. The numerical relation of AGB with radar backscatter is the following, where a and b are determined by any user:

$$\sigma^0(\mathrm{dB}) = a + b \times \log(\mathrm{AGB}).$$

For wet tropical forests, the values of a and b are about −22.5 and 3, respectively, for L-band. There are other equations which may require even more complex calculations (Yu and Saatchi, 2016):

$$\sigma^0(\mathrm{linear}) = Ax^a + \left(1 - e^{-Bx}\right) + C$$

where x is the AGB, and A, B, C, and a are coefficients that can be fitted empirically through iteration until x results in the observed σ^0 (linear).

For wet tropical forest, generally, the coefficient values are $\alpha = 0.013682$, $A = 0.21116$, $B = 0.051846$, $C = 0.02192$. The coefficient A should be fitted locally. These parameters vary depending on the species types and inundation levels.

There also exist limitations of following the above approach. Multiple researchers have encountered the AGB saturation (Carreiras, Vasconcelos and Lucas, 2012; Cohen et al., 2013). It can be expected that σ^0 to increase with forest AGB up to a saturation value that depends on the radar frequency (i.e. P, L, C, X, K) before it decreases. The longer the wavelength, the better performance is expected, that is, the larger the biomass saturation point (Mougin et al., 1999; Proisy et al., 2002). However, mangrove AGB estimation from backscatter alone is extremely difficult and is strongly site-dependent and these polarimetric trends can also be used to classify mangrove type and structure besides AGB estimation (Hong et al., 2015; Brown, Mwansasu and Westerberg, 2016). It is tough to determine forest mangrove extent by using radar alone, and as mentioned previously, this can also be difficult with optical sensors, as "color" (e.g., greenness) may not suffice to distinguish mangroves from other vegetation types. Thus, it is recommended to understand the mangrove extent by using a combination of different sensing technologies and their data sets (Nascimento et al., 2013). It is also possible to extract the mangrove forest extent from the radar data from the existing global maps of mangrove forests (Giri et al., 2011). Land Cover Classification can generally be performed with the radar backscatter as one of the layers along with data from optical instruments such as Landsat or Sentinel 2. Land Cover Classification can be performed with supervised methods (e.g., maximum likelihood, decision trees, neural network) or unsupervised methods (e.g., ISODATA) to delineate the forest cover. These algorithms could be implemented in all commercial geospatial software (e.g. ArcGIS, ENVI, etc.), open source software like QGIS and python libraries and even utilizing cloud platforms like GEE. It is crucial that the training sets must represent the entire range of spectral signature observed with remote sensing, otherwise pixels with untrained spectral signatures in radar and optical data will not be classified correctly and wrong classes will potentially be classified as mangroves and hence lead to errors in results. In order to avoid these problems, an initial unsupervised classification method can be used by manual merging of "unsupervised classes" into relevant mangroves classes.

Data from the Shuttle Radar Topography Mission also known as SRTM from February 2000 were the first data set that has facilitated the measurement of forest canopy height. SRTM was designed to measure elevation; however, due to radar microwave contact with canopy volume, the SRTM can show biased outcomes, for example, a forested hilltop might appear higher than it really is. A big assumption is considered while relating SRTM elevation measurement with mangrove canopy height (i.e. mangroves are located at mean sea level with negligible topography). There have been significant advances in the use of RADAR interferometry to map mangrove canopy height thus can lead to calculate AGB and BGB by using allometric equations (Simard et al., 2006; Lagomasino et al., 2015). Considering another assumption in mangroves with high tides, this method allows to determine mangrove height patterns which can also be adopted into training classes and validate the Land Cover Classification of the mangrove structure more accurately.

1.2.2.2 Forest Stand Height

Forest Stand Height (FSH) or the average height of trees in a forest stand of a mangrove or other forest types is known as an indicator of the forest age. The FSH also allows characterizing the history of land use, the amount of AGB, and plant and animal habitats.

FSH can be determined by the difference between digital elevation model (DEM) and digital terrain model (Treuhaft and Siqueira, 2000; Papathanassiou and Cloude, 2001). It could also be determined by the differences of two DEM produced by C and P/L band SAR image pair as longer wavelength SAR has more capability to penetrate the leaves and get backscattered from the earth surface. However, temporal and spatial decorrelation is always a challenge in implementing this approach. The best estimate of FSH made from repeat-pass interferometry is made from the combined properties of SAR backscatter power and InSAR coherence and thus leading to combined process of SAR/InSAR techniques (Papathanassiou and Cloude, 2001). FSH ground validation data is a crucial component of the data processing methodology necessary for converting interferometric SAR data into FSH estimates.

1.2.2.2.1 Relationship of Backscatter to Forest Stand Height

After topographic and geometric corrections, the backscatter power is written as

$$\gamma^0 = A\left(1 - e^{-Bh_v^C}\right).$$

where γ^0 is the terrain-corrected form of radar cross section (Small, 2011), h_v is the vegetation height, and the coefficients A, B, and C are determined in the FSH algorithm using a least-squares fit between the backscatter power and the vegetation height provided by the ground validation and/or overlap data between scenes. Sample values for these coefficients that have been automatically determined by the FSH algorithm are $A = 0.11$, $B = 0.0622$ and $C = 1.0143$.

A very general issue determining the relationship of backscatter to vegetation characteristics is that for a certain threshold of biomass, a sensitivity of increasing γ^0 to increasing biomass exists no longer and this saturation effect is wavelength-dependent. There is also a similar saturation effect that occurs between γ^0 and h_v while estimating AGB from FSH. For the equation above, the backscatter power is denoted as γ^0. This backscatter power is stored in radar sensor and other satellite platforms. The data are then processed into ground coordinates followed by aperture synthesis where these backscatter values are converted by the processor either as Digital Numbers (DN values) or in terms of calibrated radar backscatter power, either in units of σ^0 or γ^0, depending on the level of processing employed to the users.

Calibration is required in order to correct the radar power returns for gains and making all measurements proportional to the transmitted power. The σ^0 values are calibrated according to the range co-ordinate of the radar system and then normalized for the size of the area reflecting the energy back to the system (larger areas will reflect more energy). The units of σ^0 are in m^2/m^2. When the radar cross-section is adjusted to account for surface area interception in the direction of the radar view using DEM, this value is termed as γ^0 and this γ^0 is the most appropriate for quantitative analysis of the forest classification (Small, 2011).

1.2.2.2.2 Relationship of Interferometric Coherence to FSH

The interferometric coherence can be derived from the interferometric correlation, which is the normalized geometric average between two complex radar images. Arithmetically, the interferometric correlation γ is defined as

$$\gamma = \frac{E_1 E_2^*}{\sqrt{|E_1|^2 |E_2|^2}}.$$

where E_1 and E_2 are the radar cross-sections' complex values that are observed by the SAR satellite and delivered as Single Look Complexes (SCLs). The brackets in the equation indicate averaging over multiple looks, and * indicates a complex conjugation. However, the γ denoted here in the interferometric correlation expression above is not the same as the γ^0 specified for the terrain-corrected value of radar cross-section mentioned previously. It is common to indicate terrain-corrected value of γ when an image is referred to interferogram. This complex-valued correlation magnitude varies between 0 and 1 and the phase varies between 0 and 2π. This explains that a signal with low correlation will have coherence close to 0 with random poor phase and a high correlation will have coherence close to 1 with good phase.

Sampling design and statistical modeling/estimation is necessary to address the issue of uncertainty and reducing the number of field and remotely sensed data in biomass estimation. In addition, these approaches provide the flexibility between filed plot configurations and remote sensing data characteristics. The seventh chapter of the SAR handbook (Flores et al., 2019) explained several important considerations in addressing the multi-level sampling uncertainty including both the model-based and model-assisted inferential framework and the application of simulation to illustrate the estimators associated with these designs. The interested learners are encouraged to go through this book chapter while conducting field data collection and during model validation for biomass estimation.

1.2.2.2.3 Other Remote Sensing Approaches for Monitoring Forest Biomass

Other remote sensing techniques, that is, optical and hyperspectral, are also useful for monitoring forest biomass. The application of hyper-spectral remote sensing on mangrove mapping and monitoring is broadly explained in Chapter 2 of this book (Flores et al., 2019). Chapters 8 and 12 have covered the multispectral remote sensing application in the context of LAI, canopy cover and plant diversity. Moreover, a detail overview of optical and hyperspectral remote sensing technique in mangrove biomass estimation can also be found in Pham et al. (2019).

Briefly, high and very high spatial resolution optical data from multiple sensors like SPOT, IKONOS, RapidEye, WorldView-2 and Quick-BirdGeoeye-1 may play a crucial role in assessing AGB of mangrove forests. These data can be used to estimate AGB utilizing either parametric approaches including multiple regression model or non-parametric machine learning approaches support vector machine and neural networks (Pham et al., 2019). Recently launched Sentinel-2A, 2B optical sensors are also providing much opportunities as the data are open source. Similarly, hyperspectral data due to its rich spectral information with a large number of spectral bands (from visible to NIR or SWIR range) have the potentiality of vegetation classification and estimation of forest AGB. However, the only constraint lies on its air-borne sensing mechanism which captures limited area of investigation. On the other hand, space-borne hyper-spectral sensor covers a large area but the data are not freely accessible in most cases.

1.3 FOREST HEALTH AND STRESS MAPPING USING GEOINFORMATICS

Forest health and stress mapping can be carried out by many approaches like in-situ forest inventory and experimental studies and remote sensing techniques based on several indicators (Lausch et al., 2016). The remote sensing technologies are gaining an advantage over these in-situ techniques despite of its progress in the field of science which helps to obtain detailed information with high accuracy (Lausch et al., 2017). On the other hand, the use of remote sensing technologies in the study of forest health and stress mapping had increased greatly over the past years due to wide availability of earth observation data, increased degree of competency, large aerial coverage and high temporal resolution (White et al., 2016). This also allows to monitor the forest health over a long period in order to observe the trend, identify any short-term abrupt changes and infer the reasons behind it (Lausch et al., 2017). This section includes comprehensive description on how different traits respond to remote sensing technologies, various stress indicators that are used in different types of remote sensing techniques as well as the methods to map forest health.

1.3.1 Forest Health and Stress Indicators

A set of indicators are usually used to observe the condition of the forest over time and space in order to determine the forest health. Apart from field observations, forest health can be monitored with active and passive remote sensing technologies due to its ability to record physical characteristics or biotic traits such as biochemical, physiological, morphological, structural, phonological and functional traits of plants, populations or communities (Abelleira Martínez et al., 2016). Ustin and Gamon (2010) discussed the principles of image spectroscopy, which includes broad ranges of the electromagnetic spectrum that is from visible to microwave band (spectral range of 400–2500 nm), to identify and know the details of these traits of biotic and abiotic earth surfaces based on the amount of radiance received at the sensor. Later, this information is used to assess and predict forest health by developing an understanding of the distribution of forest plants and communities and their responses to different stress factors. The forests are susceptible to several intrinsic and extrinsic factors such as drought, climate change, anthropogenic activities like deforestation, pest and pathogenic attack, etc., which likely affects the functioning of the species and populations in the forest. Therefore, the species and populations develop resilience and adapt to these disturbances (Wulder et al., 2006). The characteristics of plants, populations and communities and the changes in these features that can be recorded directly and indirectly using RS techniques are referred to as Spectral Traits (ST) and Spectral Trait Variations (STVs) whereas the traits that cannot be recorded by either way are termed as Non-Spectral Traits (N-STs) (Homolová et al., 2013; Lausch et al., 2016, 2017). According to Lausch et al. (2016), ST is anatomical, morphological, biochemical, biophysical, physiological, structural, phenological or functional characteristics whereas STV is the changes to STs in terms of physiology, senescence and phenology along with different stress factors. On the other hand, N-ST refers to the same characteristics of plants, populations and communities that of ST but it cannot be detected by the sensors due to its location (forest undergrowth or roots) or present shortcomings of RS characteristics, thereby, mostly depends on in-situ techniques. This, however, gives an advantage to the ST/STV approach as the forest health can be effectively monitored over a large area in a cost-effective way by using indicators. Some of the indicators of different ST/STV categories along with the N-ST categories are listed in Table 1.2, which shows the variation in the indicator types and explains why RS cannot detect N-ST.

TABLE 1.2
Different Categories of STs and N-STs and Their Related Indicators to Assess Forest Health

Traits		Parameters	
STs	Biochemical-Biophysical ST such as pigment content, nitrogen, phosphorous, cellulose, carbon content	Physiognomic-Morphological ST Leaf size, form, type, leaf anatomy, specific leaf area, leaf carbon content, wood/stem anatomy, etc.	Phenotypical ST Plant growth form, age structure, height, crown size, lifespan, plant surface roughness, pollination mode, etc.
	Physiological and Functional ST e.g. photosynthetic, carbon sequestration, evapotranspiration.	Phenology and Senescence ST Leaf phenology type, flowering, plant, and land surface phenology.	Aggregated ST Net Primary Productivity, Fraction of Photosynthetically Active Radiation and Biomass.
	Structural ST such as spatial distribution, configuration pattern, area, size, shape, density.	8. Stress, Adaptation, and Disturbance ST For example, Plant tolerance to stress, damage, urbanity, naturalness, vulnerability.	Chorology, Distribution and Dispersal ST Dispersal mode, floristic zones, floristic areas, and gradient traits.
N-STs	Physiognomic-Morphological N-ST Seed size, seed mass, seed longevity, seed morphology, stem bark thickness, stem specific density, vessel area, root nitrogen content, xylem conductivity etc.		Phenotypical N-ST Wood porosity, woodness, wood anatomy, wood density, modification of storage, mycorrhiza type, rooting depth etc.

Modified after Lausch et al. (2016).

STs and their variations are identified through a multitude of remote sensors as shown in Figure 1.4 (details will be described in the latter part). They can be close ranged sensors such as wireless sensor networks (WSN), towers with remote sensing capabilities and field spectrometers. These methods are commonly used as a reference for identifying different forest STs. For delineating STs on a wide area, far ranged remote sensors are used. They include ground-based and air-borne LiDAR and laser-scanner as well as air-borne and space-borne optical and SAR sensors. These sensors have a wide range of applications on forest health including but not limited to the identification of ST, STV and external stressors like natural and man-made hazards. However, the identification of forest health indicators through remote sensing is based on several factors. The spectral information reflected to the sensor contains detailed information on the characteristics and content, proportion, density, shape, spatial distribution, the texture of biotic and abiotic traits and on trait variations. Lausch et al. (2016) also identified that different modeling methods and geographic data representation along with the spatial, spectral, radiometric, angular and temporal resolution of the RS techniques affect the results of forest health monitoring. Besides, a suitable algorithm and relevant assumptions should be made during the analysis.

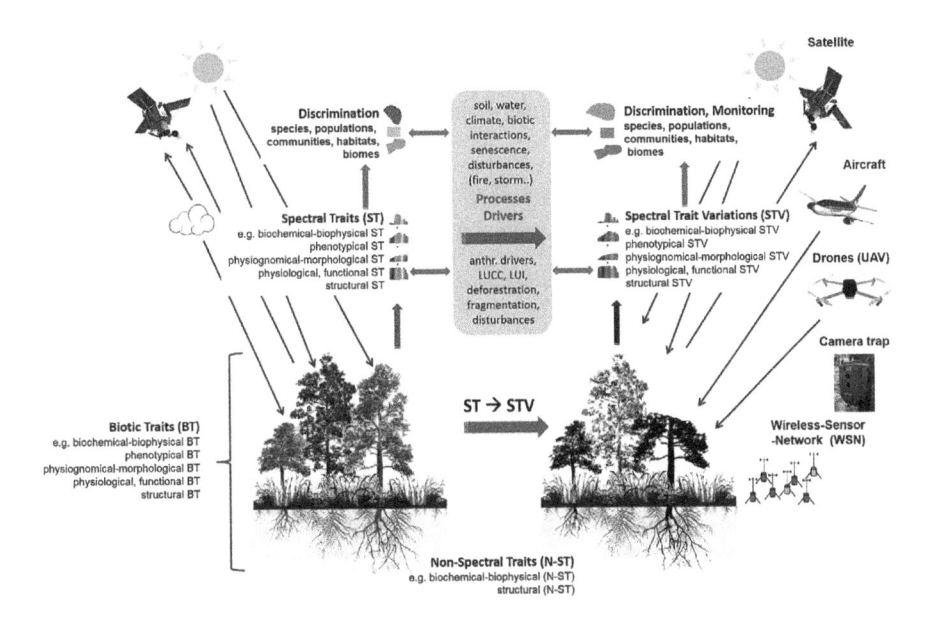

FIGURE 1.4 STs and N-ST observing and assessing forest health through different approaches. (From Lausch et al., 2016.)

1.3.2 Forest Health and Stress Monitoring Using Close-Range RS Approach

Close-range remote sensing is mainly performed to identify the spectral responses of forest health traits. These traits are further investigated to formulate practical models that can be used on a more massive scale. Different forest health traits respond to stress differently and manifest after a different length of time (Müller, 2009). This close-range remote sensing allows for widespread and long-term monitoring of stress. Furthermore, they incorporate vegetation phenological cycle information as well as within year disparities (Chen and Cihlar, 1996). By sampling from the forest areas, these methods ascertain chlorophyll, water content, leaf area and other leaf properties as well as functional traits such as photosynthesis. These methods and their functionality are briefly described in this section.

Field spectrometer analysis is the basis for spectral characteristics research. Taxonomic, phylogenic and morphological functional traits are investigated based on this as well. However, there is no standardized method for analysis and this method does not consider population effects (Asner and Martin, 2009). Spectral laboratory and plant phenomics facilities analyze time-variant biochemical, biophysical and structural traits. They also investigate plant stress and development. However, only up to 5 years of plant age can be recorded (Brosinsky et al., 2014).

Towers with remote sensing capabilities can be used for long-term monitoring. Even the multisensory approach for determining biotic and abiotic traits is possible with this. Furthermore, international networking between them is also present. However, the results found on a local scale have no meaningful way to translate to a global scale

(Baldocchi et al., 2001). Similarly, WSNs can monitor for a long period using multiple sensors. Multiple variant biochemical, biophysical and structural traits are investigated using this method. Though, these methods are mainly nonimaging (Lloret et al., 2009).

1.3.3 FOREST HEALTH AND STRESS MONITORING USING FAR-RANGE RS APPROACH

In recent years, the science of space-borne and air-borne, far-reaching remote sensing techniques have flourished. This is true for the application in forest health monitoring as well. This section reviews the most prominent of these techniques.

1.3.3.1 LiDAR

LiDAR is an air-borne or ground-based sensor that emits pulses of laser light to the subject and captures the reflection as a point cloud. Later these point clouds are used to construct the three-dimensional virtual structure of the subject. By nature, this technique is very useful in constructing a three-dimensional model of forests. However, it is even more useful in investigating the under-canopy parameters which are usually inaccessible to the traditional passive remote sensing (Maltamo et al., 2014). Thus, the LiDAR works best as part of remote sensing technique when canopy and tree structure related forest health STs are investigated.

LiDAR can precisely recreate the digital terrain model or digital surface model. It can even create a DTM with an error margin as low as 0.15 m (Andersen et al., 2005). Hence, it can produce tree height and other biophysical properties (Nilsson, 1996). Commonly LiDAR-based methods are used to define fine-resolution carbon stock (Asner et al., 2014). These methods are either area-based or individual tree-based. In area-based methods, the reflections are calculated for a specific area (Næsset, 2002) and in the individual tree method (Yao et al., 2012), LiDAR can extract the properties of a single tree based on the reflected point cloud model. The LiDAR techniques commonly use empirical models to infer the biophysical properties. Therefore, after identifying and extracting the properties of a tree, it is compared with a reference tree. This allows the model to produce information about the biomass, height, crown parameter, species, volume of tree and health parameters (Morsdorf et al., 2004). In area-based approach, the tree height can be determined with a precision of 6% with 10% for basal area and diameter as well as 20% for stem density (Holmgren, 2004). Furthermore, it can produce the AGB with a precision of 12% the AGB (Latifi et al., 2010). Individual tree approach sees 80%–97% detection rate for upper canopy trees (Latifi et al. 2015). Furthermore, coarse woody debris can be detected (Morsdorf et al., 2006; Pesonen et al., 2008). In addition, this technique allows vertical stratification of forest canopy which is useful for delineating species diversity (Zimble et al., 2003). For an in-depth review of the LiDAR system, readers are directed to Wehr and Lohr (1999).

1.3.3.2 Synthetic Aperture RADAR

SAR data emit microwave pulses and capture the backscatters in different polarizations. Much like LiDAR technique, it is an active remote sensing method that has the ability to pierce the forest canopy. This ability allows the technique to measure tree height, biomass and canopy coverage (Balzter et al., 2007). Apart from these, SAR is extensively used for detecting deforestation, forest fire and inundation. The polarization characteristics can detect tree species as well (Ulaby et al., 1987). Forest changes cause the correlation between the phases of two images to decline. Thus, this can be used as a means of forest degradation monitoring.

Currently, different space-borne SAR satellites use different wavelengths which have different usage. The L-band and P-band can penetrate deeper into the forest canopy and determine AGB. Cross-polarization allows for improved biomass models. Forest density can be mapped utilizing polarization information as well (Varghese et al., 2016). C-band can determine the LAI and it is used for forest identification (Manninen et al., 2013). SAR data set can delineate different types of forest degradation such as fire damage. The polarimetric information, phase and decomposition parameters can properly distinguish fire-damaged section of forest (Bourgeau-Chavez et al., 1994). Moreover, flooded forest area is determined from HH and VV polarization as they return large phase difference. In addition, the high double bounce off based on dead trees indicates to submerged forest area (Rignot et al., 1997). The incident angle of the SAR data has also been seen to be used parameter in identifying the inundated area. Deforestation area identification has been performed through SAR data set as well (Thapa et al., 2013). SAR information is used to determine soil moisture and evapotranspiration which are indicative of forest health. Passive radiometer produces are also used to produce this information.

1.3.3.3 Optical Sensor

Optical sensors are passive sensors that photograph or scan the earth's surface that is illuminated by sunlight. The most common usage of multispectral optical images is the vegetation index which works as the ground for determining many biophysical properties. Centimeter-level spatial resolution optical sensors such as APEX, AISA and HySPEX can be used for leaf trait identification (Ali et al., 2016). Furthermore, high-resolution optical imagery can be used for canopy coverage and LAI (Durbha et al., 2007). Some optical sensors like MODIS have a high temporal resolution which is used for rapid forest monitoring. The high spectral resolution of hyperspectral images has a multitude of usage in forest health, foliar biochemistry, biomass and aboveground carbon (Goodenough and Bannon, 2014). The nutrient information such as soil pH, fertility and water supply can be derived from hyperspectral remote sensing as well (Schmidtlein, 2005). MODIS-based model ALEXI and other models such as SEBS, MOD16 and CMRSET provide excellent information on evapotranspiration, a physiological trait for forest health (da Motta Paca et al., 2019). Moreover, SMAP satellite provided soil moisture and evapotranspiration information based on a passive radiometer. Forest fragmentation occurs primarily due to development and it has an adverse effect on forest plant species. Optical remote sensing is primarily used for determining this process (Boentje and Blinnikov, 2007). Another important ST is the fraction of photosynthetically absorbed radiation. This is globally measured through the optical sensor as well (Claverie et al., 2016; Hu et al., 2020).

1.3.3.4 Multi-Sensor Approach

All the above-mentioned sensors can be used in conjunction with one another. This results in an even better product that can be used to further improve the understanding of forest health (Pause et al., 2016). Canopy and forest STs can be inferred by combining optical hyperspectral images and LiDAR data (Joshi et al., 2016). Even integration of thermal information allows for the creation of 3D thermal canopy model (Giuliani et al., 2001). The LiDAR and SAR sensor can overcome the limitations of optical sensors. Similarly, the high temporal resolution of the optical image can be useful for overcoming limitations for other sensors. Multi-sensor approach is not only useful for combining different types of sensors but different optical sensors can also be calibrated to provide even denser observation of forest area. On the airborne system, there can be multiple sensors installed. One such system is Goddard LiDAR hyperspectral and thermal (Lausch et al., 2016). Moreover, HyspIRI mission containing an imaging spectrometer and a multispectral imager with the capability to capture mid to long-wave infrared images is launched in 2020. This sensor combination is especially useful for assessing canopy biochemistry measurement as well as the biogeochemical cycles in the forest ecosystem. Similarly, the Hyperspectral Imager Suite is a space-borne multi-sensor platform that has a combination of 185 bands hyperspectral sensor and a 5-band multispectral sensor. This is expected to be seated on-board of ALOS-3 satellite of Japan Aerospace Exploration Agency (Lee et al., 2015). Curious readers are advised to take a look at Lausch et al. (2016, 2017, 2018). It is worth mentioning about the newly launched sensor ECOSTRESS (The Ecosystem Space Borne Thermal Radiometer Experiment on Space Station) which is designed to measure the temperature of plants and use that information to better understand how plants respond to heat and water stress (http://ecostress.jpl. nasa.gov). This information is available for an individual farmer field which will facilitate managing water resources and monitoring droughts. The evapotranspiration product of ECOSTRESS is enabling us to identify stressed vegetation even way before detecting the physical degradation of plants with a naked eye.

1.3.4 Modeling Approach

1.3.4.1 Empirical Models

Empirical transfer functions are established by correlating ground information with remotely sensed reflectance values. These methods are more commonly used in forest health as a plethora of established empirical methods are currently present (Verrelst et al., 2015).

1.3.4.1.1 Parametric Models

Parametric models utilize specific spectral band combinations and correlate them with bio-geophysical variables by applying a specific fitting function. In multispectral optical remote sensing, vegetation indices are a common example of this method. In hyperspectral or quasi-hyperspectral space, red edge position, derivative-based indices methods are used to determine forest health indicators such as leaf area content or leaf chlorophyll content (Delegido et al., 2011; Clevers and Kooistra, 2012).

1.3.4.1.2   Non-Parametric Models

Non-parametric regression analysis relies on the training dataset. They include a training phase that trains the models to properly interpret the remotely sensed data sets to infer the biophysical properties that produce minimal estimation errors. Linear non-parametric such as principal component regression and partial least square regression are more simpler and are massively used in multi-spectral datasets (Geladi and Kowalski, 1986). The non-linear non-parametric methods, commonly known as machine learning methods, assume that the relationship between reflectance and bio-geophysical properties are implicit and thus these methods do not require prior knowledge about the data sets although they offer to incorporate a priory information in Bayesian methods (Verrelst et al., 2012). The most common approaches are the decision tree method (i.e. Random forest (Mutanga et al., 2012)), Kernel-based (i.e. Support vector machine (Vapnik et al., 1997)), artificial neural network (Verrelst et al., 2012) and Bayesian networks (Kalacska et al., 2005).

1.3.4.2   Physical Models

The physical models are based on the relationship between causal factors and their effects. They are based on physical laws. Radiative transfer function and coherence conversion are the two most common types of physical methods that are used in assessing forest health (Lausch et al., 2017).

1.3.4.2.1   Radiative Transfer Theory Models

Radiative transfer theory is the characterization of wave propagation through a medium by a random distribution of scatterers (Njoku, 2014). This model inversion is a robust approach to obtain biophysical properties (Jacquemoud et al., 1995; Combal et al., 2003). Commonly used Radiative Transfer Model (RTM) techniques are numerical optimization where the objective is to minimize the cost function which is used to describe the difference between estimated and measured variables (Zarco-Tejada et al., 2001). In the Lookup Table (LUT)-based approach, the reflectance model is compared with the measured one. This is done by using the values obtained from LUT and applying an appropriate cost function (Liang, 2007). Another common RTM, geometric optical model, considers the geometric shapes and assumes the optical properties are known. This model is more commonly used to determine the LAI (Li, Song and Wang, 2015). Other RTM theory such as Kuusk–Nilson forest reflectance model is also used to obtain LAI (Rautiainen et al., 2003).

1.3.4.2.2   Coherence Conversion Models

Coherence conversion methods use the coherence-amplitude or more robustly phase and coherence information of SAR imagery and apply inversion algorithm to obtain phenotypical parameters such as forest height. They are called as such as these methods use coherence information as inputs. Random volume over ground is a commonly used algorithm where surface and volumetric coherence are used as a ground layer and canopy layer. This method has been employed on X-band and L-band SAR images (Garestier et al., 2008; Praks et al., 2007). The three-stage inversion method uses a single coherence value to obtain tree height and underlying topography

(Tan and Yang, 2008). On the other hand, water cloud model uses the dielectric constant of the canopy water to describe the upper layer of the forest. The dielectric constant of canopy water is smaller than water but higher than air. This allows for pinpointing of canopy size, hence AGB through the radar backscatter inversion (Behera et al., 2016).

1.3.4.3 Hybrid Models

These methods are mainly non-parametric in nature, but the training data set is obtained through physical models. Hence, they are universally applicable like physical models but flexible and efficient in computation like non-parametric models (Verrelst et al., 2015). Commonly used methods are artificial neural networks that are trained through RTM-generated data (Danson et al., 2003). The kernel-based machine learning regression algorithms trained with RTM data have been used to obtain forest health indicators as well (Durbha et al., 2007). MERIS and CYCLOPS data sets are produced following this same principle (Zarco-Tejada et al., 2001; Baret et al., 2007).

REFERENCES

Abelleira Martínez, O. J. et al. (2016) 'Scaling up functional traits for ecosystem services with remote sensing: concepts and methods', *Ecology and Evolution*, 6(13), pp. 4359–4371. doi: 10.1002/ece3.2201.

Ali, A. M. et al. (2016) 'Retrieval of forest leaf functional traits from HySpex imagery using radiative transfer models and continuous wavelet analysis', *ISPRS Journal of Photogrammetry and Remote Sensing*, 122, pp. 68–80. doi: 10.1016/j.isprsjprs.2016.09.015.

Alongi, D. M. (2002) 'Present state and future of the world's mangrove forests', *Environmental Conservation*, 29(3), pp. 331–349. doi:10.1017/S0376892902000231.

Andersen, H.-E., Reutebuch, S. E. and McGaughey, R. J. (2005) 'Accuracy of an IFSAR-derived digital terrain model under a conifer forest canopy', *Canadian Journal of Remote Sensing*, 31(4), pp. 283–288. doi: 10.5589/m05-016.

Asner, G. P. et al. (2014) 'Targeted carbon conservation at national scales with high-resolution monitoring', *Proceedings of the National Academy of Sciences*, 111(47), pp. E5016–E5022. doi: 10.1073/pnas.1419550111.

Asner, G. P. and Martin, R. E. (2009) 'Airborne spectranomics: mapping canopy chemical and taxonomic diversity in tropical forests', *Frontiers in Ecology and the Environment*, 7(5), pp. 269–276. doi: 10.1890/070152.

Asner, G. P. and Mascaro, J. (2014) 'Mapping tropical forest carbon: Calibrating plot estimates to a simple LiDAR metric', *Remote Sensing of Environment*, 140, pp. 614–624. doi: 10.1016/j.rse.2013.09.023. Elsevier Inc.,

Avitabile, V., et al. (2016) 'An integrated pan-tropical biomass map using multiple reference datasets', *Global Change Biology*, 22(4), pp. 1406–1420. doi: 10.1111/gcb.13139.

Baccini, A., et al. (2012) 'Estimated carbon dioxide emissions from tropical deforestation improved by carbon-density maps', *Nature Climate Change*, 2(3), pp. 182–185. doi: 10.1038/nclimate1354.

Baldocchi, D., et al. (2001) 'FLUXNET: A new tool to study the temporal and spatial variability of ecosystem–scale carbon dioxide, water vapor, and energy flux densities', *Bulletin of the American Meteorological Society*, 82(11), pp. 2415–2434. doi: 10.1175/1520-0477(2001)082<2415:FANTTS>2.3.CO;2.

Baloloy, A. B., et al. (2020) 'Development and application of a new mangrove vegetation index (MVI) for rapid and accurate mangrove mapping', *ISPRS Journal of Photogrammetry and Remote Sensing*. 166, pp. 95–117. doi: 10.1016/j.isprsjprs.2020.06.001. Elsevier.

Balzter, H., Rowland, C. and Saich, P. (2007) 'Forest canopy height and carbon estimation at Monks Wood National Nature Reserve, UK, using dual-wavelength SAR interferometry', *Remote Sensing of Environment*, 108(3), pp. 224–239. doi: 10.1016/j. rse.2006.11.014.

Barbier, E.B., et al. (2011) 'The value of estuarine and coastal ecosystem services', *Ecological Monographs*, 81(2), pp. 169–193.

Baret, F., et al. (2007) 'LAI, fAPAR and fCover CYCLOPES global products derived from Vegetation', *Remote Sensing of Environment*, 110(3), pp. 275–286. doi: 10.1016/j. rse.2007.02.018.

Behera, M. D., et al. (2016) 'Above-ground biomass and carbon estimates of Shorea robusta and Tectona grandis forests using QuadPOL ALOS PALSAR data', *Advances in Space Research*. COSPAR, 57(2), pp. 552–561. doi: 10.1016/j.asr.2015.11.010.

Boentje, J. P. and Blinnikov, M. S. (2007) 'Post-Soviet forest fragmentation and loss in the Green Belt around Moscow, Russia (1991–2001): a remote sensing perspective', *Landscape and Urban Planning*, 82(4), pp. 208–221. doi: 10.1016/j.landurbplan.2007.02.009.

Bourgeau-Chavez, L. L., et al. (1994) 'Using ERS-1 SAR imagery to monitor variations in burn severity in an Alaskan fire-disturbed boreal forest ecosystem', in *Proceedings of IGARSS '94-1994 IEEE International Geoscience and Remote Sensing Symposium*. IEEE, pp. 243–245. doi: 10.1109/IGARSS.1994.399093.

Brosinsky, A., et al. (2014) 'Analysis of spectral vegetation signal characteristics as a function of soil moisture conditions using hyperspectral remote sensing', *Journal of the Indian Society of Remote Sensing*, 42(2), pp. 311–324. doi: 10.1007/s12524-013-0298-8.

Brown, I., Mwansasu, S. and Westerberg, L. O. (2016) 'L-band polarimetric target decomposition of mangroves of the rufiji delta, Tanzania', *Remote Sensing*, 8(2). doi: 10.3390/rs8020140.

Brown, S. (2002) 'Measuring carbon in forests: Current status and future challenges', *Environmental Pollution*, 116(3), pp. 363–372. doi: 10.1016/S0269-7491(01)00212-3.

Carreiras, J. M. B., Vasconcelos, M. J. and Lucas, R. M. (2012) 'Understanding the relationship between aboveground biomass and ALOS PALSAR data in the forests of Guinea-Bissau (West Africa)', *Remote Sensing of Environment*, 121, pp. 426–442. doi: 10.1016/j.rse.2012.02.012. Elsevier Inc.

Chave, J., et al. (2004) 'Error propagation and sealing for tropical forest biomass estimates', *Philosophical Transactions of the Royal Society B: Biological Sciences*, 359(1443), pp. 409–420. doi: 10.1098/rstb.2003.1425.

Chave, J., et al. (2005) 'Tree allometry and improved estimation of carbon stocks and balance in tropical forests', *Oecologia*, 145(1), pp. 87–99. doi: 10.1007/s00442-005-0100-x.

Chen, J. M. and Cihlar, J. (1996) 'Retrieving leaf area index of boreal conifer forests using landsat TM images', *Remote Sensing of Environment*, 55(2), pp. 153–162. doi: 10.1016/0034-4257(95)00195-6.

Claverie, M., et al. (2016) 'A 30+ Year AVHRR LAI and FAPAR Climate Data Record: Algorithm Description and Validation', *Remote Sensing*, 8(3), p. 263. doi: 10.3390/rs8030263.

Clevers, J. G. P. W. and Kooistra, L. (2012) 'Using Hyperspectral Remote Sensing Data for Retrieving Canopy Chlorophyll and Nitrogen Content', *IEEE Journal of Selected Topics in Applied Earth Observations and Remote Sensing*, 5(2), pp. 574–583. doi: 10.1109/JSTARS.2011.2176468.

Cohen, R., et al. (2013) 'Propagating uncertainty to estimates of above-ground biomass for Kenyan mangroves: A scaling procedure from tree to landscape level', *Forest Ecology and Management*, 310, pp. 968–982. doi: 10.1016/j.foreco.2013.09.047. Elsevier B.V.

Combal, B., et al. (2003) 'Retrieval of canopy biophysical variables from bidirectional reflectance', *Remote Sensing of Environment*, 84(1), pp. 1–15. doi: 10.1016/S0034-4257(02)00035-4.

Coppin, P., et al. (2004) 'Digital change detection methods in ecosystem monitoring: A review', *International Journal of Remote Sensing*, 25(9), pp. 1565–1596. doi:10.1080/0143116031000101675.

Danson, F. M., Rowland, C. S. and Baret, F. (2003) 'Training a neural network with a canopy reflectance model to estimate crop leaf area index', *International Journal of Remote Sensing*, 24(23), pp. 4891–4905. doi:10.1080/0143116031000070319.

Delegido, J., et al. (2011) 'Evaluation of sentinel-2 red-edge bands for empirical estimation of green LAI and chlorophyll content', *Sensors*, 11(7), pp. 7063–7081. doi: 10.3390/s110707063.

Donato, D. C., et al. (2011) 'Mangroves among the most carbon-rich forests in the tropics', *Nature Geoscience*, 4(5), pp. 293–297. doi: 10.1038/ngeo1123. Nature Publishing Group.

Drake, J. B., et al. (2002) 'Sensitivity of large-footprint lidar to canopy structure and biomass in a neotropical rainforest', *Remote Sensing of Environment*, 81(2–3), pp. 378–392. doi: 10.1016/S0034-4257(02)00013-5.

Durbha, S. S., King, R. L. and Younan, N. H. (2007) 'Support vector machines regression for retrieval of leaf area index from multiangle imaging spectroradiometer', *Remote Sensing of Environment*, 107(1–2), pp. 348–361. doi: 10.1016/j.rse.2006.09.031.

Ferraz, A., et al. (2016) 'Lidar detection of individual tree size in tropical forests', *Remote Sensing of Environment*, 183, pp. 318–333. doi: 10.1016/j.rse.2016.05.028. Elsevier Inc.

Flores, A., et al. (2019) 'SAR Handbook: Comprehensive Methodologies for Forest Monitoring and Biomass Estimation'. pp. 1–307. doi: 10.25966/nr2c-s697.

Garestier, F., Dubois-Fernandez, P. C. and Papathanassiou, K. P. (2008) 'Pine forest height inversion using single-pass X-band PolInSAR data', *IEEE Transactions on Geoscience and Remote Sensing*, 46(1), pp. 59–68. doi: 10.1109/TGRS.2007.907602.

Geladi, P. and Kowalski, B. R. (1986) 'Partial least-squares regression: a tutorial', *Analytica Chimica Acta*, 185, pp. 1–17. doi: 10.1016/0003-2670(86)80028-9.

Giri, C., et al. (2011) 'Status and distribution of mangrove forests of the world using earth observation satellite data', *Global Ecology and Biogeography*, 20(1), pp. 154–159. doi: 10.1111/j.1466-8238.2010.00584.x.

Giuliani, R., Magnanini, E. and Flore, J. A. (2001) 'potential use of infrared thermometry for the detection of water deficit in apple and peach orchards', *Acta Horticulturae*, 557, pp. 399–406. doi: 10.17660/ActaHortic.2001.557.53.

Goodenough, D. G. and Bannon, D. (2014) 'Hyperspectral forest monitoring and imaging implications', in Bannon, D. P. (ed.), p. 910402. doi:10.1117/12.2057637.

Gupta, K., et al. (2018) 'An index for discrimination of mangroves from non-mangroves using LANDSAT 8 OLI imagery', *MethodsX*, 5, pp. 1129–1139. doi: 10.1016/j.mex.2018.09.011.

Hayes, D. J. and Sader, S. A. (2001) 'Comparison of change-detection techniques for monitoring tropical forest clearing and vegetation regrowth in a time series', *Photogrammetric Engineering and Remote Sensing*, 67(9), pp. 1067–1075.

Healey, S. P., et al. (2006) 'Application of two regression-based methods to estimate the effects of partial harvest on forest structure using Landsat data', *Remote Sensing of Environment*, 101(1), pp. 115–126. doi: 10.1016/j.rse.2005.12.006.

Holmgren, J. (2004) 'Prediction of tree height, basal area and stem volume in forest stands using airborne laser scanning', *Scandinavian Journal of Forest Research*, 19(6), pp. 543–553. doi: 10.1080/02827580410019472.

Homolová, L., et al. (2013) 'Review of optical-based remote sensing for plant trait mapping', *Ecological Complexity*, 15, pp. 1–16. doi: 10.1016/j.ecocom.2013.06.003.

Hong, S. H., et al. (2015) 'Evaluation of polarimetric SAR decomposition for classifying wetland vegetation types', *Remote Sensing*, 7(7), pp. 8563–8585. doi: 10.3390/rs70708563.

Hu, Q., et al. (2020) 'Evaluation of global decametric-resolution LAI, FAPAR and FVC estimates derived from sentinel-2 imagery', *Remote Sensing*, 12(6). doi: 10.3390/rs12060912.

Huete, A. (1988) 'A soil-adjusted vegetation index (SAVI)', *Remote Sensing of Environment*, 25(3), pp. 295–309. doi: 10.1016/0034-4257(88)90106-X.

Jacquemoud, S., et al. (1995) 'Extraction of vegetation biophysical parameters by inversion of the PROSPECT + SAIL models on sugar beet canopy reflectance data. Application to TM and AVIRIS sensors', *Remote Sensing of Environment*, 52(3), pp. 163–172. doi: 10.1016/0034-4257(95)00018-V.

Jensen, J. R. (1981) 'Urban change detection mapping using landsat digital data', *The American Cartographer*, 8(2), pp. 127–147. doi: 10.1559/152304081784447318.

Jerath, M. (2012) *An Economic Analysis of Carbon Sequestration and Storage Service by Mangrove Forests in Everglades National Park, Florida.* Florida International University. doi: 10.25148/etd.FI12080626.

Jia, M., et al. (2019) 'A new vegetation index to detect periodically submerged mangrove forest using single-tide Sentinel-2 imagery', *Remote Sensing*, 11(17), pp. 1–17. doi: 10.3390/rs11172043.

Jordan, C. F. (1969) 'Derivation of Leaf-Area Index from Quality of Light on the Forest Floor', *Ecology*, 50(4), pp. 663–666. doi: 10.2307/1936256.

Joshi, N., et al. (2016) 'A review of the application of optical and radar remote sensing data fusion to land use mapping and monitoring', *Remote Sensing*, 8(1), p. 70. doi: 10.3390/rs8010070.

Kalacska, M., et al. (2005) 'Estimating leaf area index from satellite imagery using Bayesian networks', *IEEE Transactions on Geoscience and Remote Sensing*, 43(8), pp. 1866–1873. doi: 10.1109/TGRS.2005.848412.

Kaufman, Y. J. and Tanre, D. (1992) 'Atmospherically resistant vegetation index (ARVI) for EOS-MODIS', *IEEE Transactions on Geoscience and Remote Sensing*, 30(2), pp. 261–270. doi: 10.1109/36.134076.

Kumar, L. and Mutanga, O. (2017) 'Remote sensing of above-ground biomass', *Remote Sensing*, 9(9), p. 935. doi: 10.3390/rs9090935.

Lagomasino, D., et al. (2015) 'High-resolution forest canopy height estimation in an African blue carbon ecosystem', *Remote Sensing in Ecology and Conservation*, 1(1), pp. 51–60. doi: 10.1002/rse2.3.

Latifi, H., et al. (2015) 'Stratified aboveground forest biomass estimation by remote sensing data', *International Journal of Applied Earth Observation and Geoinformation*, 38, pp. 229–241. doi: 10.1016/j.jag.2015.01.016.

Latifi, H., Nothdurft, A. and Koch, B. (2010) 'Non-parametric prediction and mapping of standing timber volume and biomass in a temperate forest: application of multiple optical/LiDAR-derived predictors', *Forestry*, 83(4), pp. 395–407. doi: 10.1093/forestry/cpq022.

Lausch, A., et al. (2016) 'Linking Earth Observation and taxonomic, structural and functional biodiversity: Local to ecosystem perspectives', *Ecological Indicators*, 70, pp. 317–339. doi: 10.1016/j.ecolind.2016.06.022.

Lausch, A., et al. (2016) 'Understanding forest health with remote sensing-Part I-A review of spectral traits, processes and remote-sensing characteristics', *Remote Sensing*, 8(12), pp. 1–44. doi: 10.3390/rs8121029.

Lausch, A., et al. (2017) 'Understanding forest health with Remote sensing-Part II-A review of approaches and data models', *Remote Sensing*, 9(2), pp. 1–33. doi: 10.3390/rs9020129.

Lausch, A., et al. (2018) 'Understanding forest health with remote sensing, Part III: Requirements for a scalable multi-source forest health monitoring network based on data science approaches', *Remote Sensing*, 10(7). doi: 10.3390/rs10071120.

Lee, C. M., et al. (2015) 'An introduction to the NASA Hyperspectral InfraRed Imager (HyspIRI) mission and preparatory activities', *Remote Sensing of Environment*, 167, pp. 6–19. doi: 10.1016/j.rse.2015.06.012.

Lefsky, M. A. (2010) 'A global forest canopy height map from the moderate resolution imaging spectroradiometer and the geoscience laser altimeter system', *Geophysical Research Letters*, 37(15), pp. 1–5. doi: 10.1029/2010GL043622.

Li, C., Song, J. and Wang, J. (2015) 'Modifying geometric-optical bidirectional reflectance model for direct inversion of forest canopy leaf area index', *Remote Sensing*, 7(9), pp. 11083–11104. doi: 10.3390/rs70911083.

Liang, S. (2007) 'Recent developments in estimating land surface biogeophysical variables from optical remote sensing', *Progress in Physical Geography: Earth and Environment*, 31(5), pp. 501–516. doi: 10.1177/0309133307084626.

Liu, H. Q. and Huete, A. (1995) 'A feedback based modification of the NDVI to minimize canopy background and atmospheric noise', *IEEE Transactions on Geoscience and Remote Sensing*, 33(2), pp. 457–465. doi: 10.1109/TGRS.1995.8746027.

Lloret, J., et al. (2009) 'A wireless sensor network deployment for rural and forest fire detection and verification', *Sensors*, 9(11), pp. 8722–8747. doi: 10.3390/s91108722.

Lyon, J. G., et al. (1998) 'A change detection experiment using vegetation indices', *Photogrammetric Engineering and Remote Sensing*, 64(2), pp. 143–150.

Maltamo, M., Næsset, E. and Vauhkonen, J. (eds) (2014) *Forestry Applications of Airborne Laser Scanning*. Dordrecht: Springer Netherlands (Managing Forest Ecosystems). doi: 10.1007/978-94-017-8663-8.

Manninen, T., et al. (2013) 'Leaf area index estimation of boreal and subarctic forests using VV/HH ENVISAT/ASAR data of various swaths', *IEEE Transactions on Geoscience and Remote Sensing*, 51(7), pp. 3899–3909. doi: 10.1109/TGRS.2012.2227327.

Manogaran, G. and Lopez, D. (2018) 'Spatial cumulative sum algorithm with big data analytics for climate change detection', *Computers and Electrical Engineering*, 65, pp. 207–221. doi: 10.1016/j.compeleceng.2017.04.006. Elsevier Ltd.

Meyer, V., et al. (2013) 'Detecting tropical forest biomass dynamics from repeated airborne lidar measurements', *Biogeosciences*, 10(8), pp. 5421–5438. doi: 10.5194/bg-10-5421-2013.

Morsdorf, F., et al. (2004) 'LIDAR-based geometric reconstruction of boreal type forest stands at single tree level for forest and wildland fire management', *Remote Sensing of Environment*, 92(3), pp. 353–362. doi: 10.1016/j.rse.2004.05.013.

Morsdorf, F., et al. (2006) 'Estimation of LAI and fractional cover from small footprint airborne laser scanning data based on gap fraction', *Remote Sensing of Environment*, 104(1), pp. 50–61. doi: 10.1016/j.rse.2006.04.019.

Mougin, E., et al. (1999) 'Multifrequency and multipolarization radar backscattering from mangrove forests', *IEEE Transactions on Geoscience and Remote Sensing*, 37(1 PART 1), pp. 94–102. doi:10.1109/36.739128.

Müller, J. (2009) 'Forestry and water budget of the lowlands in northeast Germany — consequences for the choice of tree species and for forest management', *Journal of Water and Land Development*, 13a(1), pp. 133–148. doi: 10.2478/v10025-010-0024-7.

Mutanga, O., Adam, E. and Cho, M. A. (2012) 'High density biomass estimation for wetland vegetation using WorldView-2 imagery and random forest regression algorithm', *International Journal of Applied Earth Observation and Geoinformation*, 18, pp. 399–406. doi: 10.1016/j.jag.2012.03.012.

Næsset, E. (2002) 'Predicting forest stand characteristics with airborne scanning laser using a practical two-stage procedure and field data', *Remote Sensing of Environment*, 80(1), pp. 88–99. doi: 10.1016/S0034-4257(01)00290-5.

Næsset, E., et al. (2011) 'Model-assisted regional forest biomass estimation using LiDAR and InSAR as auxiliary data: A case study from a boreal forest area', *Remote Sensing of Environment*, 115(12), pp. 3599–3614. doi: 10.1016/j.rse.2011.08.021. Elsevier Inc.

Nascimento, W. R. et al. (2013) 'Mapping changes in the largest continuous Amazonian mangrove belt using object-based classification of multisensor satellite imagery', *Estuarine, Coastal and Shelf Science*, 117, pp. 83–93. doi: 10.1016/j.ecss.2012.10.005.

Nelson, R., et al. (2009) 'Estimating Siberian timber volume using MODIS and ICESat/GLAS', *Remote Sensing of Environment*, 113(3), pp. 691–701. doi: 10.1016/j.rse.2008.11.010. Elsevier B.V.

Neukermans, G. et al. (2008) 'Mangrove species and stand mapping in gazi bay (kenya) using quickbird satellite imagery', *Journal of Spatial Science*, 53(1), pp. 75–86. doi: 10.1080/14498596.2008.9635137.

Nielsen, A. A. (2007) 'The regularized iteratively reweighted MAD method for change detection in multi- and hyperspectral data', *IEEE Transactions on Image Processing*, 16(2), pp. 463–478. doi: 10.1109/TIP.2006.888195.

Nilsson, M. (1996) 'Estimation of tree heights and stand volume using an airborne lidar system', *Remote Sensing of Environment*, 56(1), pp. 1–7. doi: 10.1016/0034-4257(95)00224-3.

Njoku, E. G. (ed.) (2014) *Encyclopedia of Remote Sensing*. New York: Springer (Encyclopedia of Earth Sciences Series). doi: 10.1007/978-0-387-36699-9.

da Motta Paca, V. H., et al. (2019) 'The spatial variability of actual evapotranspiration across the Amazon River Basin based on remote sensing products validated with flux towers', *Ecological Processes*, 8(1), p. 6. doi: 10.1186/s13717-019-0158-8.

Papathanassiou, K. P. and Cloude, S. R. (2001) 'Single-baseline polarimetric SAR interferometry', *IEEE Transactions on Geoscience and Remote Sensing*, 39(11), pp. 2352–2363.

Pause, M., et al. (2016) 'In Situ/Remote Sensing Integration to Assess Forest Health—A Review', *Remote Sensing*, 8(6), p. 471. doi: 10.3390/rs8060471.

Pendleton, L., et al. (2012) 'Estimating global "Blue Carbon" emissions from conversion and degradation of vegetated coastal ecosystems', *PLoS ONE*, 7(9). doi: 10.1371/journal.pone.0043542.

Pesonen, A., et al. (2008) 'Airborne laser scanning-based prediction of coarse woody debris volumes in a conservation area', *Forest Ecology and Management*, 255(8–9), pp. 3288–3296. doi: 10.1016/j.foreco.2008.02.017.

Pham, T. D., et al. (2019) 'Remote sensing approaches for monitoring mangrove species, structure, and biomass: Opportunities and challenges', *Remote Sensing*, 11(3), pp. 1–24. doi: 10.3390/rs11030230.

Praks, J., et al. (2007) 'Height estimation of boreal forest: Interferometric model-based inversion at L- and X-Band Versus HUTSCAT profiling scatterometer', *IEEE Geoscience and Remote Sensing Letters*, 4(3), pp. 466–470. doi: 10.1109/LGRS.2007.898083.

Proisy, C., et al. (2002) 'On the influence of canopy structure on the radar backscattering of mangrove forests', *International Journal of Remote Sensing*, 23(20), pp. 4197–4210. doi: 10.1080/01431160110107725.

Rautiainen, M., et al. (2003) 'Application of a forest reflectance model in estimating leaf area index of Scots pine stands using Landsat-7 ETM reflectance data', *Canadian Journal of Remote Sensing*, 29(3), pp. 314–323. doi: 10.5589/m03-002.

Reiche, J., et al. (2015) 'Fusing Landsat and SAR time series to detect deforestation in the tropics', *Remote Sensing of Environment*, 156, pp. 276–293. doi: 10.1016/j.rse.2014.10.001. Elsevier Inc.

Rignot, E., Salas, W. A. and Skole, D. L. (1997) 'Mapping deforestation and secondary growth in Rondonia, Brazil, using imaging radar and thematic mapper data', *Remote Sensing of Environment*, 59(2), pp. 167–179. doi: 10.1016/S0034-4257(96)00150-2.

Rouse, J. W., et al. (1974) 'Monitoring vegetation systems in the Great Plains with ERTS', in *Proceedings of the Third Earth Resources Technology Satellite-1 Symposium*. Greenbelt: NASA, pp. 301–317.

Saatchi, S. (2019) 'SAR methods for mapping and monitoring forest biomass', in *SAR Handbook: Comprehensive Methodologies for Forest Monitoring and Biomass Estimation*. Berlin, Boston: NASA. doi: 10.25966/hbm1-ej07.

Saatchi, S. S., et al. (2011) 'Benchmark map of forest carbon stocks in tropical regions across three continents', *Proceedings of the National Academy of Sciences*, 108(24), pp. 9899–9904. doi: 10.1073/pnas.1019576108.

Schmidtlein, S. (2005) 'Imaging spectroscopy as a tool for mapping Ellenberg indicator values', *Journal of Applied Ecology*, 42(5), pp. 966–974. doi: 10.1111/j.1365-2664.2005.01064.x.

Shimu, S. A., et al. (2019) 'NDVI based change detection in sundarban mangrove forest using remote sensing data', *2019 4th International Conference on Electrical Information and Communication Technology, EICT 2019*. IEEE, (December), pp. 1–5. doi: 10.1109/EICT48899.2019.9068819.

Simard, M., et al. (2006) 'Mapping height and biomass of mangrove forests in Everglades National Park with SRTM elevation data', *Photogrammetric Engineering and Remote Sensing*, 72(3), pp. 299–311. doi: 10.14358/PERS.72.3.299.

Singh, A. (1989) 'Review Articlel: Digital change detection techniques using remotely-sensed data', *International Journal of Remote Sensing*, 10(6), pp. 989–1003. doi: 10.1080/01431168908903939.

Small, D. (2011) 'Flattening gamma: Radiometric terrain correction for SAR imagery', *IEEE Transactions on Geoscience and Remote Sensing*, 49(8), pp. 3081–3093. doi: 10.1109/TGRS.2011.2120616.

Tan, L. and Yang, R. (2008) 'Investigation on tree height retrieval with polarimetric SAR interferometry', in *IGARSS 2008-2008 IEEE International Geoscience and Remote Sensing Symposium*. IEEE, pp. V-546-V-549. doi: 10.1109/IGARSS.2008.4780150.

Thapa, R. B., et al. (2013) 'The tropical forest in south east Asia: Monitoring and scenario modeling using synthetic aperture radar data', *Applied Geography*, 41, pp. 168–178. doi: 10.1016/j.apgeog.2013.04.009.

Thomas, N., et al. (2017) 'Distribution and drivers of global mangrove forest change, 1996–2010', *PLoS ONE*, 12(6), pp. 1–14. doi: 10.1371/journal.pone.0179302.

Treuhaft, R. N. and Siqueira, P. R. (2000) 'Vertical structure of vegetated land surfaces from interferometric and polarimetric radar', *Radio Science*, 35(1), pp. 141–177. doi: 10.1029/1999RS900108.

Ulaby, F. T., et al. (1987) 'Relating polaization phase difference of SAR signals to scene properties', *IEEE Transactions on Geoscience and Remote Sensing*, GE-25(1), pp. 83–92. doi: 10.1109/TGRS.1987.289784.

Ulaby, F. T. and Dobson, M. C. (1989) *Handbook of Radar Scattering Statistics for Terrain, Artech House*. Norwood, MA: Artech House.

Ustin, S. L. and Gamon, J. A. (2010) 'Remote sensing of plant functional types', *New Phytologist*, 186(4), pp. 795–816. doi: 10.1111/j.1469-8137.2010.03284.x.

Vapnik, V., Golowich, S. E. and Smola, A. J. (1997) 'Support vector method for function approximation, regression estimation and signal processing', in Mozer, M. C., Jordan, M. I., and Petsche, T. (eds.) *Advances in Neural Information Processing Systems 9*. MIT Press, pp. 281–287. Available at: http://papers.nips.cc/paper/1187-support-vector-method-for-function-approximation-regression-estimation-and-signal-processing.pdf.

Varghese, A. O., Suryavanshi, A. and Joshi, A. K. (2016) 'Analysis of different polarimetric target decomposition methods in forest density classification using C band SAR data', *International Journal of Remote Sensing*, 37(3), pp. 694–709. doi: 10.1080/01431161.2015.1136448.

Verrelst, J., et al. (2012) 'Machine learning regression algorithms for biophysical parameter retrieval: Opportunities for Sentinel-2 and -3', *Remote Sensing of Environment*, 118, pp. 127–139. doi: 10.1016/j.rse.2011.11.002.

Verrelst, J., et al. (2015) 'Optical remote sensing and the retrieval of terrestrial vegetation bio-geophysical properties - A review', *ISPRS Journal of Photogrammetry and Remote Sensing*, pp. 273–290. doi: 10.1016/j.isprsjprs.2015.05.005. International Society for Photogrammetry and Remote Sensing, Inc. (ISPRS).

Vogelmann, J. E., Tolk, B. and Zhu, Z. (2009) 'Monitoring forest changes in the southwestern United States using multitemporal Landsat data', *Remote Sensing of Environment*, 113(8), pp. 1739–1748. doi: 10.1016/j.rse.2009.04.014.

Wan, L., et al. (2019) 'A small-patched convolutional neural network for mangrove mapping at species level using high-resolution remote-sensing image', *Annals of GIS*, 25(1), pp. 45–55. doi: 10.1080/19475683.2018.1564791. Taylor & Francis.

Wang, L., Silván-Cárdenas, J. L. and Sousa, W. P. (2008) 'Neural network classification of mangrove species from multi-seasonal Ikonos imagery', *Photogrammetric Engineering and Remote Sensing*, 74(7), pp. 921–927. doi: 10.14358/PERS.74.7.921.

Wang, L., Sousa, W. P. and Gong, P. (2004) 'Integration of object-based and pixel-based classification for mapping mangroves with IKONOS imagery', *International Journal of Remote Sensing*, 25(24), pp. 5655–5668. doi: 10.1080/014311602331291215.

Wehr, A. and Lohr, U. (1999) 'Airborne laser scanning—an introduction and overview', *ISPRS Journal of Photogrammetry and Remote Sensing*, 54(2–3), pp. 68–82. doi: 10.1016/S0924-2716(99)00011-8.

White, J. C., et al. (2016) 'Remote sensing technologies for enhancing forest inventories: A review', *Canadian Journal of Remote Sensing*, 42(5), pp. 619–641. doi: 10.1080/07038992.2016.1207484.

Wulder, M. A., et al. (2006) 'Surveying mountain pine beetle damage of forests: A review of remote sensing opportunities', *Forest Ecology and Management*, 221(1–3), pp. 27–41. doi: 10.1016/j.foreco.2005.09.021.

Yao, W., Krzystek, P. and Heurich, M. (2012) 'Tree species classification and estimation of stem volume and DBH based on single tree extraction by exploiting airborne full-waveform LiDAR data', *Remote Sensing of Environment*, 123, pp. 368–380. doi: 10.1016/j.rse.2012.03.027.

Yu, Y. and Saatchi, S. (2016) 'Sensitivity of L-Band SAR backscatter to aboveground biomass of global forests', *Remote Sensing*, 8(6), p. 522. doi: 10.3390/rs8060522.

Zarco-Tejada, P. J., et al. (2001) 'Scaling-up and model inversion methods with narrowband optical indices for chlorophyll content estimation in closed forest canopies with hyperspectral data', *IEEE Transactions on Geoscience and Remote Sensing*, 39(7), pp. 1491–1507. doi: 10.1109/36.934080.

Zhang, X. and Tian, Q. (2013) 'A mangrove recognition index for remote sensing of mangrove forest from space', *Current Science*, 105, pp. 1149–1154.

Zimble, D. A., et al. (2003) 'Characterizing vertical forest structure using small-footprint airborne LiDAR', *Remote Sensing of Environment*, 87(2–3), pp. 171–182. doi: 10.1016/S0034-4257(03)00139-1.

2 Does Mid-Resolution Landsat Data Provide Sufficient Accuracy for Image Classification and Mapping at Species Level? A Case of the Bangladesh Sundarbans, Bay of Bengal

Manoj Kumer Ghosh
University of Rajshahi

Lalit Kumar
University of New England

CONTENTS

2.1 INTRODUCTION

Mangrove forests are one of the most dominant and productive ecosystems of the coastal environment (Kirui et al. 2013) and are considered as an important resource for their ability to store and sequester carbon (Suratman 2008, Donato et al. 2011, Kirui et al. 2013). The Sundarbans mangrove forests occupy the landscape of both Bangladesh and India, encompassing an area of almost one million hectares, and are the largest contiguous mangrove forest in the world (Ghosh et al. 2015, Rahman and Asaduzzaman 2013, Spalding 2010). In the year 1992, the Sundarbans was designated as a Ramsar site under the Ramsar Convention while UNESCO declared it as a World Heritage Site in the year 1997 (Siddiqi 2001), citing its rich biological diversities and conservation values.

Mangrove forests play a crucial role in the maintenance of the local coastal ecosystem and balancing the global environment (Biswas et al. 2007) by providing a range of renewable resources such as sources of food and medicine. The landscape is home to a diverse range of biota (Giri et al. 2007) and is a reliable source of forest and non-wood forest products (NWFPs) for the local people in the vicinity (Rahman 2000). For instance, the endangered Royal Bengal Tiger (*Panthera tigris*) along with other fauna such as *Batagur baska, Pelochelys bibroni,* and *Chelonia mydas* are found in the Sundarbans. The landscape is ecologically endangered, highly populated, fragile but economically valuable (Danda 2007). Approximately 3 million people living in the villages surrounding the Sundarbans depend on mangrove forest resources for sustaining their livelihood and economic activities (Giri et al. 2007). An increase in the use of such forest resources in the Sundarbans, however, could have detrimental effects on mangrove biodiversity where climate change is further exacerbating the extent of such impacts (Danda 2007). Cost- and time-effective mapping with acceptable accuracy levels is vital to monitor the state of such valuable resources in the Sundarbans.

Remote sensing techniques have the potential for identifying and mapping mangrove species composition effectively and at a low cost. The data from different satellite sensors that have been used in the identification of mangrove forests include imagery from Landsat Thematic Mapper (TM)/Enhanced Thematic Mapper (Giri et al. 2014, Long and Skewes 1996), SPOT (Franklin 1993, Pasqualini et al. 1999), China Brazil Earth Resource Satellite (CBERS) (Li, Wang, and Jiang 2003), Space-borne Imaging Radar (SIR) (Pasqualini et al. 1999), Advanced Space-borne Thermal Emission and Reflection Radiometer (ASTER) (Vaiphasa, Skidmore, and de Boer 2006), IKONOS and Quick Bird (Wang et al. 2004), and Compact Airborne Spectrographic Imager (CASI) (Green et al. 1998). Among them, IKONOS, Quick Bird and CASI are the high-resolution sensors while Landsat, SPOT, ASTER, and CBERS are the mid- and low-resolution sensors. According to Wang, Sousa, and Gong (2004) the application of Landsat and SPOT data in mapping mangrove species has produced mixed results. They also argued that spectral resolution is less effective in discriminating different mangrove species than spatial resolution. It has also been reported that air-borne sensors like CASI are more successful in accurate discrimination among mangrove species than conventional satellite data (Green et al. 1998). However, acquiring most of the mid- and high-resolution satellite images apart from Landsat data will result in substantial costs and is difficult for researchers and managers in developing countries, especially where large areas need to be mapped.

From the assessed literature, it appears that using public domain medium-resolution image data, such as Landsat TM, ASTER, SPOT, etc., the identification of different mangrove species remains a challenging task. In many less developed countries, the high cost of purchasing fine resolution satellite imagery prohibits its extensive use (Kirui et al. 2013). The use of freely available open source satellite image, such as Landsat, is the option for those who do not have financial access to purchase fine resolution images. However, the accuracy is always a concern in feature identification and subsequent thematic mapping in remote sensing. Therefore, taking Sundarbans as a case study, this work aims to (i) evaluate if mid-resolution Landsat satellite image combined with traditional classification algorithms produces an acceptable accuracy at the species level, and (ii) identify and map major mangrove cover at the species level. We anticipate that demonstration of the use of Landsat satellite data to successfully map mangrove cover at species level with an acceptable accuracy will motivate other researchers and relevant stakeholders to develop a reliable time-series database that can be used in the continuous forest monitoring activities.

2.2 METHOD

2.2.1 STUDY AREA

The present work was conducted in the Bangladesh territory Sundarbans mangrove forest. This mangrove forest is located on the Ganges Delta that is created by the three mighty rivers Ganges, Brahmaputra, and Meghna in the Bay of Bengal (Giri et al. 2007, Ghosh, Kumar, and Roy 2017). It extends approximately between 21°32′N to 22°30′N latitude and 89°00′E to 89°51′E longitude (Figure 2.1) and covers an area of around 6017 km^2 (Sarker et al. 2016) which is approximately three-fifths of the entire Sundarbans that is located in the Bay of Bengal. This forest comprises several mudflats and islands created because of the sedimentation process from the Ganges and its tributaries. This study site is characterized by a tropical climate with a dry season (November to April) and a monsoonal period (May to October) (Ghosh et al. 2015). The annual average precipitation is 192 cm, of which 75% occurs between June and September; the annual maximum temperature is 35°C and average humidity is 82% (Islam et al. 2014). The elevation within the Sundarbans varies from 0.9 to 2.11 m above mean sea level (Rahman 2000). The amplitude of the tide within the estuaries is between 3.5 and 4 m, but varies seasonally between 1 and 6 m (Ghosh et al. 2015, Islam et al. 2014).

2.2.2 IMAGE USED

Landsat image of the entire Sundarbans was acquired from the US Geological Survey, Center for Earth Resources Observation and Science (EROS) website (www.glovis. usgs.gov). In this study, Landsat Operational Land Imager (OLI) was applied to map mangrove cover at the species level in the Sundarbans. The OLI image has 11 spectral bands that cover the visible, near-infrared, short-wave infrared, and thermal infrared regions of the electromagnetic spectrum (Table 2.1).

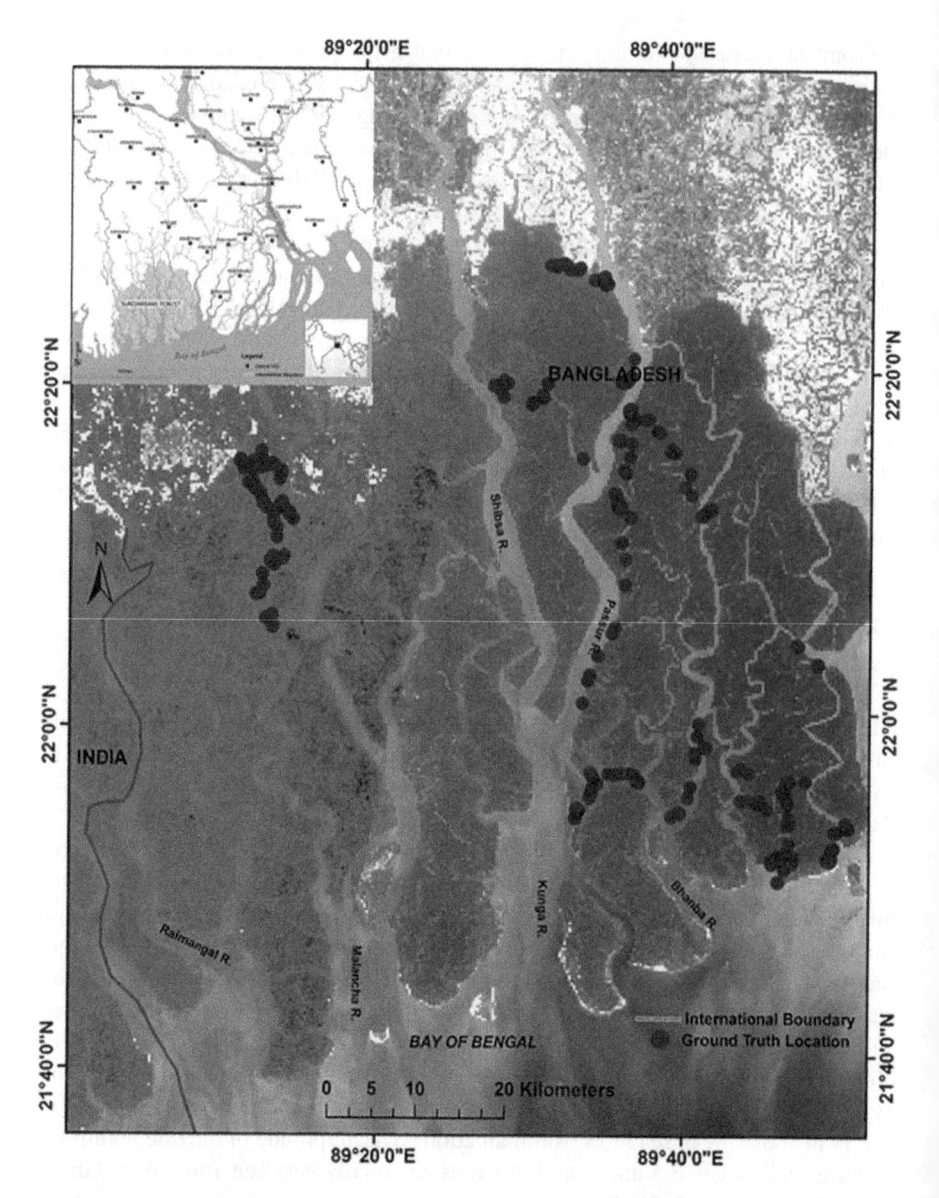

FIGURE 2.1 Study area in false color composite image showing ground truth points in different localities of the Sundarbans.

2.2.3 IMAGE PRE-PROCESSING

First, radiometric corrections were carried out using sun azimuth and sun elevation data extracted from the image's header file to remove the influence of the atmosphere, and then image registration was carried out using 51 ground control points (GCPs) with a root mean square error (RMSE) of 0.004 pixels. GCPs used for the registration were collected from prominent geographic features and river channel

TABLE 2.1
Information of Landsat Data Used in This Study

Sensor	Path/Row	Resolution	Acquisition Date	Band	Wavelength (in µm)
OLI	137–138/45	30 M	4th February 2015	Coastal/Aerosol	0.435–0.451
				Blue	0.452–0.512
				Green	0.533–0.590
				Red	0.636–0.673
				Near infrared	0.851–0.879
				Short-wave infrared-1	1.566–1.651
				Thermal infrared-1	10.60–11.19
				Thermal infrared-2	11.50–12.51
				Short-wave infrared-2	2.107–2.294
				Cirrus	1.363–1.384

junctions. The image frames were then mosaicked to get the study site as it is covered by two image frames. Afterward, the mangrove areas were extracted from the images using the on-screen digitization process as they had a clear boundary created by river channels. Thereafter water areas were masked from the images so the results would not be affected by water turbidity in the classified image.

2.2.4 IMAGE ANALYSIS

In the field of remote sensing, the maximum likelihood classification (MLC) technique is considered one of the most effective methods (Green et al. 1998, Gao 1999, Bischof, Schneider, and Pinz 1992) used for supervised classification. The algorithm is used for computing the weighted distance or likelihood of unknown measurement vectors belonging to one of the known classes that are based on the Bayesian equation (Giri et al. 2014). In the MLC algorithm, spectral information in each class meets the normal distribution criteria (Bischof, Schneider, and Pinz 1992) and also is a very effective method in classifying mangroves with traditional remote sensing data. Therefore, this method was adopted in this study to identify mangrove species composition using medium resolution Landsat data.

One of the authors (M.K. Ghosh) has made several visits to the Sundarbans in the last several years and mapped different mangrove species with GPS. Based on this extensive field experience and the information obtained from previously published literature (Giri et al. 2007, Chaffey, Miller, and Sandom 1985, Emch and Peterson 2006, Treygo and Dean 1989) and species level map of the Sundarbans, five major mangrove species out of 19 known species (Sarker et al. 2016) were chosen for the classification, viz., *Heritiera fomes, Excoecaria agallocha, Ceriops decandra, Sonneratia apetala,* and *Xylocarpus mekongensis.* The majority of the remaining species are distributed in lower density, spreading over wide areas in small patches, and do not form large enough mono-specific patches for training site selection and therefore cannot be detected through medium-resolution satellite data. Hence, we purposively mapped five major mangrove species of the Sundarbans that constitute higher density cover and other remaining species were classified under the major five mangrove species based on dominance. Training sites were selected for those five species visually on the satellite image prior to the fieldwork. Afterward, detailed fieldwork for training and validation sampling of each representative mangrove species was completed during February and March 2016. Based on representation and distribution, training samples between 53 and 144 for each mangrove species were selected randomly and stratified by land cover class. Care was taken to ensure the inclusion of all the targeted species in sample selection, with samples well distributed and fulfiling the minimum number required to validate the accuracy evaluation process (Congalton and Green 2008, Sinha et al. 2014). The training and validation sample sets for different mangrove species were collected from the field using handheld GPS. Therefore, collected GPS locations of various mangrove species were overlaid on the 2015 satellite image to extract the spectral signatures for each of the species. The absorption and peak point of reflectance characteristics were identified from the spectral profile which defined the identification characteristics of the species. Finally, those characteristics were applied for image classification. Spectral profiles of those identified mangrove species are shown in Figure 2.2. The spectral reflectance of all mangrove species showed very similar reflectance patterns in the lower wavelengths of the spectrum. In contrast, in the higher wavelengths, reflectance patterns were different from each other. Figure 2.3 shows the workflow of different stages for image processing and classification. The whole image processing was carried out using ENVI 5.1 software (Exelis Visual Information Solutions, Inc., Boulder, UT, USA).

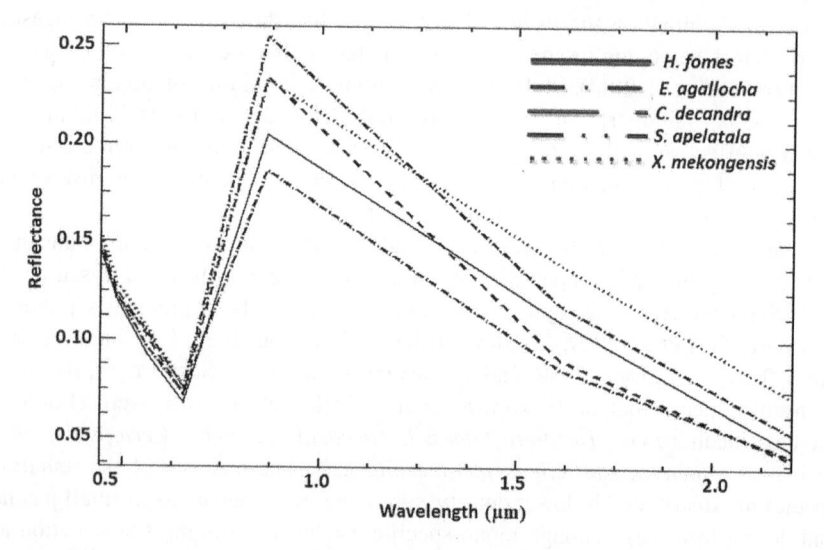

FIGURE 2.2   Spectral profiles of the five mangrove species.

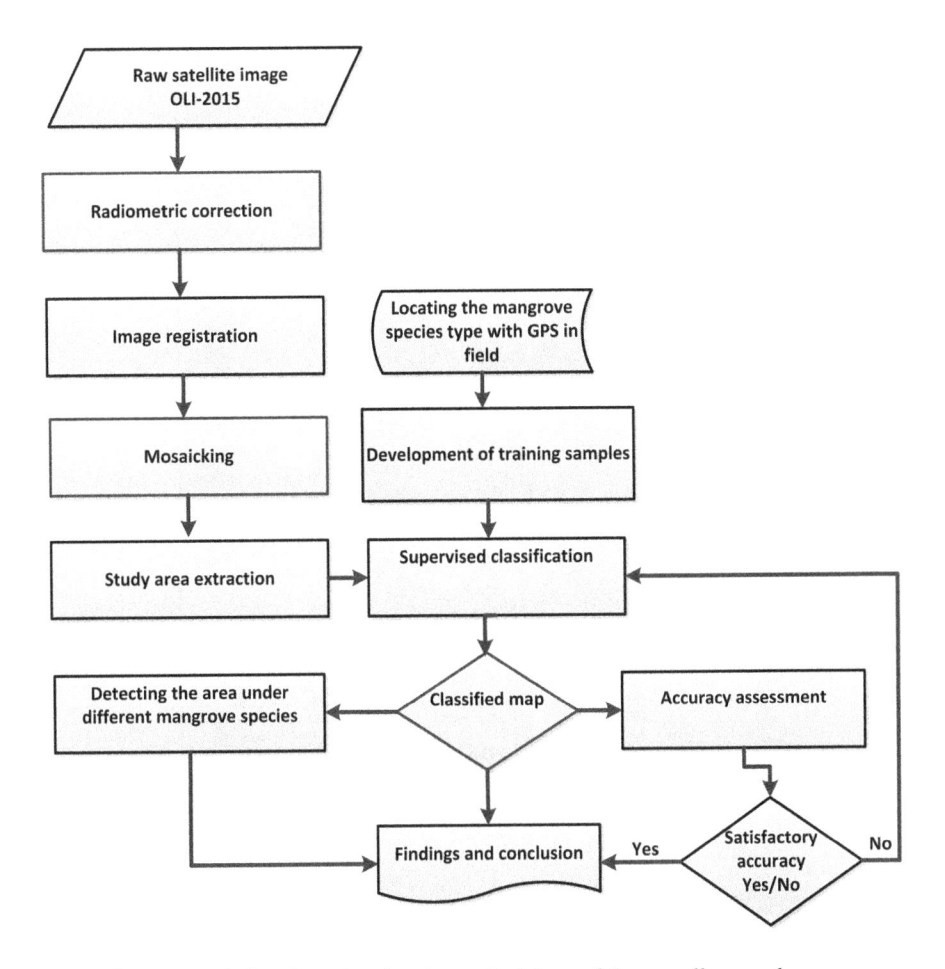

FIGURE 2.3 Work flowchart showing the methodology of the overall research.

2.2.5 CLASS ACCURACY

Validation or accuracy assessment of the classification is an important step in any image classification process (Sinha et al. 2014). To evaluate the efficacy of the classification, a confusion matrix was developed through available ground truth data. Confusion matrix provides information on the number of pixels that are classified according to the reference data set and the number of pixels that are classified otherwise (Bhowmik and Cabral 2013). Three different primary accuracies, that is, overall accuracy, producer's accuracy, user's accuracy, and the kappa coefficient were calculated to quantify the assessment of satellite image analysis.

2.3 RESULTS

This study identified and mapped five mangrove species composition, viz., *H. fomes,* *E. agallocha, C. decandra, S. apetala,* and *X. mekongensis* and their distribution at the species level in the Sundarbans based on Landsat satellite image classification

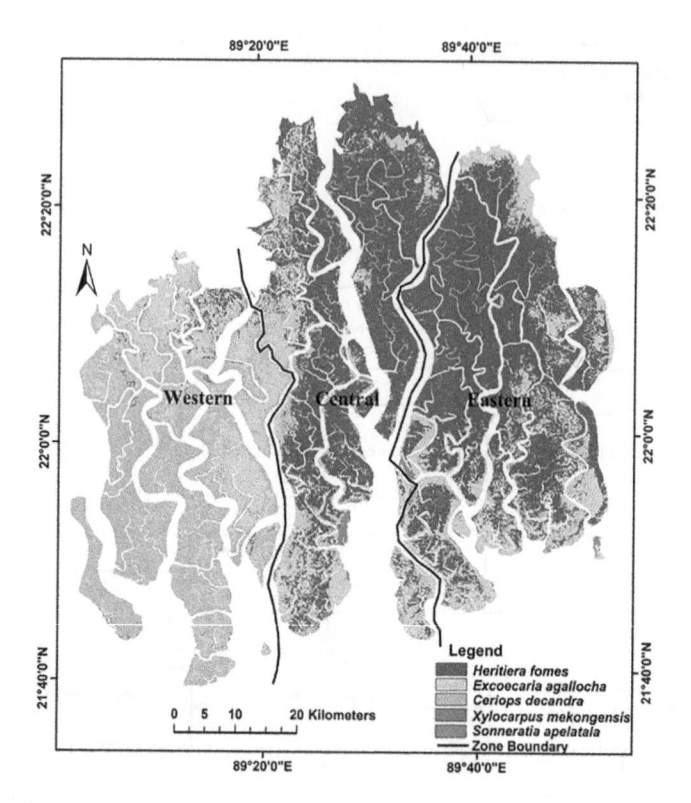

FIGURE 2.4 Classified image showing major mangrove species composition in the Bangladesh Sundarbans for the year 2015.

(Figure 2.4). Error matrix of the image classification is shown in Table 2.2. The results indicate that the Landsat classification for mangrove species composition for the year 2015 had an overall accuracy of 89.10% and a kappa coefficient of 0.87. The results also revealed that the lowest producer's and user's accuracy were obtained for *X. mekongensis* (73%) and *E. agallocha* (84%), respectively. In contrast, *H. fomes* and *C. decandra* produced a very high producer's accuracy whereas *X. mekongensis* produced the highest user accuracy. The producer's accuracy for *H. fomes* and *C. decandra* was 94% and 95%, respectively, while the user's accuracy was 94% for *X. mekongensis* (Table 2.2).

Overall accuracy: 89.10% and kappa coefficient: 0.87

Supply of freshwater, salinity level in water, erosion, and sedimentation process, and variation in drainage are the forces that substantially influence the spatial distribution of mangrove species in the Sundarbans (Rahman 2000). The spatial composition and distribution of the five studied mangrove species are presented in Figure 2.4. *H. fomes* is distributed mostly in the central and eastern region of the Sundarbans, especially in the areas, which are frequently flooded and less affected by water salinity. *E. agallocha* is the dominant woody species and is sparsely distributed all over the Sundarbans with dense patches in the northwest and south-central and east regions, where there is highest seasonal variation of salinity levels with relatively

TABLE 2.2
Confusion Matrix for 2015 Landsat OLI Image, Sundarbans

Class	*H. fomes*	*E. agallocha*	*C. decandra*	*S. apetala*	*X. mekongensis*	Total	User's Accuracy (%)
H. fomes	**359**	18	2	8	13	400	90
E. agallocha	21	**546**	17	7	59	650	84
C. decandra	0	12	**476**	5	39	532	89
S. apetala	0	4	2	**220**	12	238	92
X. mekongensis	0	16	2	0	**336**	354	94
Total	380	596	499	240	459	**2174**	
Producer's accuracy (%)	94	92	95	92	73		

Note: Bold shows the number of correctly classified pixels for each class.

longer duration of moderate salinity. They are often mixed with *H. fomes* in the eastern and *C. decandra* in the western parts of the Sundarbans. Dense stands of *C. decandra* are located in the western region of the Sundarbans that is characterized by higher salinity and shares its habitat with *E. agallocha* and *H. fomes*. *S. apetala* is an indicator species for newly accreted mud banks and is an important species for wildlife (Ghosh, Kumar, and Roy 2016). Our analysis showed that *S. apetala* is scantily distributed, mainly in the southeastern part of the eastern region, northern, and southern parts of the central region and northwestern parts of the western region of the forest. Similarly, *X. mekongensis* is distributed mostly in the western region of the study area and some parts of the central region. Overall, *H. fomes*, *E. agallocha*, *C. decandra*, *S. apetala*, and *X. mekongensis* occupy 196,090 (49.5%), 119,304 (30.1%), 70,610 (17.8%), 8,621 (2.2%), and 1,438 (0.4%) of the Sundarbans, respectively. Our results demonstrate that among the five species, *H. fomes* is the most dominant species followed by *E. agallocha*, *C. decandra*, *S. apetala*, and *X. mekongensis*, respectively.

2.4 DISCUSSION AND CONCLUSION

Wetlands, including mangrove forest ecosystems, are some of the most important as well as vulnerable ecosystems in the world (Guo et al. 2017). Mangrove species are also projected to shift its niche under a changing climate. For instance, Record et al. (2013) modeled some major mangrove species at a global scale and found that mangrove species will shift their range by at least 2° of latitude, with Central America and the Caribbean regions to be the highest sufferers. Mukhopadhyay et al. (2015) projected a loss of 17% mangrove cover of Bangladesh by 2105 that could result in a significant decline in freshwater loving species and ecosystem services they provide. This necessitates the protection and conservation of this precious ecosystem but with judicial management without hampering the livelihood rights of the low-income communities that specifically depend on this mangrove forest ecosystem services. For instance, Abdullah et al. (2016) observed a heavy dependence of local people on the forest resources of the

Sundarbans in Bangladesh, where forest income represents almost 74%, 48%, and 23% of the household income for lower-, middle-, and higher-income households.

The quality of thematic maps produced from image classification techniques is evaluated mostly based on the criteria called accuracy assessment, which is historically set at ≥85% (Foody 2008). Liu, Frazier, and Kumar (2007) found the user's accuracy, producer's accuracy, and overall accuracy as consistent among 14 different measures commonly used and suggested them to be used in accuracy assessment of classified imagery in remote sensing work. Many studies are available that achieved satisfactory accuracy level in the mangrove forest discrimination using fine resolution satellite images. Heenkenda et al. (2014) reported an accuracy level of 89% in discriminating five mangrove species at species level using high-resolution (0.5 m) World View 2 image in Northern Australia while Heumann (2011) obtained 94% of accuracy using the same sensor with the application of object-based image analysis and support vector machine in discriminating mangrove species at Isabela Island in the Galapagos Archipelago, Ecuador. Roslani et al. (2014) found an overall accuracy of 87.5% and kappa statistic of 0.85 for delineating mangrove species in Matang Mangrove forest, Malaysia, using RapidEye image of 5-m resolution integrated with NDVI. Koedsin and Vaiphasa (2013) found an overall accuracy of 86% (all spectral band combination), 87% (SFS feature selector) and 92% (genetic search algorithm) for five mangrove species at the species level in Thailand using the EO-1 Hyperion image having 220 spectral bands and 30 m spatial resolution. Gao (1999) reported an overall accuracy of 81.4% for mapping mangrove species using SPOT multispectral data with three bands in Waitemata Harbor of New Zealand. Vo et al. (2013) obtained an overall accuracy of 75% for mapping mangrove stands using SPOT satellite image with an object-based approach at the Mekong Delta in Vietnam. Wang et al. (2004) found an accuracy of 73% and 74% for IKONOS and QuickBird images, respectively, both are very high-resolution images, for mapping the mangroves of the Caribbean Coast of Panama using object-based classification. Neukermans et al. (2008) used high-resolution QuickBird imagery employing fuzzy per pixel classification technique to map and recognize mangrove stands at Gazi Bay of Southern Kenya and obtained an overall accuracy of only 72%.

Our analysis with 30 m resolution Landsat OLI data produced an overall accuracy of 89.10% with a Kappa coefficient of 0.87 (Table 2.2). The accuracy was well over the recommended value >85% as suggested by Foody (2008). This is also highly comparable with the recent work of Giri et al. (2014) who obtained an overall accuracy and Kappa value for the year 1999 and 2010 at 80% and 0.77 and 85.71% and 0.81, respectively, for discriminating mangrove species at Indian Sundarbans using Landsat with Hyperion technique. Our analysis with mid-resolution Landsat data also produced very high user and producer accuracy for all the five identified mangrove species; this indicates the efficacy and reliability of the classification. The user's accuracy is a measure of the reliability of the map. It informs how well the map represents the real picture on the ground. On the other hand, the producer's accuracy measures how well certain areas have been classified (Banko 1998). This present analysis yielded the user's accuracy well over 80% for all the five species and producer's accuracy of more than 90% for all the identified mangrove species except *X. mekongensis* that yielded 73% producer's accuracy. It, therefore, indicates high confidence level in this mapping.

The resultant thematic map (Figure 2.4) prepared with this accuracy level discriminates all the five mangrove species of the Bangladesh Sundarbans with enough accuracy to be used as a baseline data for the future resource planning. The analysis further depicts that almost half of the Sundarbans is currently occupied by *H. fomes*, especially in the central and eastern parts and is in line with the finding of Sarker et al. (2016). Similarly, *E. agallocha* and *C. decandra* occupy 30% and 18% of the total area. *E. agallocha* is sparsely distributed all over the Sundarbans while *C. decandra* are found densely in the western part and in a mixed stand with other species, aligning with the results of Sarker et al. (2016). *S. apetala* and *X. mekongensis* constitute the least coverage in the Bangladesh Sundarbans and is supported by the findings of Sarker et al. (2016). Valderrama-Landeros et al. (2018) obtained an overall accuracy of 64% using the same coarse spatial resolution sensor in the arid regions. Our findings compared to the findings of Valderrama-Landeros et al. (2018) suggest that mangroves in tropical regions are more capable of being discriminated at the species level with mid-resolution space borne data, compared to arid regions where pore-water hyper-salinity causes an increase of stress conditions, and consequently mangrove stands form dense clusters that are not feasible to discriminate. Despite using Landsat image data with conventional image processing techniques, we found an overall acceptable level of accuracy that signifies the applicability of conventional mid-resolution image for the classification and identification of mangrove forests at the species level. Thus, our work clearly shows that even with the mid-resolution images and traditional classification algorithm of conventional Landsat sensor, high accuracy output with finer details at the species level can be achieved. The availability of images such as Landsat at low or even no direct cost to the end user could enable greater use of such satellite imagery for making better informed policy decisions (Hansen et al. 2013), especially for the management of resources such as the mangrove forests in Bangladesh and other developing countries. The high accuracy obtained here gives confidence in the use of such images in regions where high spatial resolution imagery is not available or cannot be afforded.

REFERENCES

Abdullah, A.N.M., Stacey, N., Garnett, S.T. and Myers, B., 2016. Economic dependence on mangrove forest resources for livelihoods in the Sundarbans, Bangladesh. *Forest Policy and Economics*, 64, pp. 15–24.

Banko, G., 1998. *A Review of Assessing the Accuracy of Classifications of Remotely Sensed Data and of Methods Including Remote Sensing Data in Forest Inventory*. International Institution for Applied System Analysis, Luxemberg.

Bhowmik, A.K. and Cabral, P., 2013. Cyclone Sidr impacts on the Sundarbans floristic diversity. *Earth Science Research*, 2(2), p. 62.

Bischof, H., Schneider, W. and Pinz, A.J., 1992. Multispectral classification of Landsat-images using neural networks. *IEEE transactions on Geoscience and Remote Sensing*, 30(3), pp. 482–490.

Biswas, S.R., Choudhury, J.K., Nishat, A. and Rahman, M.M., 2007. Do invasive plants threaten the Sundarbans mangrove forest of Bangladesh? *Forest Ecology and Management*, 245(1–3), pp.1–9.

Chaffey, D.R., Miller, F.R. and Sandom, J.H., 1985. *A forest inventory of the Sundarbans, Bangladesh*. Land Resources Development Centre, Survey, UK.

Congalton, R.G. and Green, K., 2008. *Assessing the Accuracy of Remotely Sensed Data: Principles and Practices*. CRC Press, New York.

Danda, A.A., 2007. Surviving in the Sundarbans: Threats and Responses-An Analytical description of life in an Indian Riparian commons. PhD Thesis, University of Twente, Enschede, The Neherland, pp. 14–23.

Donato, D.C., Kauffman, J.B., Murdiyarso, D., Kurnianto, S., Stidham, M. and Kanninen, M., 2011. Mangroves among the most carbon-rich forests in the tropics. *Nature Geoscience*, 4(5), pp. 293–297.

Emch, M. and Peterson, M., 2006. Mangrove forest cover change in the Bangladesh Sundarbans from 1989–2000: A remote sensing approach. *Geocarto International*, 21(1), pp. 5–12.

Foody, G.M., 2008. Harshness in image classification accuracy assessment. *International Journal of Remote Sensing*, 29(11), pp. 3137–3158.

Franklin, J., 1993. Discrimination of tropical vegetation types using SPOT multispectral data. *Geocarto International*, 8(2), pp. 57–63.

Gao, J., 1999. A comparative study on spatial and spectral resolutions of satellite data in mapping mangrove forests. *International Journal of Remote Sensing*, 20(14), pp. 2823–2833.

Ghosh, A., Schmidt, S., Fickert, T. and Nüsser, M., 2015. The Indian Sundarban mangrove forests: history, utilization, conservation strategies and local perception. *Diversity*, 7(2), pp. 149–169.

Ghosh, M.K., Kumar, L. and Roy, C., 2016. Mapping long-term changes in mangrove species composition and distribution in the Sundarbans. *Forests*, 7(12), p. 305.

Ghosh, M.K., Kumar, L. and Roy, C., 2017. Climate variability and mangrove cover dynamics at species level in the Sundarbans, Bangladesh. *Sustainability*, 9(5), p. 805.

Giri, C., Pengra, B., Zhu, Z., Singh, A. and Tieszen, L.L., 2007. Monitoring mangrove forest dynamics of the Sundarbans in Bangladesh and India using multi-temporal satellite data from 1973 to 2000. *Estuarine, Coastal and Shelf Science*, 73(1–2), pp. 91–100.

Giri, S., Mukhopadhyay, A., Hazra, S., Mukherjee, S., Roy, D., Ghosh, S., Ghosh, T. and Mitra, D., 2014. A study on abundance and distribution of mangrove species in Indian Sundarban using remote sensing technique. *Journal of Coastal Conservation*, 18(4), pp. 359–367.

Green, E.P., Clark, C.D., Mumby, P.J., Edwards, A.J. and Ellis, A.C., 1998. Remote sensing techniques for mangrove mapping. *International Journal of Remote Sensing*, 19(5), pp. 935–956.

Guo, M., Li, J., Sheng, C., Xu, J. and Wu, L., 2017. A review of wetland remote sensing. *Sensors*, 17(4), p. 777.

Hansen, M.C., Potapov, P.V., Moore, R., Hancher, M., Turubanova, S.A., Tyukavina, A., Thau, D., Stehman, S.V., Goetz, S.J., Loveland, T.R. and Kommareddy, A., 2013. High-resolution global maps of 21st-century forest cover change. *Science*, 342(6160), pp. 850–853.

Heenkenda, M.K., Joyce, K.E., Maier, S.W. and Bartolo, R., 2014. Mangrove species identification: Comparing WorldView-2 with aerial photographs. *Remote Sensing*, 6(7), pp. 6064–6088.

Heumann, B.W., 2011. An object-based classification of mangroves using a hybrid decision tree—Support vector machine approach. *Remote Sensing*, 3(11), pp. 2440–2460.

Islam, Md. T., Broström, G., Christensen, K.H., Drivdal, M., Weber, J.E.H., Shendryk, I., Hellstrom, M., Klemedtsson, L., Kljun, N., and Alwmark, C. 2014. Vegetation changes of Sundarbans based on Landsat Imagery analysis between 1975 and 2006. *Landscape Environment* 8(1), pp. 1–9.

Kirui, K.B., Kairo, J.G., Bosire, J., Viergever, K.M., Rudra, S., Huxham, M. and Briers, R.A., 2013. Mapping of mangrove forest land cover change along the Kenya coastline using Landsat imagery. *Ocean & Coastal Management*, 83, pp. 19–24.

Koedsin, W. and Vaiphasa, C., 2013. Discrimination of tropical mangroves at the species level with EO-1 Hyperion data. *Remote Sensing*, 5(7), pp. 3562–3582.

Li, S., Wang, H. and Jiang, X., 2003. Application of CBERS-1 CCD in the mangrove Remote Sensing Survey. *Marine Science Bulletin-Tianjin-Chinese Edition*, 22(6), pp. 30–35.

Liu, C., Frazier, P. and Kumar, L., 2007. Comparative assessment of the measures of thematic classification accuracy. *Remote Sensing of Environment*, 107(4), pp. 606–616.

Long, B.G. and Skewes, T.D., 1996. A technique for mapping mangroves with Landsat TM satellite data and geographic information system. *Estuarine, Coastal and Shelf Science*, 43(3), pp. 373–381.

Mukhopadhyay, A., Mondal, P., Barik, J., Chowdhury, S.M., Ghosh, T. and Hazra, S., 2015. Changes in mangrove species assemblages and future prediction of the Bangladesh Sundarbans using Markov chain model and cellular automata. *Environmental Science: Processes & Impacts*, 17(6), pp. 1111–1117.

Neukermans, G., Dahdouh-Guebas, F.J.G.K., Kairo, J.G. and Koedam, N., 2008. Mangrove species and stand mapping in Gazi Bay (Kenya) using Quickbird satellite imagery. *Journal of Spatial Science*, 53(1), pp. 75–86.

Pasqualini, V., Iltis, J., Dessay, N., Lointier, M., Guelorget, O. and Polidori, L., 1999. Mangrove mapping in North-Western Madagascar using SPOT-XS and SIR-C radar data. *Hydrobiologia*, 413, pp. 127–133.

Rahman, M.L., 2000. The Sundarbans: A unique wilderness of the world. In McCool, Stephen, F., Cole, David, N., Borrie, William, T. and O'Loughlin, Jennifer, Comps (eds.). *Wilderness Science in a Time of Change Conference* (Vol. 2, pp. 23–27).

Rahman, M.R., and M. Asaduzzaman. 2013. Ecology of Sundarban, Bangladesh. *Journal of Science Foundation* 8 (1–2), pp. 35–47.

Record, S., Charney, N.D., Zakaria, R.M. and Ellison, A.M., 2013. Projecting global mangrove species and community distributions under climate change. *Ecosphere*, 4(3), pp. 1–23.

Roslani, M.A., Mustapha, M.A., Lihan, T. and Wan, J., 2014. Applicability of RapidEye satellite imagery in mapping mangrove vegetation species at Matang mangrove forest reserve, Perak, Malaysia. *Journal of Environmental Science and Technology*, 7(2), pp. 123–136.

Sarker, S.K., Reeve, R., Thompson, J., Paul, N.K. and Matthiopoulos, J., 2016. Are we failing to protect threatened mangroves in the Sundarbans world heritage ecosystem? *Scientific Reports*, 6(1), pp. 1–12.

Siddiqi, N.A., 2001. Mangrove forestry in Bangladesh. Institute of Forestry and Environmental Sciences, University of Chittagong. Nibedon Press Limited, Chittagong.

Sinha, P., Kumar, L., Drielsma, M. and Barrett, T., 2014. Time-series effective habitat area (EHA) modeling using cost-benefit raster based technique. *Ecological Informatics*, 19, pp. 16–25.

Spalding, M., 2010. *World Atlas of Mangroves*. Earthscan, London, UK, p. 319.

Suratman, M.N., 2008. Carbon sequestration potential of mangroves in Southeast Asia. In *Managing Forest Ecosystems: The Challenge of Climate Change* (pp. 297–315). Springer, Dordrecht, Netherlands.

Treygo, W. and Dean, P.B., 1989. *The Environment and Development in Bangladesh: An Overview and Strategy for the Future*. Canadian International Development Agency, Ottowa, ON.

Vaiphasa, C., Skidmore, A.K. and de Boer, W.F., 2006. A post-classifier for mangrove mapping using ecological data. *ISPRS Journal of Photogrammetry and Remote Sensing*, 61(1), pp. 1–10.

Valderrama-Landeros, L., Flores-de-Santiago, F., Kovacs, J.M. and Flores-Verdugo, F., 2018. An assessment of commonly employed satellite-based remote sensors for mapping mangrove species in Mexico using an NDVI-based classification scheme. *Environmental Monitoring and Assessment*, 190(1), p. 23.

Vo, Q.T., Oppelt, N., Leinenkugel, P. and Kuenzer, C., 2013. Remote sensing in mapping mangrove ecosystems—An object-based approach. *Remote Sensing*, 5(1), pp. 183–201.

Wang, L., Sousa, W.P. and Gong, P., 2004. Integration of object-based and pixel-based classification for mapping mangroves with IKONOS imagery. *International Journal of Remote Sensing*, 25(24), pp. 5655–5668.

Wang, L., Sousa, W.P., Gong, P. and Biging, G.S., 2004. Comparison of IKONOS and QuickBird images for mapping mangrove species on the Caribbean coast of Panama. *Remote Sensing of Environment*, 91(3–4), pp. 432–440.

3 Effect of Statistical Relative Radiometric Normalization on Spectral Response of Mangrove Vegetation of Indian Sundarbans – A Comparative Performance Evaluation on Sentinel 2A Multi-Spectral Data

Abhisek Santra
Haldia Institute of Technology

Debashis Mitra
Indian Institute of Remote Sensing (IIRS), Indian
Space Research Organization (ISRO)

Shreyashi S. Mitra
Haldia Institute of Technology

CONTENTS

3.1   INTRODUCTION

The launch of various satellite sensors helps to deliver a series of earth observation data at varying spatial and spectral resolution. However, for every case, the surface radiation interacts with the atmosphere before reaching those sensors. The resultant satellite images show the brightness values that deviated from the actual ones due to such atmospheric interactions. It is more pronounced in areas of non-bright earth surface objects such as vegetation and water bodies (Hadjimitsis et al., 2010). Varying atmospheric conditions over those areas, such as attenuation due to absorption and scattering; sensor-target illumination condition; Instantaneous Field Of View and other related calibration procedures influence the spectral characteristics of an area (Teillet, 1986). Radiometric normalization helps to estimate the atmospheric and radiometric noise caused by non-surface factors and rectifies the radiometric differences between different images (Wu et al., 2018). It is not only indispensable to detect the actual change in surface reflectance from the multi-temporal images but also necessary to compare those images with respect to one another (Lo and Yang, 2000, El Hajj et al., 2008). In this regard, two possible calibration approaches were developed based on the brightness value transformation to physical signals. These are absolute and relative radiometric normalization (RRN) (Chen et al., 2005). The absolute calibration involves on-site ground measurement for atmospheric correction and sensor calibration at the time of image acquisition (Santra et al., 2019). The method establishes a relationship between the sensor-measured brightness values and the actual object reflectance at the real time of image acquisition to eliminate the radiometric distortion between images. Therefore, it requires certain real-time atmospheric parameters to frame 'atmosphere-surface-sensor' interaction model (Vermote et al., 1997). These models capture Top of Canopy reflectance from the Top of Atmosphere reflectance that is obtained from the radiance measured by the sensor (El Hajj et al., 2008). Besides, certain time-invariant calibration sites are needed to calibrate the on-orbit sensor parameters (Helder et al., 2013). Some other approaches based on the Dark Object Subtraction (Teillet and Fedosejevs, 1995, Chavez, 1996, Song et al., 2001) concept require radiative transfer codes for making absolute correction. However, not all archived time-series historical data were captured with all the necessary information required for absolute radiometric correction. This raises a big question mark about the practical usability of absolute correction specifically for time-series digital change detection (Schroeder et al., 2006, Canty et al., 2004). On the contrary, the RRN does not require any sort of in-situ atmospheric information at the time of data acquisition and it is not depended on the radiative transfer models discussed earlier. It is time-efficient and cost-effective as well. The fundamental concept behind any of such RRN approach is the application of linear relationship between the brightness values of the bi-temporal images (Canty and Nielsen, 2008, Sun et al., 2014, Rahman et al., 2015, Zhang et al., 2016, Santra et al., 2019). These methods normalize the band-wise brightness values of the target image based on the parameters retrieved from the reference image. However, the bi-temporal images must be of the same sensor. The resultant normalized image would reveal the same atmospheric and illumination condition as the reference image (Lo and Yang, 2000). Several comparative studies have been reported using the traditional RRN approaches on multispectral and thermal images (Bao et al., 2012, Xu et al., 2012, Rahman et al., 2015).

The effect of normalization is beneficial for multi-directional applications that depend upon the spectral reflectance of the earth surface features. In the context of monitoring mangrove vegetation, radiometric normalization plays an important role. It stretches the vegetation index values for better identification of features (Santra et al., 2019). Even it provides better results while forming new spectral indices (Santra et al., 2020, Sinha et al., 2020, Santra et al., 2021). Therefore, in this chapter, six statistical relative radiometric normalization (SRRN) techniques were applied on the bi-temporal Sentinel 2A images over the Indian Sundarbans region. The target image was normalized using these techniques and error statistic was used to see the performance of these selected SRRN methods. Finally, the response of vegetation indices was evaluated on the best selected normalized image.

3.1.1 STUDY AREA

The Sundarbans is located at the confluence of three rivers of Ganga, Brahmaputra and Meghna. It is considered as the world's largest active delta with halophytic plant species formation (Mondal et al., 2019). The Indian Sundarbans (Figure 3.1) covers approximately 2100 km^2 of area in the humid tropical climatic zone (Gupta et al., 2018). The delta was originated in the last phase of Miocene period (Hazra et al., 2002). It is highly sensitive to the variability of climate and population dynamics (DasGupta and Shaw, 2013). The temperature ranges from 9°C to 35°C in between winter to summer seasons with average annual precipitation of 144 cm. However, nearly 3/4th of the total annual rainfall is reported in the rainy season from June to September (Mondal et al., 2019). The dominant mangrove species found in the delta are *Avicennia officinalis* (Bain), *Bruguiera gymnorhiza* (Kankra), *Ceriops tagal* (Passur), *Excoecaria agallocha* (Gewa), *Heritiera* (Sundari), *Phoenix paludosa* (Hental), *Sonneratia apetala* (Keora) and *Xylocarpus granatum* (Dhundul) (Giri et al., 2014, Samanta and Hazra, 2017). People of the delta are earning their livelihood mainly from primary activities like fishing and collecting forest products. However, more than half of the population is also engaged in agricultural activities (Mondal et al., 2019). It also attracts global population for tourism purposes.

3.2.1 DATABASE AND METHODOLOGY

3.2.1.1 Database and Software Used

The study was conducted using Sentinel 2A multispectral imagery covering the areas of Indian Sundarbans. The open access cloud-free Sentinel 2A data were procured from the USGS Earth explorer data dissemination portal. The bi-temporal period

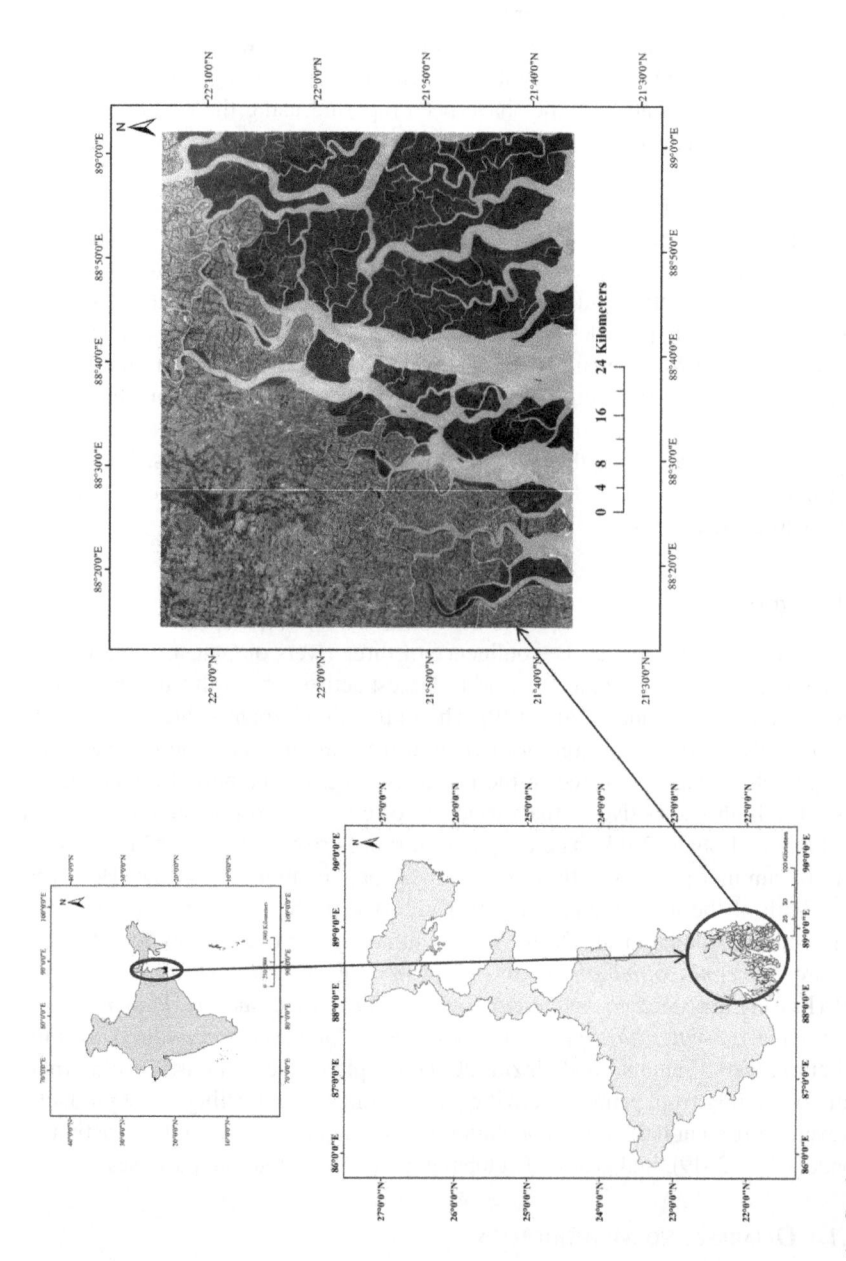

FIGURE 3.1 Study area.

TABLE 3.1
Details of the Database

Sensor	Date of Acquisition	Bands	
		Spatial Resolution (m)	Spectral Resolution (µm)
Sentinel 2A	Reference Image. 16th December, 2016	10	B2. 0.490
			B3. 0.560
			B4. 0.665
			B8. 0.842
		20	B5. 0.705
			B6. 0.740
			B7. 0.783
	Target Image. 31st December, 2019		B8A. 0.865
			B11. 1.610
			B12. 2.190
		60	B1. 0.443
			B9. 0.940
			B10. 1.375

was selected as 2016 and 2019. The images of these 2 years were considered as the reference and target images, respectively. To avoid the seasonal difference in surface reflectance, the similar month images were considered. The data consist of 13 spectral bands at three different spatial resolutions of 10 m, 20 m and 60 m. Both the data sets were co-registered to a common reference system of UTM Zone 45 with WGS 84 datum. Details of the data are given in the following Table 3.1. Pre-processing and processing of the Sentinel 2A data were done in QGIS 3.8, Erdas Imagine 9.2, ENVI 4.7 and ArcGIS 10.5. The statistical analyses were conducted in the IDL, R and SPSS software environment.

3.2.1.2 Methods

The present research was designed keeping in mind the logical understanding and selection of bi-temporal image based SRRN. In this study, based on the statistical information present in the image were considered and applied six different SRRN algorithms to normalize the target image of the year 2019. However, before the application of these SRRN techniques, both the target images were converted to the at-sensor radiance images using the following Eq. 4.1.

$$radiance = \left[\frac{\text{pixel value of band } i \times \cos\left(\text{incidence angle}\right) \times \text{solar irradance of band } i}{\pi d^2} \right]$$

$$(3.1)$$

where $d^2 = 1.0/U$.

The six applied SRRN techniques were – Haze Correction (HC), Min-Max (MM) normalization, Mean-Standard (MS) deviation normalization, Simple Image

Regression (SIR) normalization, Dark-set and Bright-set (DB) normalization and Iteratively Re-weighted Multi-Variate Alteration Detection (IR-MAD) normalization (Jensen, 1983, Hall et al., 1991, Yuan and Elvidge, 1996, Santra et al., 2019). The objective of the application of the SRRN techniques is the rectification of the target image from the reference image using a linear transformation process. This linear transformation can be expressed by a straight line moving through the central part of the scatter diagram of the bi-temporal reference and target images, apparently represented by the true time-invariant objects in terms of their radiometric properties (Yuan and Elvidge, 1996). The target image is therefore expressed as

$$t_{n_k} = \left(a_k r_k - b_k \right) \tag{3.2}$$

where t_{n_k} and r_k are the normalized target and reference images of band k, respectively, and a_k and b_k are the normalization coefficients of different SRRN algorithm for band k.

The first five SRRN methods mentioned above use Eq. (3.2) for normalizing the target images. The HC normalization method is the simplest of all the methods selected for this study. It assumes the pixels having zero reflectance on both reference and target images should have the same minimum brightness values. It is an offset correction with gain value 1 (Yuan and Elvidge, 1996). The HC normalization coefficients are

$$a_k = 1 \tag{3.3}$$

$$b_k = \left(r_{k_{\min}} - t_{k_{\min}} \right) \tag{3.4}$$

where $r_{k_{\min}}$ and $t_{k_{\min}}$ are the haze values in band k of the reference and target images, respectively.

The MM normalization technique assumes the similar minimum and maximum pixel values in both reference and target images in all bands (Yuan and Elvidge, 1996). The target image is normalized using the following Eqs. 3.5 and 3.6.

$$a_k = \frac{\left(r_{k_{\max}} - r_{k_{\min}} \right)}{\left(t_{k_{\max}} - t_{k_{\min}} \right)} \tag{3.5}$$

$$b_k = \left(r_{k_{\min}} - a_k t_{k_{\min}} \right) \tag{3.6}$$

where $r_{k_{\max}}$ and $r_{k_{\min}}$ are the maximum and minimum pixel values of band k of the reference image, respectively, and $t_{k_{\max}}$ and $t_{k_{\min}}$ are the maximum and minimum values of band k of the target image, respectively.

The MS normalization method normalizes the target-image keeping the band-wise mean and standard deviation values of the reference image (Santra et al., 2019). The normalization coefficients used for this method are

$$a_k = \frac{S_{r_k}}{S_{t_k}} \tag{3.7}$$

$$b_k = r_k \text{ mean} - a_k t_k \text{mean} \tag{3.8}$$

where S_{r_k} and r_k mean are the standard deviation and mean values of band k of the reference image respectively. S_{t_k} and t_k mean are the standard deviation and mean values of band k of the target image, respectively.

The SIR normalization method assumes that the pixels sampled at one time period are linearly related to the co-registered pixels in another time (Du et al., 2002). This linear relationship is achieved by the least-square regression method. The target image is regressed against the reference image using this least-square method (Jensen, 1983). So that,

$$Q = \sum \left(r_k - a_k t_k - b_k \right)^2 \tag{3.9}$$

For the entire scene the following equations (Eqs. 3.10 and 3.11) are applied to retrieve the normalization coefficients.

$$a_k = \frac{S_{t_k r_k}}{S_{t_k t_k}} \tag{3.10}$$

$$b_k = \left(r_{k \text{ mean}} - a_k t_{k \text{ mean}} \right) \tag{3.11}$$

where $S_{t_k t_k} = \dfrac{\sum \left(t_k - t_{k \text{ mean}} \right)^2}{n}$ and $S_{t_k r_k} = \dfrac{\sum \left(t_k - t_{k \text{ mean}} \right)\left(r_k - r_{k \text{ mean}} \right)}{n}$; $n = $ total number of pixels in the band k of the images.

In contrast with the single minimum and maximum pixel values of band k in the MM normalization technique, the DB normalization considers the mean values of two sets of dark and bright radiometric control points (Hall et al., 1991, Santra et al., 2019). However, identification of such dark and bright sets of pixels in an image is rather challenging. The most conventional approach in this regard is the Tasselled Cap (TC) or Kauth-Thomas transformation (Kauth and Thomas, 1976, Crist and Cicone, 1984, Hall et al., 1991). The TC transformation identifies the moist and dry vegetation surface well (Santra and Mitra, 2014). In this study, the TC transformed images were considered to identify the band specific dark and bright radiometric control points for both the target and reference images. The normalized coefficients of this method were estimated using the following equations (Eqs. 3.12 and 3.13).

$$a_k = \frac{\left(r_{kb \text{ mean}} - r_{kd \text{ mean}} \right)}{\left(t_{kb \text{ mean}} - t_{kd \text{ mean}} \right)} \tag{3.12}$$

$$b_k = \left(r_{kd \text{ mean}} - a_k t_{kd \text{ mean}} \right) \tag{3.13}$$

where $r_{kb \text{ mean}}$ and $r_{kd \text{ mean}}$ are the mean values of bright and dark sets of pixels of band k in reference image, respectively, and $t_{kb \text{ mean}}$ and $t_{kd \text{ mean}}$ are the mean values of bright and dark sets of pixels of band k in target images, respectively.

Canty, Nielsen, and Schmidt (2004) introduced Multivariate Alteration Detection (MAD), an automated method of selecting time-invariant pixels within the multispectral image. These pseudo-invariant pixels are used to normalize the target image in time-efficient manner. These no-change pixels form linear combinations of the intensities for all the channels in bi-temporal images (Santra et al., 2019). This method identifies large number of pixels out of which some are used for testing and rest are applied to perform the linear regression (Canty et al., 2004). To improve the sensitivity of the MAD approach Nielsen (2007) introduced the IR-MAD. However, in both the cases the statistical Canonical Component Analysis (CCA) (Hotelling, 1936) was used. However, the CCA in IR-MAD is Iterative CCA. These methods identify changes from the canonical difference of the multivariate images (Nielsen et al., 1998). Suppose two N-band multispectral bi-temporal images F and G are acquired over a study area at two different time periods t_1 and t_2, MAD variates are calculated simply from this difference (Falco et al., 2016).

$$D = a^T F - b^T G \tag{3.14}$$

where $a^T F = a_1 F_1 + a_2 F_2 + \cdots + a_n F_n$, $b^T G = b_1 G_1 + b_2 G_2 + \cdots + b_n G_n$ and $= (F_1, F_2, F_3, \ldots, F_n)^T$ and $G = (G_1, G_2, G_3, \ldots, G_n)^T$.

To find out the suitable vectors a and b that maximize the variance $V\{a^T F - b^T G\}$ two generalized eigenvector problems coupled with the parameter ρ need to be solved using the following equations (Eqs. 3.15 and 3.16) (Marpu et al., 2011, Falco et al., 2012).

$$\sum_{FG} \sum_{GG}^{-1} \sum_{FG}^{T} a = \rho^2 \sum_{FF} a \tag{3.15}$$

$$\sum_{FG} \sum_{FF}^{-1} \sum_{FG}^{T} a = \rho^2 \sum_{GG} b \tag{3.16}$$

where $\sum_{FF}$ and $\sum_{GG}$ are the covariance matrices of F and G, respectively, and $\sum_{FG}$ is the cross-covariance matrix between F and G, ρ is the canonical correlation between $a_i^T F$ and $b_i^T G$.

Now, the MAD transformation can be achieved through the following Eq. 3.17.

$$M_i = a_i^T F - b_i^T G \tag{3.17}$$

where $a_i^T F$ and $b_i^T G$ are canonical variates (CVs) and M_i is the ith MAD variate. CVs and MAD variates are mutually uncorrelated.

Now, to get the change probability $P_r(\text{change})$, first a random variable 'Z' be calculated to represent the sum of squares of standardized MAD variates (Eq. 3.18).

$$Z = \sum_{i=1}^{N} \left(\frac{M_i}{\sigma_{M_i}^{nc}} \right)^2 \tag{3.18}$$

where $\sigma_{M_i}^{nc}$ is the variance of no-change distribution.

The no-change observations are uncorrelated and their distribution pattern is Gaussian in nature. The realizations z of the random variable Z must be chi-square (χ^2) distributed with K degrees of freedom (Falco et al., 2012). Finally, the P_r (change) be obtained from the following equations (Eq. 3.19).

$$P_r(\text{change}) = P_{\chi^2, \, K(Z)} \qquad (3.19)$$

In IR-MAD approach, high weights are given on observations that exhibit little change to increase the detection of observations whose status over temporal span is uncertain (Nielsen, 2007). This is achieved by applying probabilities of no-change as weights that are used to calculate covariance matrix in successive iteration. In each of the iterations, a better no-change background is identified that represent better separability between the change and no-change objects (Falco et al., 2016). Iterations are applied until the largest absolute change in canonical correlations becomes lower than the predefined small value, e.g. 10^{-6} (Nielsen, 2007).

In this study, the IR-MAD was applied to the bi-temporal Sentinel 2A images at three different resolution levels in ENVI-IDL environment. However, before applying the algorithm the images were converted to the sensor radiance images using Eq. (3.1) mentioned earlier.

After the application of all the above-mentioned SRRN techniques, the Root Mean Square Error (RMSE) statistic was applied on the entire scene to estimate the visual closeness and statistical robustness between the normalized target and the reference images for each band (Santra et al., 2019) using the following equation (Eq. 3.20).

$$\text{RMSE} = \sqrt{\frac{1}{|\text{Scene}|} \sum \left(t_{n_k} - r_k \right)^2} \qquad (3.20)$$

The lower the RMSE value, better the correspondence between the normalized target and reference images.

The objective behind such a comparison of the selected SRRN algorithms is to observe the effect of radiometric normalization on the spectral response of the mangrove vegetation of the Indian Sundarbans. The best identified method generated normalized image was used to estimate Normalized Difference Vegetation Index (NDVI) (Rouse et al., 1974) and Combined Mangrove Recognition Index (CMRI) (Gupta et al., 2018) using the following equations (Eq. 3.21 and 3.22).

$$\text{NDVI} = \frac{\text{NIR} - R}{\text{NIR} + R} \qquad (3.21)$$

$$\text{CMRI} = (\text{NDVI} - \text{NDWI}) \qquad (3.22)$$

where Normalized Difference Water Index (NDWI) (Gao, 1996) is $\text{NDWI} = \dfrac{G - \text{NIR}}{G + \text{NIR}}$ and G, R and NIR are the pixel values of green, red and near infra-red spectral bands.

The NDVI was chosen to see the effect of normalization on mangrove vegetation health and CMRI was considered for observing the response of such normalization on the capability of differentiating mangrove vegetation from the rest of the vegetation classes. Random sample locations were identified from the original and normalized target images. Regression analysis and Analysis of Variance (ANOVA) test were conducted to see the effect of SRRN on target image.

3.2 RESULTS AND DISCUSSION

3.2.1 COMPARATIVE RESULTS OF DIFFERENT SRRN TECHNIQUES

The normalized target images of the year 2019 were obtained applying the six aforesaid SRRN algorithms (Figure 3.2). However, it is very difficult to assess visually the effect of different normalization techniques on the target images. Therefore, band-wise statistical comparison is required to find out the best normalized image. The computed normalization coefficients for the first five methods are presented in Table 3.2. These coefficients helped to generate different normalized target images using the linear relationship between the reference and the normalized target images mentioned in Eq. (3.2).

The HC normalized image (Figure 3.2a) retained only the difference between the minimum pixel values of the reference and target images. The MM normalized image (Figure 3.2b) on the other hand retained the minimum and maximum pixel values of the reference image. Similarly, the MS normalized image (Figure 3.2c) kept the mean and standard deviation values of the reference image. The SIR normalization method (Figure 3.2d) considers the variance and covariance of target and reference images for each band to rectify the target image. Therefore, instead of the specific values mentioned in the previous three methods, this normalization technique considers the entire master and slave images. This method proves to be very effective in similar seasonal and cloudless bi-temporal data. Unlike the SIR rectified image, the DB normalized image (Figure 3.2e) takes into account only the dark and bright patches of the reference and target images. The mean pixel values of those patches were applied to generate the normalized target image. The TC transformation images helped to identify the dark moist and bright dry patches in the study area. For this purpose, more than 10% of the total number of pixels was considered. However, in this study, this method did not respond effectively.

Finally, the IR-MAD normalized image (Figure 3.2f) depends on the no-change sample set using iterative canonical component analysis. Figure 3.3 shows the canonical correlations vs. the number of iterations for all the 13 spectral bands under three different spatial resolution contexts. From Figure 3.3a it is evident that the high-resolution (10 m) comprising bands 2, 3, 4 and 8 stabilize after 12 iterations, whereas moderate 20 m resolution images with 5, 6, 7, 8A, 11 and 12 bands (Figure 3.3b) and coarse 60 m resolution images with bands 1, 9 and 10 converge after 9 and 18 iterations, respectively (Figure 3.3c). The observation period, atmospheric diversity and radiometric differences between the reference and target images may affect the number of iteration in such convergence. Based on these statistics, the chi-square images and chi-square histograms at 95% threshold level were generated. The chi-square histograms of the three spatial resolution help to identify the change and no-change pixels (Figure 3.4).

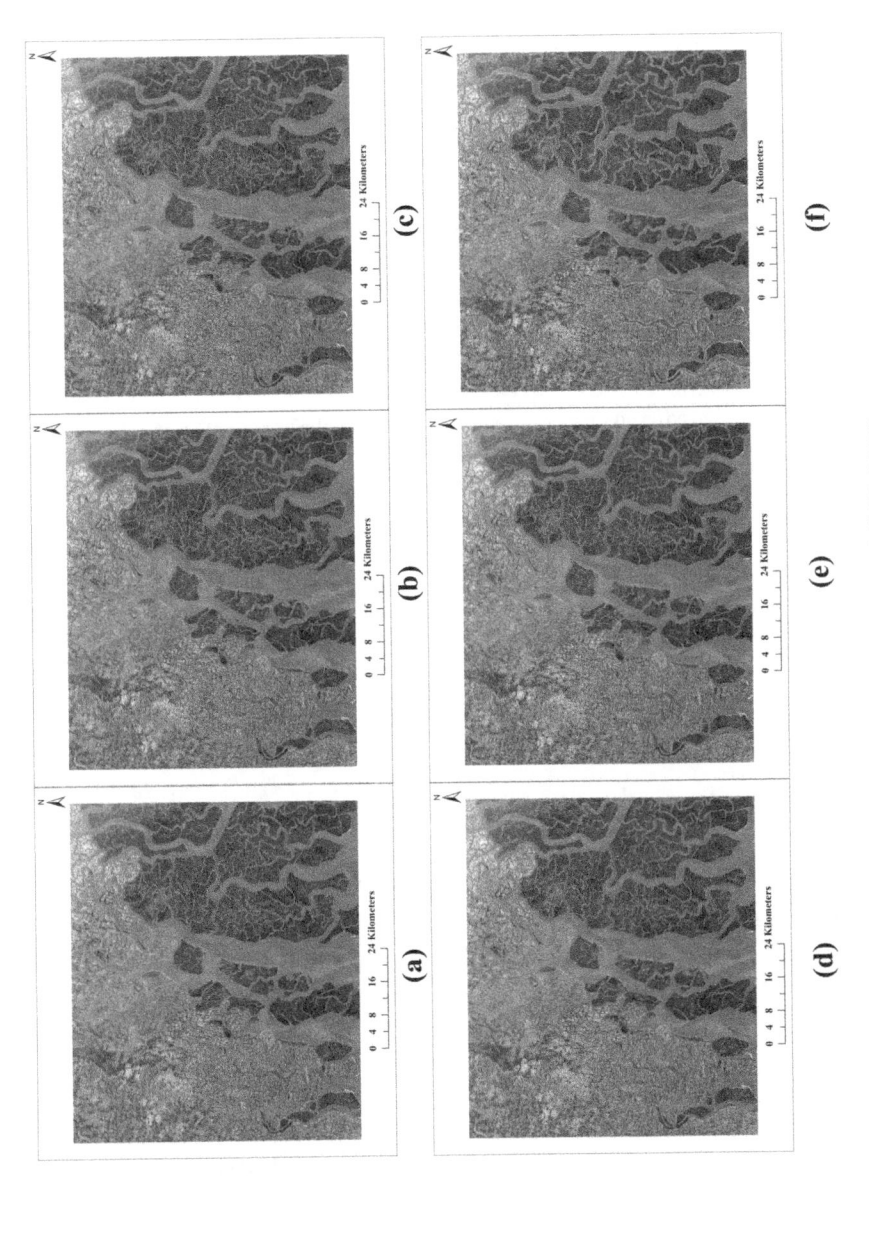

FIGURE 3.2 Normalized images: (a) HC, (b) MM, (c) MS, (d) SIR, (e) DB and (f) IR-MAD.

TABLE 3.2
Normalization Coefficients

	Statistical Normalization Techniques								
	HC		MM		MS		SIR		DB	
Bands	A	b	a	B	a	b	A	b	a	B
Band 1	1.00	−6.39	1.56	−39.48	1.9	−60.8	1.94	−60.79	1.62	−41.63
Band 2	1.00	0.23	1.74	−17.29	2.4	−32.5	2.35	−32.52	3.12	−68.94
Band 3	1.00	−0.77	1.67	−14.71	2.5	−34.2	2.51	−34.16	2.13	−30.75
Band 4	1.00	−1.08	1.53	−8.40	1.4	0.8	1.38	0.76	1.71	−15.46
Band 5	1.00	−0.97	1.50	−7.76	1.1	2.7	1.10	2.69	1.71	−15.62
Band 6	1.00	−1.05	1.23	−3.62	1.2	−0.3	1.21	−0.30	1.31	−6.43
Band 7	1.00	−1.26	1.31	−4.52	1.2	−0.1	1.20	−0.12	1.48	−14.33
Band 8	1.00	0.02	1.59	−2.90	1.4	−0.7	1.43	−0.70	2.09	−10.94
Band 8A	1.00	−1.82	1.25	−3.68	1.2	−2.4	1.22	−2.36	1.36	−8.00
Band 9	1.00	−0.02	2.05	−2.08	1.9	−2.4	1.88	−2.35	4.40	−28.02
Band 10	1.00	0.00	20.94	−0.10	1.7	−0.1	1.67	−0.02	34.75	−0.71
Band 11	1.00	−0.22	1.24	−0.34	1.5	−1.2	1.46	−1.16	1.28	−0.77
Band 12	1.00	−0.04	1.31	−0.06	1.5	−0.3	1.48	−0.28	1.38	−0.03

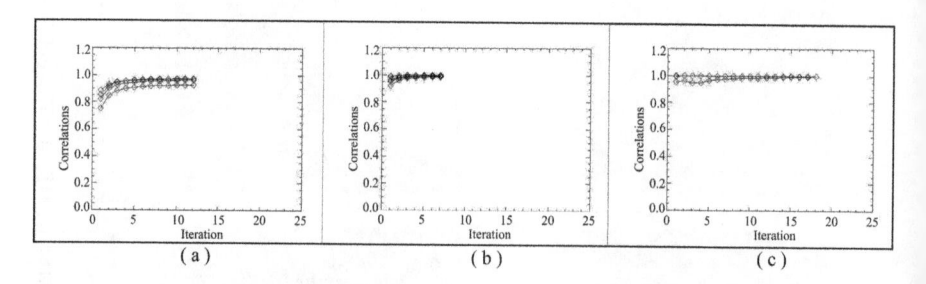

FIGURE 3.3 Canonical correlations: (a) Bands 5, 6, 7, 8; (b) bands 5, 6, 7, 8A, 11, 12; (c) bands 1, 9, 10.

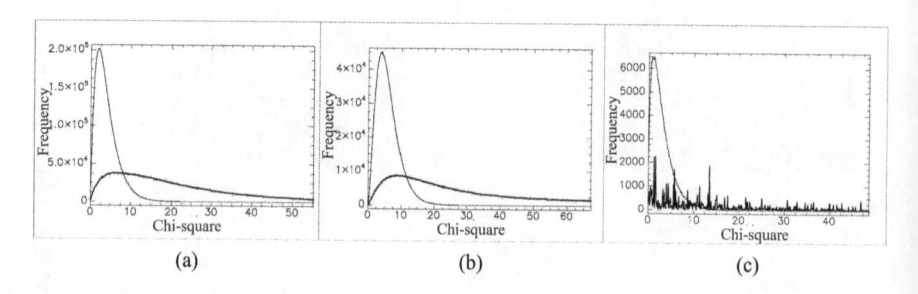

FIGURE 3.4 Chi-square histograms: (a) Bands 5, 6, 7, 8; (b) bands 5, 6, 7, 8A, 11, 12; (c) bands 1, 9, 10.

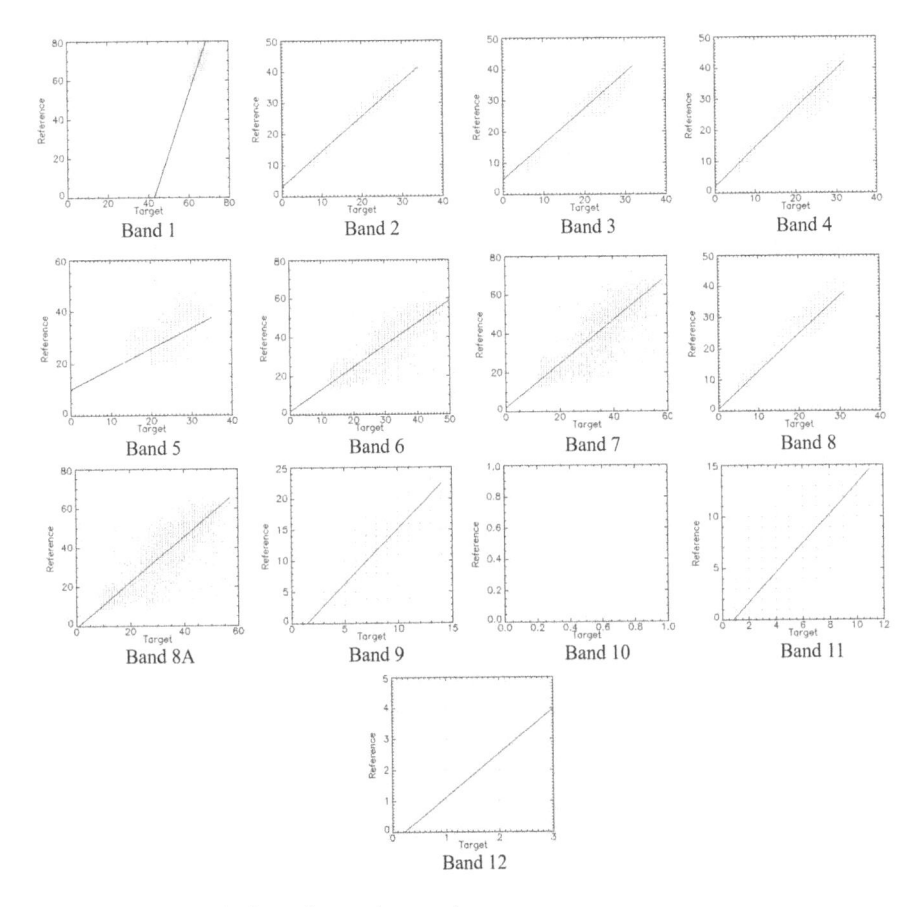

FIGURE 3.5 Band-wise orthogonal regression.

Figure 3.5 represents the comparison of the band-by-band orthogonal regression analysis using the pseudo-invariant features obtained from the images. However, sample size may exert differential effects in normalization (Yang and Lo, 2000). Larger sample set may produce better results statistically. However, this may also require robust system that supports such rigorous calculations.

Overall, from the measure of accuracy, it was observed that the IR-MAD method offers the best among the possible solution provided by the selected methods applied here in normalizing the target image (Table 3.3). After that the simple HC provided better results. The remaining four methods, in this case, did not provide acceptable statistical results in comparison with the two techniques mentioned before. Also, the IR-MAD specifically provided better normalization effect on the higher resolution bands. Therefore, this time-efficient automated canonical component based iterative method in this study proved to be advantageous, as it identifies the time-invariant pixels in the background of time-uncertain objects more accurately with less human intervention.

TABLE 3.3
Band-Wise RMSE

			RMSE			
Bands	HC	MM	MS	SIR	DB	IRMAD
Band 1	2.636634	3.544446	4.296789	4.849088	2.912501	6.387001
Band 2	15.88546	8.241571	15.60297	14.00535	9.141931	0.224999
Band 3	3.423399	6.775439	14.22057	14.53863	6.845019	0.766
Band 4	3.066267	6.927201	12.25006	11.63927	5.79693	1.083
Band 5	2.975894	13.34934	10.14931	11.0624	19.57164	0.016899
Band 6	4.730747	6.905157	5.55326	5.584372	5.728323	0.97
Band 7	6.757649	5.587762	7.582107	8.086274	6.554415	1.054001
Band 8	3.851681	9.144722	8.503295	8.362188	8.748899	1.2593
Band 8A	3.974254	6.622953	5.865969	6.84438	7.876528	1.824299
Band 9	1.125742	13.49611	10.98146	10.67338	26.71062	0.0212
Band 10	0.0094	1.139142	0.024379	0.026937	1.410001	0.0013
Band 11	1.035438	1.145325	2.040938	1.860791	1.08181	0.21719
Band 12	0.437813	0.422331	0.554752	0.529871	0.54369	0.0379
Average	**3.83926**	**6.407807**	**7.509682**	**7.543302**	**7.9171**	**1.066392**

3.2.2 Effect of SRRN on the Spectral Response of the Mangrove Vegetation

The objective of the study was to assess the effect of SRRN on the spectral response pattern of mangrove vegetation of the Indian Sundarbans. Two spectral vegetation indices were chosen for this purpose, namely, the widely accepted NDVI and the newly developed CMRI that address efficiently the mangrove vegetation in this region (Gupta et al., 2018). Figure 3.6 depicts the index images generated from the best identified IR-MAD normalization technique. Thirty sample points were randomly chosen from the different categories of mangrove vegetation. Regression analysis of the original image based indices and the IR-MAD normalized image based indices were conducted. The regression analysis shows the correlation for both the indices (Figure 3.7) (Table 3.4).

The one-way ANOVA test supports the subtle change of the effect radiometric correction on the median values of both the NDVI and CMRI index images (Table 3.5). Table 3.5 also shows that the differences in the mean values among the two NDVI sample values are not great enough to exclude the possibility that the difference is due to random sampling variability; there is not a statistically significant difference ($P = 0.147$). On the other hand, the differences in the median values among the treatment sample values of the CMRI images are greater than would be expected by chance; there is a statistically significant difference ($P = <0.001$).

Figure 3.8 shows the box-plot comparison of the two index images generated from the original target image and the IR-MAD normalized target image. The figure also shows the stretching effect of data values due to radiometric normalization. The stretching of NDVI values may help to identify the vegetation health in a better way.

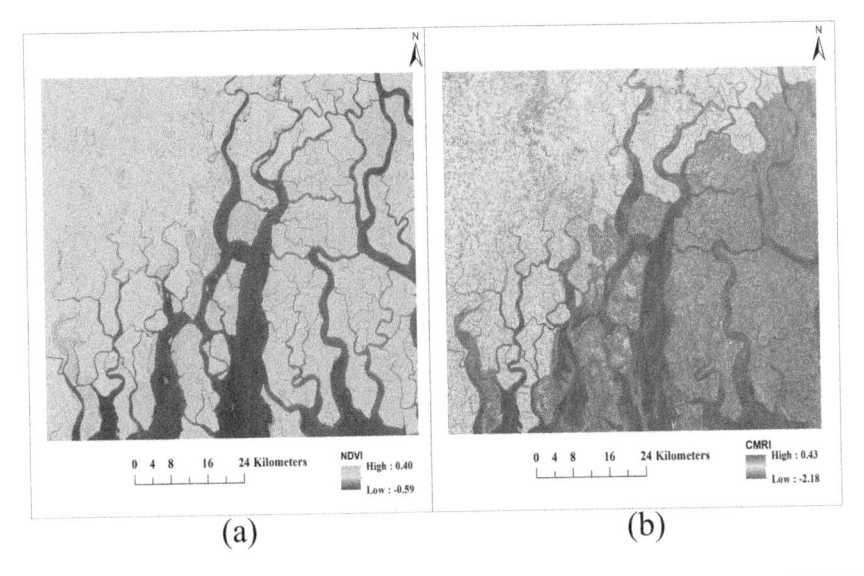

FIGURE 3.6 Index images generated from the IR-MAD normalized target image: (a) NDVI and (b) CMRI.

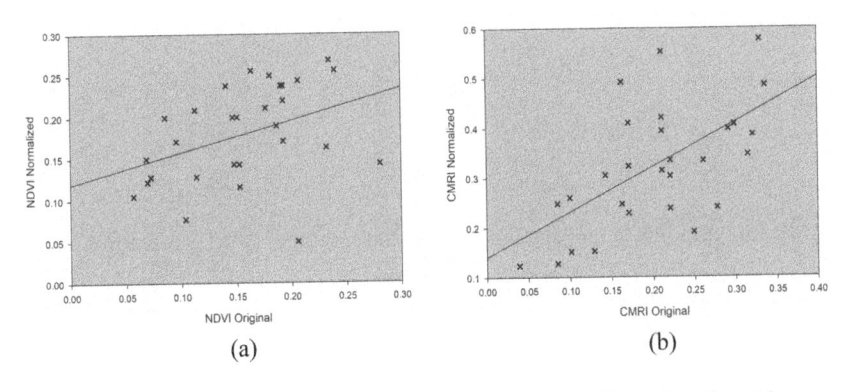

FIGURE 3.7 Correlation between vegetation indices generated from the original image and the IR-MAD normalized image (a) NDVI and (b) CMRI (x axis shows the vegetation index sample values from original image and y axis shows the vegetation index sample values from IR-MAD normalized image).

TABLE 3.4
Regression Analysis

Vegetation Indices	R	R Square	Adjusted R Square	Standard Error of Estimate
NDVI	0.387	0.149	0.119	0.0538106
CMRI	0.608	0.37	0.348	0.0970221

TABLE 3.5

One-Way ANOVA

Vegetation Index		Sum of Squares	Df	Mean Square	F	Sig.
NDVI	Between groups	.070	19	.004	1.421	.289
	Within groups	.026	10	.003		
	Total	.095	29			
CMRI	Between groups	.324	18	.018	2.107	.104
	Within groups	.094	11	.009		
	Total	.419	29			

Normality Test (NDVI) (Shapiro-Wilk). Passed ($P = 0.481$)

Normality Test (CMRI) (Shapiro-Wilk). Passed ($P = 0.766$)

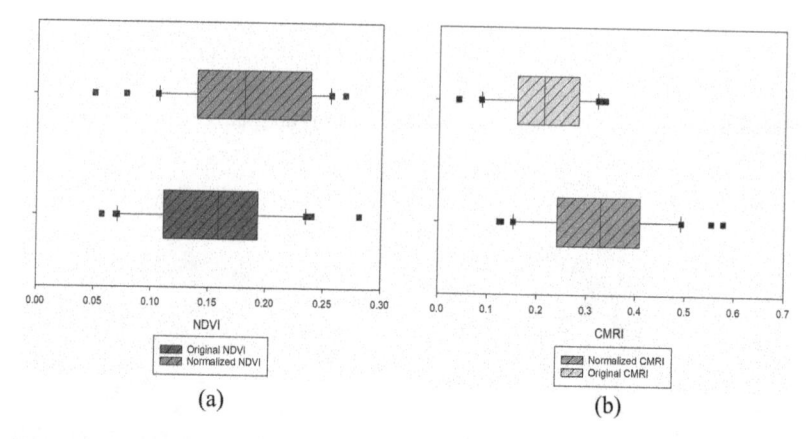

(a)   (b)

FIGURE 3.8   Box plot of the sample vegetation indices values (a) NDVI and (b) CMRI.

However, the stretching of the CMRI values has some other significance. As this index is applied to discriminate the mangrove vegetation, the stretched values may improvise the discrimination capability of the mangroves from the other vegetation.

3.3   CONCLUSION

RRN is useful as well as necessary in land cover change analysis. The time-series satellite images carry inherent radiometric differences due to sensor and atmospheric conditions at the time of image acquisition. In this chapter, six relative image-statistics-dependent radiometric normalization techniques were applied. Most of these approaches apply the conventional SRRN methods that often reduce the radiometric differences caused by the actual land use and land cover change. These methods use either single or a set of pixel values in deriving the normalization coefficients using which the target image is normalized. The simple HC in this respect performed significantly well in reducing the differences. However, this sometimes affects the actual change studies in time-series analysis. Comparatively complex statistical concept-based IR-MAD in this regard performed best in capturing the actual spectral response of the mangrove vegetation in this study area. The IR-MAD method is depended upon the statistical iterative canonical component analysis. This transformation process helped to minimize the inherent target image noise in the change components in the background of potential change and higher auto-correlation. The method was proved advantageous as it is fast, automated and requires limited user-end information to generate the chi-square and CV images to identify the time-invariant pixels. Therefore, it has a strong edge over the other pseudo-invariant feature-based normalization techniques. However, results derived from such an iterative pseudo-invariant feature-based normalization technique may be compared with the other complex iterative principal component analysis and wavelet transformation based results. As a consequence of such image-based SRRN, spectral response of the mangrove vegetation in the study area was enhanced. Results showed that in both the NDVI and CMRI images generated from the IR-MAD normalized target image, the values over the mangrove vegetative surface were stretched. This stretching effect may not only help to identify the mangrove vegetation health in a better way but also to improvise the discrimination capability of mangroves from other types of vegetation in the Indian Sundarbans. However, this capability may be validated with the ground truth data.

ACKNOWLEDGEMENTS

The authors are thankful to the Geoinformatics Cell, Department of Civil Engineering, Haldia Institute of Technology for providing the infrastructure to conduct the research work.

REFERENCES

Bao, N., Alex, M. L., Andrew, F., David, M., Andrew, M. & Zhongke, B. 2012. Comparison of relative radiometric normalization methods using pseudo-invariant features for change detection studies in rural and urban landscapes. *Journal of Applied Remote Sensing*, 6, 1–18.

Canty, M. J. & Nielsen, A. A. 2008. Automatic radiometric normalization of multitemporal satellite imagery with the iteratively re-weighted MAD transformation. *Remote Sensing of Environment*, 112, 1025–1036.

Canty, M. J., Nielsen, A. A. & Schmidt, M. 2004. Automatic radiometric normalization of multitemporal satellite imagery. *Remote Sensing of Environment*, 91, 441–451.

Chavez, P. S. 1996. Image-based atmospheric corrections - revisited and improved. *Photogrammetric Engineering and Remote Sensing*, 62, 1025–1036.

Chen, X., Vierling, L. & Deering, D. 2005. A simple and effective radiometric correction method to improve landscape change detection across sensors and across time. *Remote Sensing of Environment*, 98, 63–79.

Crist, E. P. & Cicone, R. C. 1984. Application of the tasselled cap concept to simulated Thematic Mapper data. *Photogrammetric Engineering and Remote Sensing*, 50, 343–352.

Dasgupta, R. & Shaw, R. 2013. Changing perspectives of mangrove management in India – An analytical overview. *Ocean & Coastal Management*, 80, 107–118.

Du, Y., Teillet, P. M. & Cihlar, J. 2002. Radiometric normalization of multitemporal high-resolution satellite images with quality control for land cover change detection. *Remote Sensing of Environment*, 82, 123–134.

El Hajj, M., Bégué, A., Lafrance, B., Hagolle, O., Dedieu, G. & Rumeau, M. 2008. Relative radiometric normalization and atmospheric correction of a SPOT 5 time series. *Sensors*, 8, 2774–2791.

Falco, N., Marpu, P. R. & Benediktsson, J. A. Comparison of ITPCA and IRMAD for automatic change detection using initial change mask. *2012 IEEE International Geoscience and Remote Sensing Symposium*, 22–27 July 2012 2012. 6769–6772.

Falco, N., Marpu, P. R. & Benediktsson, J. A. 2016. A toolbox for unsupervised change detection analysis. *International Journal of Remote Sensing*, 37, 1505–1526.

Gao, B.C. 1996. NDWI—A normalized difference water index for remote sensing of vegetation liquid water from space. *Remote Sensing of Environment*, 58, 257–266.

Giri, S., Mukhopadhyay, A., Hazra, S., Mukherjee, S., Roy, D., Ghosh, S., Ghosh, T. & Mitra, D. 2014. A study on abundance and distribution of mangrove species in Indian Sundarban using remote sensing technique. *Journal of Coastal Conservation*, 18, 359–367.

Gupta, K., Mukhopadhyay, A., Giri, S., Chanda, A., Datta Majumdar, S., Samanta, S., Mitra, D., Samal, R. N., Pattnaik, A. K. & Hazra, S. 2018. An index for discrimination of mangroves from non-mangroves using LANDSAT 8 OLI imagery. *MethodsX*, 5, 1129–1139.

Hadjimitsis, D. G., Papadavid, G., Agapiou, A., Themistocleous, K., Hadjimitsis, M. G., Retalis, A., Michaelides, S., Chrysoulakis, N., Toulios, L. & Clayton, C. R. I. 2010. Atmospheric correction for satellite remotely sensed data intended for agricultural applications: Impact on vegetation indices. *Natural Hazards and Earth System Sciences*, 10, 89–95.

Hall, F. G., Strebel, D. E., Nickeson, J. E. & Goetz, S. J. 1991. Radiometric rectification: Toward a common radiometric response among multidate, multisensor images. *Remote Sensing of Environment*, 35, 11–27.

Hazra, S., Ghosh, T., Dasgupta, R. & Sen, G. 2002. Sea level and associated changes in the Sundarbans. *Science and Culture*, 68, 309–321.

Helder, D., Thome, K. J., Mishra, N., Chander, G., Xiong, X., Angal, A. & Choi, T. 2013. Absolute radiometric calibration of landsat using a pseudo invariant calibration site. *IEEE Transactions on Geoscience and Remote Sensing*, 51, 1360–1369.

Hotelling, H. 1936. Relations between two sets of variates. *Biometrika*, 28, 321–377.

Jensen, J. R. 1983. Urban/suburban landuse analysis. In: Colwell, R. N. (ed.) *Manual of Remote Sensing*. 2nd ed. Falls Church, Virginia: American Society of Photogrammetry.

Kauth, R. J. & Thomas, G. S. 1976. The tasselled cap - A graphic description of the spectral-temporal development of agricultural crops as seen by Landsat.

Lo, C. P. & Yang, X. 2000. Relative radiometric normalization performance for change detection from multi-date satellite images. *Photogrammetric Engineering and Remote Sensing*, 66, 967–980.

Marpu, P. R., Gamba, P. & Canty, M. J. 2011. Improving change detection results of IR-MAD by eliminating strong changes. *IEEE Geoscience and Remote Sensing Letters*, 8, 799–803.

Mondal, B., Saha, A. K. & Roy, A. 2019. Mapping mangroves using LISS-IV and Hyperion data in part of the Indian Sundarban. *International Journal of Remote Sensing*, 40, 9380–9400.

Nielsen, A. A. 2007. The regularized iteratively reweighted MAD method for change detection in multi- and hyperspectral data. *IEEE Transactions on Image Processing*, 16, 463–478.

Nielsen, A. A., Conradsen, K. & Simpson, J. J. 1998. Multivariate alteration detection (MAD) and MAF postprocessing in multispectral, bitemporal image data: New approaches to change detection studies. *Remote Sensing of Environment*, 64, 1–19.

Rahman, M. M., Hay, G. J., Couloigner, I., Hemachandran, B. & Bailin, J. 2015. A comparison of four relative radiometric normalization (RRN) techniques for mosaicing H-res multi-temporal thermal infrared (TIR) flight-lines of a complex urban scene. *ISPRS Journal of Photogrammetry and Remote Sensing*, 106, 82–94.

Rouse, J.W., Haas, R.H., Schell, J.A., & Deering, D.W. 1974. Monitoring vegetation systems in the Great Plains with ERTS, In: Freden, S.C., Mercanti, E.P. & Becker, M. (eds.) *Third Earth Resources Technology Satellite–1 Syposium. Volume I: Technical Presentations*, NASA SP-351, NASA, Washington, D.C., pp. 309-317.

Samanta, K. & Hazra, S. 2017. Mangrove forest cover changes in Indian Sundarban (1986–2012) using remote sensing and GIS. In: Hazra, S., Mukhopadhyay, A., Ghosh, A. R., Mitra, D. & Dadhwal, V. K. (eds.) *Environment and Earth Observation: Case Studies in India*. Cham: Springer International Publishing.

Santra, A. & Mitra, S. S. 2014. A comparative study of tasselled cap transformation of DMC and ETM+ images and their application in forest classification. *Journal of the Indian Society of Remote Sensing*, 42, 373–381.

Santra, A., Mitra, S. S., Sinha, S. & Routh, S. 2020. Performance testing of selected spectral indices in automated extraction of impervious built-up surface features using Resourcesat LISS-III image. *Arabian Journal of Geosciences*, 13, 1229.

Santra, A., Mitra, S. S., Sinha, S., Routh, S. & Kumar, A. 2021. Identification of impervious built-up surface features using resourcesat-2 LISS-III-based novel optical built-up index. In: Kumar, P., Sajjad, H., Chaudhary, B. S., Rawat, J. S. & Rani, M. (eds.) *Remote Sensing and GIScience: Challenges and Future Directions*. Cham: Springer International Publishing.

Santra, A., Santra Mitra, S., Mitra, D. & Sarkar, A. 2019. Relative Radiometric Normalisation - performance testing of selected techniques and impact analysis on vegetation and water bodies. *Geocarto International*, 34, 98–113.

Schroeder, T. A., Cohen, W. B., Song, C., Canty, M. J. & Yang, Z. 2006. Radiometric correction of multi-temporal Landsat data for characterization of early successional forest patterns in western Oregon. *Remote Sensing of Environment*, 103, 16–26.

Sinha, S., Santra, A. & Mitra, S. S. 2020. Semi-automated impervious feature extraction using built-up indices developed from space-borne optical and SAR remotely sensed sensors. *Advances in Space Research*, 66, 1372–1385.

Song, C., Woodcock, C. E., Seto, K. C., Lenney, M. P. & Macomber, S. A. 2001. Classification and change detection using landsat TM data: When and how to correct atmospheric effects? *Remote Sensing of Environment*, 75, 230–244.

Sun, Y., Zhang, X., Shuai, T. & Zhuang, Z. Radiometric normalization of multitemporal hyperspectral satellite images. *2014 IEEE Geoscience and Remote Sensing Symposium*, 13–18 July 2014 2014. 4204–4207.

Teillet, P. M. 1986. Image correction for radiometric effects in remote sensing. *International Journal of Remote Sensing*, 7, 1637–1651.

Teillet, P. M. & Fedosejevs, G. 1995. On the dark target approach to atmospheric correction of remotely sensed data. *Canadian Journal of Remote Sensing*, 21, 374–387.

Vermote, E. F., Tanre, D., Deuze, J. L., Herman, M. & Morcette, J. 1997. Second Simulation of the satellite signal in the solar spectrum, 6S: An overview. *IEEE Transactions on Geoscience and Remote Sensing*, 35, 675–686.

Wu, W., Sun, X., Wang, X., Fan, J., Luo, J., Shen, Y. & Yang, Y. 2018. A long time-series radiometric normalization method for landsat images. *Sensors*, 18.

Xu, Q., Hou, Z. & Tokola, T. 2012. Relative radiometric correction of multi-temporal ALOS AVNIR-2 data for the estimation of forest attributes. *ISPRS Journal of Photogrammetry and Remote Sensing*, 68, 69–78.

Yuan, D. & Elvidge, C. D. 1996. Comparison of relative radiometric normalization techniques. *ISPRS Journal of Photogrammetry and Remote Sensing*, 51, 117–126.

Zhang, J., Mu, Q. & Huang, J. 2016. Assessing the remotely sensed Drought Severity Index for agricultural drought monitoring and impact analysis in North China. *Ecological Indicators*, 63, 296–309.

4 Remote Sensing as a Tool for Mangrove Ecosystem Mapping and Monitoring: On Sundarbans' Perspective

Kaushik Gupta
University of Manitoba
Jadavpur University

Niloy Pramanick
Jadavpur University

Anirban Mukhopadhyay
University of Manitoba
Jadavpur University

CONTENTS

4.1 INTRODUCTION

4.1.1 BACKGROUND

Mangrove forests are immensely diverse ecosystems along tropical seacoasts consisting of salt-tolerant plants with aerial roots that work as sediment entrapments and provide a microenvironment to many marine species (Upadhyay and Mishra, 2008). Coastal mangroves help regulating coastal flooding and erosion, as well as protecting inland communities like agricultural lands, livestock and homesteads and other nearshore communities, thus having a high socio-economic value (Bennett and Reynolds, 1993) as well as coordinating a source and sink system for many biochemical substances, their transformation, accumulation and remediation (Costanza et al., 1997; Birch et al., 2015; Chaudhuri et al., 2014). Most areas rich in mangrove density and diversity are predominantly inaccessible or logistically difficult to study on field and substantially time taking, hence there was a demand of a better, cost-effective and less time-consuming method of studying mangrove ecosystems (Mumby et al., 1999). Over the last few decades remote sensing studies have stood up as a response to this problem for being logistically accessible and prompt study technique, its most distinctive feature providing scope of studying areas that where otherwise inaccessible and remote to human ground-based study. Numerous studies have been carried out around the world focusing on the application of remote sensing for mangrove ecosystem mapping and monitoring (Alatorre et al., 2011; Giri et al., 2007; Ramsey et al., 1996).

Habitat loss is a major issue effecting mangrove ecosystems worldwide, anthropogenic activities, shifting of salinity gradient, natural disasters have a significant contribution to this problem (Giri et al., 2008). Mangroves have proved to be a distinctive ecosystem and economic service providers, which builds up the demand of its conservation and restoration. Monitoring and assessment of the distribution of mangrove ecosystems, species diversity, health of these ecosystems on a temporal basis is essential. Use of remote sensing in the process of mangrove ecosystem studies provides us with an array of information like overall biomass estimation, ecosystem productivity and health, change detection in land-cover and land use patterns, providing base-level data for estuarine and coastal zone management strategies, etc. (Dahdouh-Guebas, 2002; Cornejo et al., 2005; Green et al., 1996; Selvam et al., 2003).

4.1.2 AIM

The aim of this chapter is to provide an overview on the application of remote sensing techniques in monitoring and mapping mangrove vegetation. This review focuses on the use of in-situ and remotely sensed data (multispectral, hyperspectral and microwave remote sensing data) and diverse algorithms and vegetative indices (Normalized Difference Vegetation Index (NDVI), Soil Adjusted Vegetation Index (SAVI), Simple Ratio (SR), Enhanced Vegetation Index (EVI), Mangrove Recognition Index (MRI), Combined Mangrove Recognition Index (CMRI), Mangrove Vegetation Index (MVI), etc.) applied for proper feature extraction. Different methodologies, data sets, sensors and identification techniques are overviewed discussing their application. In the end we also perform a case study on the Indian Sundarbans to present the mapping capability of some selected classification indices.

4.1.3 ATTRIBUTES OF MANGROVE AND MANGROVE ECOSYSTEMS

Several authors in over numerous literatures (Hamilton and Snedaker, 1984; Aksomkoae, 1993; FAO, 1994; Kathiresan and Bingham, 2001; Lugo and Snedaker, 1974) over time have explained the subject of mangrove management and ecology. Though this study focuses to provide a review of spectral properties of mangroves, in this section, brief information on mangrove forest ecology, characteristics, distribution and services is described.

Scientists and experts over the years have used the term 'mangroves' in different ways (Tomlinson, 1986); the most common of which is used to identify trees and shrubs, even the overall ecosystems which have adapted to exist in the tidal environments through morphological adaptations (e.g. neumatophores, salt excretion glands and vivipary). In general, mangroves are classified as salt-tolerant trees and shrubs growing along the intertidal zones of lagoons, estuaries and deltas in over 124 tropical and sub-tropical countries, growing mainly on the soft silt plains (UNFAO, 2007). Mangrove ecosystems work as an ecotonal zone between the marine ecosystem and the terrestrial ecosystems influenced by the freshwater, nutrients and sediment load from the rivers with a constant action of tides and waves from the saline open sea waters. Mangrove growth, distribution and structure are highly governed by the salinity gradient, soil composition, geography and climate of the region they are found to exist. Most trees and shrubs growing in this peripheral zone of moderate to high saline content are termed as mangroves; however, only a few families of flora (e.g. *Rhizophoraceae, Avicenniaceae, Sonneratiaceae*) (Barik and Chowdhury, 2014) are found to be morphologically adapted to thrive in the saline water environment, where mangroves are generally said to be found. The exact number of species has not yet been noted; though out of almost 110 species of mangroves, there have been noted only 54 species belonging to 20 genera from 16 families that are classified as true mangroves (Tomlinson, 1986; Saenger et al., 1983; Lugo and Snedaker, 1975) with Asia having the highest species diversity followed narrowly by Africa.

Gaseous exchange, nutrient absorption and anchoring to the soil are a challenge in the saline silt-dominated coastal zones; mangroves have a signature adaptation of aerial roots which serves them this purpose and also encountered as the primary identification feature of this floral species. Structures of aerial roots vary with species. For example, stilt roots growing the lower trunk of *Rhizophora* spp. Snake-like roots found in *Ceriops* spp. are adaptations where roots run horizontally on the soil surface, while vertical outgrowths of the root systems, 'neumatophores' spreading in a considerable distance of the parent tree extending a long distance above the soil are found in the genera *Avicennia, Sonneratia*, etc. Another unique feature of mangroves is their reproduction mechanism; in order to successfully propagate themselves, mangroves have specialized adaptations, like *Vivipary*, that is, members in the *Rhizophoraceae* family do not release the fruit, and instead, the seed germinates when attached to the parental tree (Juncosa, 1982), while other genera like *Avicennia, Nypa* and *Aegiceras* exhibit cryptovivipary, where the embryo emerges from the seed instead of the fruit after its abscises from the parent tree (Carey, 1934).

4.1.4 Role and Utilization Potential of Mangrove

Mangrove and mangrove forest ecosystems are very important and they provide a broad range of ecological and economical services. Mangrove belts act as a first line of defense against most disasters; it serves as a natural barricade against wave- and tide-dominated actions along the coast hindering coastal erosion process (Dahdouh-Guebas et al., 2005; Mazda et al., 1997, 2002). Mangroves also work as an entrapment for upland river runoff, further helping in the protection of coral reefs and sea-grass beds. Scientists and experts also suggest that healthy and dense coastal vegetation like the Mangroves with proper management have the potential to act as bio-shields against storms, tsunamis and other coastal hazards (Alongi, 2008; Barbier, 2006; Danielsen et al., 2005; Othman, 1994). Mangrove ecosystems helps sustain a diverse range of biological diversity by providing habitats, nurseries, spawning grounds and nutrients to a number of marine fauna like, crabs, offshore and reef fishes and larvae, animal fauna like, birds, mammals, insects and reptiles and amphibians, even associated floral species such as some exclusive fungal and algal species (Manson et al., 2005; Nagelkerken et al., 2008; Cannicci et al., 2008; Mumby et al., 2004). They also provide habitation to many endangered species ranging from mammals (*Panthera tigris tigris*, deer, otters, and manatees) to avian species (pelicans, egrets, eagles and herons) and several reptiles (crocodiles, snakes, monitor lizards etc.) hence having a very important role in marine food chain. Mangroves are also responsible for maintaining the carbon cycle in estuarine systems (Bouillon et al., 2008; Kristensen et al., 2008), besides contributing in the accumulation, remediation and transformation of many marine pollutants.

Furthermore, mangroves are directly beneficial to humans as they contribute to the human livelihood in many ways such as, wood products (poles, posts, timber, charcoal and fuel wood) and non-wood forest products (edible items, medicines, honey, alcohol, etc.) (Alongi, 2008; Walters et al., 2008; Bandaranayake, 1998; Gaodi et al., 2010; Wells and Ravilious, 2006; Costanza et al., 1997) as well as serving as rich fishing zones. With the growing popularity of the idea of eco-tourism, areas of mangrove forests especially the ones which are easily accessible serve as good tourism places, incurring potential source of income to local populations.

4.2 STUDY SITE

The Sundarbans is one of the largest and most diverse mangrove ecosystems, globally. Being shared between India and Bangladesh, this massive delta is fed by two main rivers, namely, Ganges and Brahmaputra. The entire stretch of the Sundarbans is limited between the Hoogly River (India) in the west and Meghna River (Bangladesh) in the east. The Sundarbans support a huge spectrum of flora and fauna, with over 35 reptile species, 42 mammal species and 290 bird species (Iftekhar and Islam, 2004). The most dominant vegetation type found in the region is mangroves (27 mangrove species). The assemblages of common mangrove species found in the region are *Avicennia, Xylocarpus, Sonneratia, Bruguiera* and *Rhizophora*. The two major threats faced by this vegetation type in Sundarbans are climate induced and/or subsidence influenced sea-level rise and changing salinity regime (McLeod and Salm 2006). Change in sea level is a threat to the spatial distribution of mangroves, whereas shifting salinity gradients have been observed to alter the species assemblages. Over the past few decades the region has experienced species-level succession within the mangrove vegetation types as more salinity-tolerant species like *Avicenia* replacing more freshwater-loving species like *Heritiera* (Chaffey, 1985; Gopal, 2006). Apart from being ecologically important, the region also harbours great economic value. Economic activities like fishery, fuelwood and timber collection practices and honey collection are some of the most common occupational practices in the region. Considering the ecological and economic important of the region, the Sundarbans was declared a World Heritage Site by the UNESCO in 1987.

4.3 MANGROVE CHARACTERISTICS THROUGH REMOTELY SENSED DATA

4.3.1 CHARACTERISTICS OF MANGROVES THROUGH OPTICAL REMOTE SENSING DATA

Like most other vegetation types, remote sensing of mangroves also exploits the spectral and textural characteristics of canopy cover and leaves (Ramsey et al., 1996), though they can be separated from other vegetation communities using several other factors like leaf morphology, stand height, distribution pattern and canopy cover. Canopy cover is an important factor in terms of mangrove classification as it depends on other components like Leaf Area Index (LAI), background reflectance and leaf inclination (Díaz and Blackburn, 2003). Periodic changes in climatic conditions and soil salinity levels affect the spectral characteristics of the leaves and their morphology which is sensitive for spectral signatures (Wang et al., 2008). Spectral signatures of mangroves differ from other vegetation communities majorly due to their leaf morphology and canopy cover, though spectral signatures of new or young leaves differ from older leaves and heterogeneity in a particular area may stand as a limiting factor or a model developed for their classification.

As discussed earlier, mangroves are generally found in the ecotonal zone between water and land, hence classification of these areas in terms of remote sensing consists of differentiation into three major class types, that is, soil, water and vegetation. One of the major problems faced when using optical methods for classification of mangroves in this zone is the effect of seasonal and diurnal tidal dynamics, the most prominent problem faced when using optical remote sensing data on this zone is the difference in spectral values in low-tide and high-tide conditions in the same region (Blasco et al). Contributing to the difficulty, species diversity, mixing of mangrove and non-mangrove classes and canopy dynamics also provide an eminent challenge in mangrove discrimination. Species diversity in Asia is much greater than that of the tropical and sub-tropical areas of the new world (Ramsey et al., 1996), variation in dominant species and the difference in their spectral characters furnish a greater challenge in setting up of a robust methodology for discrimination of mangroves using optical methods.

4.3.2 Mangrove Moisture Content Properties

Studies show mangroves have tolerance to a wide range of soil salinity. Salinity has been long studied as an important factor for plant survival and growth. Mangroves, however, in contrast to most other plant species are highly soil salinity tolerant. Saline soil in comparison to non-saline soil offers a higher physiological challenge to the plants due to its highly negative water potential of soil pore water, making water acquisition a greater energy-involved process. Hence, the process of maintaining water uptake in such saline conditions is an important aspect of salt tolerance.

In order to combat high salinity and a higher energy involved water acquisition from soil mangroves have evolved a number of adaptations like alterations of leaf size and angle, succulence or water storage in leaves, suberization of roots and biomass partitioning (Reef and Lovelock, 2015). Water storage in plants facilitates advantage to a range of problems arising due to high salinity gradient in the estuarine soil; mangrove leaves play a vital role as water storage organs. In mangroves with high soil salinity increased leaf water content is observed in many mangrove species (Suárez and Sobrado, 2000; Parida et al., 2004). Succulence of leaves enables mangroves to impound large amounts of solutes to maintain turgidity at low water potential and without adversely increasing cell osmotic pressure (Parida et al., 2004). These studies across a long time imply that mangroves have higher water content in their leaves as compared to other plants, as it imparts an advantage of existing in saline conditions. This feature may impart some distinctive features of mangroves as compared to other form of non-mangrove vegetation while using different water classification indices.

4.3.3 Characteristics of Mangroves through Radar Remote Sensing Data

Unlike optical systems, radar systems are active remote sensing systems that depend majorly on the intensity of the returning signals. Synthetic Aperture Radar or SAR-based remote sensing techniques involve the estimation of backscattering coefficients '$\sigma°$' in decibels (dB). SAR systems operate under an array of configurations like wavelengths, incidence angle and polarization of transmitted and received signals, different combinations of these configurations may generate different signatures of the same target, making these systems much complex to use and interpret compared to optical systems (Kasischke et al., 1997). Though, 'all weather condition' and 'night time' imaging capability of radar systems gives it an edge over optical-based remote sensing systems. In application, SAR configurations have a huge role to play in determining the type of interaction it may produce with the target. Apart from system configuration, interaction of radar signals and targets (vegetation, specifically mangroves) depends intensely on the internal (moisture content concerning the dielectric properties of the target) and external (canopy structure, leaf structure and orientation, orientation of stilt roots and branches) components of the target (Kuenzer et al., 2011). As for mangroves, the distinct root system is expected to yield a distinct signature compared to other vegetation types.

4.4 METHODS AND STUDIES ON MANGROVE DISCRIMINATION AND MONITORING

Remote sensing techniques have been used for the assessment and monitoring of the extent, species diversity and health status of the mangrove ecosystems over the last two decades. Here, we tend to provide an overview of the different types of sensors, classification techniques and indices used for identifying and assessing mangrove habitats over the last few decades. Successful discrimination of mangroves is a challenge due to their high diversity, local adaptations, comprehensive field information, etc., which are the base-line information required prior to any image-analysis method or identification studies. Same species of mangrove may differ in their structural or morphological components based on their local environment; this hinders the development of a robust method for mangrove classification through remote sensing. Hence proper study of the local environmental setting, thorough ground survey information about species diversity, species assemblage and plant morphology is required. These informations are essential for verification and training of image analysis techniques.

4.4.1 MANGROVE IDENTIFICATION STUDIES BASED ON MULTISPECTRAL REMOTE SENSING DATA

Optical or multispectral remote sensing sensors (medium-resolution imagery) have been widely used over the years in the field of mangrove ecosystem mapping in local and regional scales. Numerous sensors and methods have been formulated and used in different geographical areas for mangrove mapping, most common of which are Landsat TM, MSS, ETM + and SPOT XS in sensors, other sensors such as IRS LISS III, THEOS (MS), MERIS, MODIS and Sentinel-2 were also used by some researchers. Spectral signatures of mangroves in remotely sensed data are very distinct and correspond to the Red, NIR, SWIR and Mid-IR region. This distinct spectral signature enhances the better separation of mangroves from other vegetation types (Giri, 2016). Information from these spectral bandwidths are commonly used in most classification techniques and indices developed or used to differentiate mangrove from non-mangrove vegetation types.

Most studies on mangrove and mangrove ecosystems been done over the Sundarbans have used medium-resolution multispectral imageries, mostly, the Landsat series. Long-term global coverage, multiple optical bands and moderate spatial resolution make the Landsat series a popular choice for these studies. Products from Sentinel-2, MODIS and LISS are some other widely used data sets (Table 4.1). Giri et al. (2007) attempted to monitor the changes in the mangrove forest of Bangladesh and Indian Sundarbans over the years 1970s–2000s. For this study, the authors heavily relied on

TABLE 4.1

Overview of Remote Sensing Techniques, Sensors and Methods for Mapping and Monitoring of Mangrove Ecosystem of Sundarbans

Data Type	Sensors	Sensor Type	Bands	Indices	References
RADAR Data	ERS-1	Satellite borne	C band		Dwivedi et al., (1999)
	JERS-1 SAR	Satellite borne	L band		Syed et al., (2001)
	ENVISAT ASAR	Satellite borne	C band		Karmaker et al., (2006)
	ALOS PAL SAR	Satellite borne	L band		Turkar et al., (2011)
	ALOS PAL SAR	Satellite borne			Cornforth et al., (2013)
	RADARSAT	Satellite borne	C band		Kumar et al., (2013)
Hyperspectral Data	Field Spectroradiometer	In-Situ	350–2500 nm		Manjunath et al., (2013)
	Hyperion EO-1	Satellite borne		ATGP	Chakravortty et al., (2013)
	Hyperion EO-1	Satellite borne		MPVI; NDWVI; SIAI; NDII; ACVI	Kumar et al., (2017)
	AVIRIS-NG and Field Spectroradiometer*	Air-borne and In-Situ		EVI; SIPI; SAM based Supervised Classification	Chaube et al., (2019)
	Hyperion EO-2 and LISS IV*	Satellite borne	VNIR (1–70); SWIR (71–242); Red, Green, NIR	NDVI	Mondal et al. (2019)
	Hyperion EO-1	Satellite borne		DNVI; ARI; MCARI; NDII; NDLI; PRI: NDNI; ARVI; CRI-1; SIPI	Ghosh et al., (2020)
	AVIRIS-NG; Sentinel-2*	Air-borne and Satellite borne			Hati et al., (2020)
	AVIRIS-NG	Air-borne		SAM, SVM	Kumar et al., (2020)

(Continued)

TABLE 4.1 *(Continued)*

Overview of Remote Sensing Techniques, Sensors and Methods for Mapping and Monitoring of Mangrove Ecosystem of Sundarbans

Data Type	Sensors	Sensor Type	Bands	Indices	References
Multispectral Data	Landsat MSS, Landsat TM, Landsat ETM+, SPOT XS, QuickBird	Satellite borne	Red, NIR	NDVI	Giri et al., (2007)
	Landsat ETM+	Satellite borne	VNIR	Supervised Classsification; MLC	Rahman et al., (2013)
	Landsat TM and Landsat ETM+	Satellite borne	VNIR	Supervised Classsification; MLC	Giri et al., (2014)
	LISS IV	Satellite borne	Green, Red, NIR	NDVI; OSAVI; TDVI	Manna et al., (2014)
	MODIS	Satellite borne	VNIR	EVI	Dutta et al., (2015)
		Satellite borne		Markov Chain and Cellular Automata	Mukhopadhyay et al., (2015)
	Landsat 8 OLI	Satellite borne	Green, Red, NIR	CMRI, NDVI, SAVI, SR	Gupta et al., (2018)
	Landsat TM, Landsat ETM+, Landsat 8 OLI	Satellite borne	VNIR	SAVI; RFC	Ranjan et al., (2018)
	Landsat 8 OLI	Satellite borne	VNIR, SWIR		Mukhopadhyay et al., (2018)
	MODIS	Satellite borne	VNIR	EVI; NDVI	Mandal et al., (2020)

* Complimentary datasets used in the study

the use of the Landsat Series (Landsat MSS, Landsat TM and Landsat ETM+). The analysis involved classification and mapping of the areal coverage of mangroves within the selected time period using supervised classification technique and NDVI. Other complimentary data sets included SPOT XS and QuickBird images. The study concluded that in the light of the changing climate, dynamic erosion and accretion rates and anthropogenic stress, the areal coverage of mangroves remains more or less the same. With some insignificant change within 1990–2000 (2.5%), which can be owed to the classification errors, there was not any serious change noticed over the 30-year time period. Another study, Rahman et al. (2013) compared classification methods for classifying mangrove vegetation type, focused specifically on the Landsat ETM+ bands. The study included mangrove forests from both the Indian and Bangladesh Sundarbans. Three classification approaches were used, unsupervised classification with k-mean clustering, supervised classification using maximum likelihood decision rule and band-ratio supervised classification. The results showed that the band-ratio supervised classification approach produced higher accuracy compared to the other two classification methods. Band ratios B4/B2 (band 4/band 2), B5/B7 and B7/B4 yielded the best discrimination of the mangrove and non-mangrove boundary, when using Landsat ETM+ data sets. Another interesting study over the Indian Sundarbans by Giri et al. (2014) attempted species-level classification of mangroves and change detection from 1999 to 2010. The study used Landsat TM and Landsat ETM+ data sets, which were subjected to supervised classification using Maximum Likelihood Classifier. Hyper-spectral imagery from Hyperion EO-1 was used as a complimentary data set and used in the study to extract species specific spectral signatures over the same study area. This information was used in the classification performed over the Landsat series datasets. Results from the study indicate a decline of total mangrove vegetation from 55.01% (of the study area) in 1999 to 50.63% in 2010. Species-level classification and change detection of the mangrove species showed a significant decrease in areal coverage of certain mangrove species like *Avicennia* spp., *Ceriops* spp. and *Bruguiera* spp. In the same year, Manna et al. (2014) used high-resolution LISS IV multispectral data to estimate above-ground biomass of *Avicennia marina* plantation of the Indian Sundarbans to assess the carbon sequestration potential of the selected patch of mangroves. The study used reflectance values from the LISS IV bands, which were further incorporated into generating outputs from three different indices, namely, NDVI, Optimized Soil Adjusted Vegetation Index (OSAVI) and Transformed Difference Vegetation Index, and correlated with the observed above-ground biomass. The OSAVI produced the best results, which can be owed to its capability to reduce the influence of background soil spectra and hence dominated by reflectance from vegetation type. Dutta et al. (2015) in this study adopted a satellite-based approach to estimate ecological disturbances caused by cyclones in the Indian and Bangladesh Sundarbans forests. The study included the use of MODIS multispectral imagery; EVI was applied on the satellite images. Changes in the EVI outputs between the pre- and post-cyclone conditions were used to estimate the damage to mangrove forests. In a study focused on the Bangladesh Sundarbans, Mukhopadhyay et al. (2015) used multispectral data from the Landsat series combined with survey informations from the Bangladesh Forest Department, attempted to study the species composition of the mangrove forests. The study included the use of species zonation maps from 1985 to 1995 which was used as the baseline data, developing on this Cellular Automata-Markov Chain model was run to estimate the mangrove species composition for the year 2005. The outputs were calibrated against the base data from 2005. Upon successful calibration the model was used in a projection study of mangrove species zonation for the years 2025, 2055 and 2105. Apart from the use of satellite imagery in assessing the health status, change detection and prediction of species dynamics of mangroves, study pertaining to development of a new improved mangrove classification technique was also accomplished focusing on the Sundarbans mangroves. Gupta et al. (2018) used Landsat 8 OLI imagery to develop a new mangrove discrimination index, CMRI.

Apart from having distinct greenness values, mangroves also exhibit succulence as a morphological adaptation to survive in the saline environment making them sensitive to water indices like NDWI. Building on that idea, the new index uses informations from the NDVI and NDWI outputs over the same region of interest to present its own range of values suitable for a more accurate discrimination of mangroves from non-mangrove vegetation classes. The accuracy of CMRI was tested against other already existing indices like NDVI, SAVI and SR, where the CMRI performed better than the others. The new index was tested on other mangrove forests across India like Bhitarkhanika and Andaman Islands to test its robustness. Ranjan and Kanga (2018) used the Landsat series data (Landsat TM, Landsat ETM+ and Landsat 8 OLI) to assess the temporal variations in the spatial patterns of mangrove forests in the Indian Sundarbans. The study included the use of SAVI and supervised classification (Random Forest Classification) to map the mangrove forest and finally estimating the forest dynamics. Mukhopadhyay et al. (2018) attempted high-resolution spatial mapping of mangrove species migration with changing aquatic salinity gradients. The study used Landsat 8 OLI images to map mangrove species distribution in the Sundarbans. This species dynamics was overlaid on the baseline salinity maps developed from in-situ field measurements, and a prediction was attempted to estimate the species distribution of mangroves with changing salinity gradient by 2050. In a more recent study, Mandal et al. (2020) attempted to estimate the time, length and periodicity of the active photosynthetic season using EVI and NDVI outputs from MODIS multispectral imagery. The study concluded that EVI performed much better compared to NDVI in predicting canopy greenness.

4.4.2 MANGROVE IDENTIFICATION USING HYPERSPECTRAL REMOTE SENSING DATA

As discussed in the earlier Section 4.1, multispectral sensors have been widely used in the mapping and monitoring of mangrove forests, though the broad bandwidths of these sensors offers deficiency in characterizing small-scale and species-level variations. To this existing situation hyperspectral remote sensing provides new venues for monitoring subtle variations within the vegetation type. Large number of narrow (<10 nm) bandwidths over the spectral range of 0.38–2.5 µm provides greater detail of the complete mangrove spectra (Green et al., 1998). Hyperspectral remote sensing provides detailed insights beyond the non-photosynthetic spectrum, thus furnishing vital informations on leaf water content, bio-physical and bio-chemical properties of the vegetation type under the influence of the local climate and its environment (Vaiphasa et al., 2005; Green et al., 1998). Hence, this remote sensing approach has a high application in monitoring the health status of the mangrove ecosystems.

Numerous studies have been performed focusing on the Sundarbans, have used hyperspectral remote sensing techniques to classify mangroves, studying species-level variations of spectra and monitoring the mangrove health status (Table 4.1). Though satellite-borne sensors (Hyperion EO-1) and air-borne sensors (AVIRIS HG) remain utilized in these studies, some studies have also used in-situ hyper-spectral remote sensing to discriminate mangroves at a species level, extracting species-specific spectral differences (Manjunath et al., 2013). A number of studies have been accomplished using Hyperion EO-1 datasets, like Chakravortty (2013), attempted to accomplish species-level distinction of dominant mangrove vegetation using sub-pixel classification-based methods. In their study they used automated target generation process to accomplish species-level discrimination of mangroves. This study was further extended to detect the changing patter of mangroves at a species level with association to saline blanks within mixed mangrove patches (Ghosh and Chakravortty, 2020). Another similar study was done by Mondal et al. (2019), where the authors used hyperspectral remote sensing data (Hyperion EO-1) to extract species-level spectral information of the Sundarbans mangroves and training sets were generated. These training sets were then used in an Object-Based Image Analysis classification technique performed over high-resolution (5 m) multispectral LISS IV satellite imagery. Kumar et al. (2019) showed five different classification indices, namely, Mangrove Probability Vegetation Index (MPVI), Normalized Difference Wetland Vegetation Index (NDWVI), Shortwave Infrared Absorption Index, Normalized Difference Infrared Index (NDII) and Atmospherically Corrected Vegetation Index (of which the three were new and two already published) to extract mangroves over a portion of the Indian Sundarbans.

Apart from satellite-borne sensors (Hyperion EO-1), several other studies used the air-borne Aerial Visible-Infrared Imaging Spectrometer – Next generation (AVIRIS-NG) for mangrove classification and assessment of mangrove health status. Chaube et al. (2019) used AVIRIS-NG hyperspectral data complimented with in-situ spectroradiometer observations to achieve species-level discrimination of mangroves using Spectral Angle Mapper classification technique. The study also involves the use of EVI, carotenoid reflectance index and structure insensitive pigment index (SIPI) to assess the health of the mangrove species found in the Indian Sundarbans. Another study by Hati et al. (2020) attempted to assess the capability of the AVIRIS-NG product in accurately mapping and monitoring mangrove health status. In their study they used nine such vegetation health indices like Anthocyanin Reflectance Index, Modified Chlorophyll Absorption Ratio Index, NDII, Normalized Difference Lignin Index, Photochemical Reflectance Index, Normalized Difference Nitrogen Index, Atmospherically Resistant Vegetation Index, CRI-1 (Carotenoid Reflectance Index) and SIPI. The results from these indices were later validated using the Discriminant Normalized Vegetation Index derived from Sentiel-2 multispectral imagery in the same location and time frame.

4.4.3 Target-Specific Classification Indices and Other Vegetation Indices

Over the years' numerous vegetation indices and mangrove classification algorithms have been used to identify and discriminate mangrove vegetation types from non-mangroves, though the two work differently. Most vegetation indices involve band rationing techniques using two bands or more which have contrasting reflectance from the same target (Table 4.2). Indices like NDVI involve the use of Red and NIR bands, where the Red band shows high absorption and NIR has a high reflectance from vegetation dominated target areas (Pearson and Miller, 1972). The contrasts in this spectral information are used in this index to provide a range of values that exclusively represent vegetation. Other simpler band ratio-based indices include SR, which basically extracts information based on the ratios of spectral values from the Red and NIR band (Huete, 1988). Such index is sensitive to any target within the image scene that produces a contrasting spectral signature within the two selected bands. Some other vegetation indices involve 'optimization', where mathematical coefficients are used to eliminate or decrease the effect of background signals and atmospheric influences. SAVI is such an index that involves the ratios of reflectance from the NIR and Red band optimized with 'L' which is a soil adjustment factor (Huete, 1988). The inclusion of L in the SAVI equation diminishes the influence of background signals of the soil producing a better extraction of vegetation. This method was a further improvement of the already existing NDVI, which only involved differential ratios between the Red and NIR bands. Another example of such 'optimized' index is the EVI. Contrasting to the other indices mentioned before, EVI utilizes information from three bands, Red, NIR and Blue. Apart from the band informations, the EVI equation involves gain factor (G), canopy background adjustment factor (L), and aerosol resistance terms C1 and C2, which is used to correct the aerosol influence in the Red band using the spectral information from the Blue band (Liu and Huete, 1995). Contrasting to NDVI which is sensitive to chlorophyll, EVI utilizes canopy structural variations like LAI and canopy type (Huete et al., 2002). The utilization of the Blue band in EVI is limited to diminishing the influence of aerosol; it does not contribute to the estimation of biophysical properties of vegetation. Hence, a modification was done for the existing 'three-band' EVI, that involved taking out the blue band out of the equation. This modification led to the development of the 'two-band' EVI or EVI-2 (Jiang et al., 2008). Without the Blue band the index will be slightly more affected by atmospheric aerosol, but with the advancement of better sensors and atmospheric correction algorithms, this difference grows more insignificant.

Vegetation indices like the ones discussed above have been used to separate mangroves from non-mangroves, though the sole purpose of these indices is to identify vegetation. Variation in canopy spectra and greenness produces different output values in these indices, which is confined within a range. In practice, ground information and biophysical properties of a particular vegetation type are used to select a value range within the output range of these indices to highlight a particular vegetation type. In contrast to this approach, target-specific classification indices aim towards classifying only a desired vegetation. Such indices are developed taking into consideration the

TABLE 4.2

An Overview on the Various Indices Used for Assessment of Mangrove Ecosystems, with Their Formulas

Classification Techniques	Indices	Formula	Data Used	References
Vegetation indices	Simple Ratio (SR)	$(\mathrm{NIR}/\mathrm{Red})$		Huete (1988)
	NDVI	$\dfrac{(\mathrm{NIR}-\mathrm{Red})}{(\mathrm{NIR}+\mathrm{Red})}$		Pearson (1972)
	SAVI	$\dfrac{(\mathrm{NIR}-\mathrm{Red})(1+L)}{(\mathrm{NIR}+\mathrm{Red}+L)}$		Huete (1988)
	EVI	$G\times\dfrac{(\mathrm{NIR}-\mathrm{Red})}{(\mathrm{NIR}+C1\times\mathrm{Red}-C2\times\mathrm{Blue}+L)}$		Liu and Huete (1995)
	Two-band Enhanced Vegetation Index (E V1_2)	$G\times\dfrac{(\mathrm{NIR}-\mathrm{Red})}{(L+\mathrm{NIR}+\mathrm{Red}\times C)}$		Jiang et al. (2008)
Target-based classification indices	Mangrove Recognition Index (MILI)	$\left|\mathrm{GVI}\,l-\mathrm{GVI}\,h\right|\times\mathrm{GVI}\,l\times\left(\mathrm{WI}\,l+\mathrm{WI}\,h\right)$	Landsat	Zhang & Tian (2013)
	Mangrove Index (MI)	$10{,}000\times\dfrac{(\mathrm{NIR}-\mathrm{SWIR})}{(\mathrm{NIR}\times\mathrm{SWIR})}$	Landsat	Winarso et al. (2014)
	Combined Mangrove Recognition Index (CM111)	$\mathrm{NDVI}-\mathrm{NDWI}$	Landsat 8	Gupta et al. (2018)
	MPVI	$\dfrac{n\sum\limits_{n-1}^{n}R_ir_i-\sum\limits_{n-1}^{n}R_i\cdot\sum\limits_{n-1}^{n}r_i}{\sqrt{n\sum\limits_{n-1}^{n}R_i^2-\left(\sum\limits_{n-1}^{n}R\right)^2}\sqrt{n\sum\limits_{n-1}^{n}r_i^2-\left(\sum\limits_{n-1}^{n}r_i^2\right)^2}}$	Hyperion E0-1	Kumar et al. (2019)
	NDWVI	$\dfrac{(R_{2203}-R_{559})}{(R_{2203}+R_{559})}$	Hyperion E0-1	Kumar et al. (2019)
	I – MVI	$\dfrac{(\mathrm{NIR}-\mathrm{Green})}{(\mathrm{SWIR}\,1-\mathrm{Green})}$	Landsat 8, Sentinel-2	Baloloy (2020)

biophysical characteristics of the target which are exclusive to that vegetation type and its environment (Table 4.2). MRI, developed by Zhang and Tian (2013), is a target-specific classification index aimed to identify mangrove forests from multispectral satellite imagery. This index takes into consideration not only the spectral characteristics of mangroves but also the background effect of tides prevalent in the region. Tidal levels have been noted to have an effect on the spectral properties of mangroves (Zhang et al., 2017); the MRI is based on this observation. The index involves GVI (Green Vegetation Index) which represents spectral properties of the vegetation type and WI (Water Index) representing spectral properties of canopy water content, which is further estimated in low-tide and high-tide conditions. Incorporating the GVI and WI estimates in low and high tidal conditions the MRI output highlights areas with mangrove vegetation cover. The limitation of this method is related to the prior knowledge of the tidal cycles in the study area and acquisition of the satellite images in the desired tidal conditions, which challenges the robustness of the method. Another such index, the CMRI, was developed to separate mangrove forest type from non-mangrove forest type using the greenness and moisture content of the mangroves (Gupta et al., 2018). Apart from greenness-sensitive spectral properties mangroves are also present a degree of sensitivity to spectral properties related to wetness or moisture content. As discussed in Section 3.3, mangrove leaves exhibit succulence as a morphological adaptation to exist in the saline estuarine environment. Developing on this observation, the CMRI takes into consideration the outputs from NDVI and Normalized Difference Water Index (NDWI) to highlight mangrove vegetation types. The method involves generating NDVI and NDWI outputs for the study area, as the outputs from the two indices are negatively correlated, further subtraction of the NDWI output from the NDVI output enhances and magnifies the range of the target specific values for the mangrove forests. The index was tested in three different types of mangrove forests in India, substantiating its robustness. Unlike the MRI, the CMRI omits the need of region-specific tidal informations. A recently developed mangrove classification technique, MVI, focuses on a similar theory to that of the CMRI, utilizing greenness and moisture content of mangrove pixels. But, unlike the CMRI, MVI administers the use of NIR, Green and SWIR1 band combinations from Sentinel-2 directly in the equation (Baloloy et al., 2020). The method focuses on the ability of SWIR1 and Green bands to highlight mangrove vegetation types. The method involved extraction of spectral properties which were used for the development of the index and further generating a MVI threshold for mangrove vegetation type. The method is fairly rapid and negates the disadvantages of user-skill bias. This index was developed using the Sentinel-2 satellite data but the authors have also tested its execution on similar Landsat band combinations, adding to the robustness of this index. The automation of this index using IDL-based MVI mapper for offline processing and online based Google Earth Engine (GEE) interface makes this a rapid and easy-to-access tool for mangrove mapping-based studies.

4.4.4 Mangrove Identification Using SAR Data

In Section 3.3, we discussed the various factors that work as a prerequisite for the application of radar remote sensing in studying mangroves and mangrove ecosystems. In this section, we elaborate on numerous studies that have used SAR data for studying mangroves in the Sundarbans mangrove forest (Table 4.1). Dwivedi et al. (1999) attempted to use ERS-1 C-band SAR data to map mangrove vegetation over the Indian Sundarbans; in this study, the authors also used IRS-IB LISS II optical data compliment the classification outputs from the SAR product. Another study over the Bangladesh Sundarbans by Syed et al. (2001) tested and compared the capability of JERS-1 L-band SAR data with JERS-1 VNIR (optical) data and Landsat TM product for accurate classification of mangroves. The results from the study show that JERS-1 SAR produces the highest accuracy among the three chosen satellite data types. A more elaborate study was done by Karmaker (2006), detailing on the role of polarization of SAR signals in estimating mangrove biomass over the Indian Sundarbans. The author used ENVISAT Advanced Synthetic Aperture Radar C-band SAR data for the study; analysis of the C-band SAR data shows that the variations of the canopy components of the mangrove forest like leaves, branches and twigs have a higher correlation with the C-band estimated outputs. Below-canopy components can be better studied using L-band or P-band SAR products. Results from this study also show that cross-polarization (HV) performs better for the estimation of total basal area, compared to HH polarization. Contrastingly, co-polarization (VV) offers a better correlation with canopy components compared to HH polarizations. Another study by Turkar et al. (2011) compared the accuracies of fully polarized and dual-polarized L-band ALOS PAL SAR data for mangrove classification. This study was done over the mangrove forests of Mumbai and Sundarbans. It was observed that fully polarized complex yields a better classification accuracy (95%) compared to dual-polarized complex (92%). Another study focusing on the use of radar data to classify and discriminate mangrove vegetation from other vegetation types was done by Kumar and Patnaik (2013). The study used C-band dual-pol RADARSAT data to classify mangrove forest using a decision rule-based classification technique. Several other such studies based on the study of mangroves using radar remote sensing over the Indian and Bangladesh Sundarbans can be found in the case study by Cornforth et al. (2013).

4.5   CASE STUDY: INDIAN SUNDARBANS

4.5.1   STATION AND GROUND INFORMATION

In this section a case study is elaborated, where a few most commonly used vegetation indices have been applied to differentiate mangrove vegetation types from non-mangrove vegetation using Landsat 8 OLI satellite imagery. The area selected for the study consists of the entire Indian Sundarbans. Numerous field-based surveys were conducted between 2017 and 2018 across the northern, western and central part of the Indian Sundarbans. Information collected from the field surveys consisted of documentation of density of mangrove patches, species diversity, identifying common mangrove assemblages and recording locations (Ground Control Points [GCPs]) at each observation site. The observation sites were carefully distributed to account for all the possible variations of mangrove forest types. Sections of these observation sites included areas close to the local communities with the presence of peripheral mangroves hugging the coastline; areas with massive mangrove monocrop assemblages as well as some places representing a higher degree of mixed mangrove patches with immense species diversity. Some of the selected study locations include Lothian Islands, Bhagabatpur, Pathar Pratima, Kshetra Mohanpur, Choto Rakhashkhali, Dhanchi, Jharkhali, Gosaba, Bally, Bonny Camp, Herobhanga, Sajhnekhali, Dobanki Camp, Netidhopani, Satjelia and Kumirmari.

4.5.2 Geospatial Analysis

For this case study we relied on the use of Landsat 8OLI multispectral imagery, available from the United States Geological Survey, Earth Explorer. The Landsat 8 OLI satellite image contains spectral information in eight bands (Coastal Aerosol, Blue, Green, Red, Near-infra Red, Short-Wave Infra-Red [SWIR] 1, SWIR2 and Cirrus) of which bands Red, Green, NIR and SWIR1 were used for this study. Pre-processing of images involved radiometric calibration, where the DN values were converted to TOA (top-of-atmosphere) spectral radiance; the requisite information for calibration are provided in the image metadata file. Finally, the spectral bands were converted to TOA planetary reflectance, using the formula illustrated in Landsat 8 User's Handbook, 2016 (USGS Landsat, 2016).

Conversion to TOA reflectance helps us estimate the spectral properties of mangroves, which was further useful in generating outputs from various vegetation indices and mangrove classification indices used in the study. Vegetation Indices like SR, NDVI, EVI and SAVI were applied on the radiometrically corrected Landsat 8 multispectral images (Table 4.2). The outputs present a range of values, from where we extract a segment of the range that provides the best representation of mangrove vegetation type. Ground information collected from the field surveys were used to estimate this value range that represents the mangrove vegetation type. Similarly, indices developed with the specific purpose of discriminating mangroves from non-mangrove vegetation types such as MRI, MVI and CMRI were also applied to the images.

4.5.3 Results and Discussion

Several remote sensing techniques, sensors, classification techniques and indices have been discussed in the earlier section. In a rough comparison, multispectral remote sensing approach using moderate resolution satellite imagery remains one of the most widely used techniques for mangrove mapping. To provide an overview on the application of these methods and approaches we have performed a case study highlighting the most commonly used classification techniques and indices over the Indian Sundarbans using Landsat 8 OLI satellite image. Several studies have been accomplished using vegetation indices for the classification of mangroves, using the distinct greenness values of the mangrove forest types. In this study we used four vegetation indices, namely, NDVI, SR, SAVI and EVI in an attempt to indicate their capability in identifying mangrove forests. SR uses a ratio between the reflectance of NIR and Red bands to generate a range of values (Figure 4.1b). It was observed that the higher values of the index (1.3 and higher) correspond to dense mangrove patches, though the index did not perform well for the identification of sparsely distributed mangroves and peripheral mangroves. As for the NDVI output the range of values distributed within the mangrove dominated areas (0.2–0.32) ranged quite low (Figure 4.1a). The output could not provide a clear distinction between the mangrove and non-mangrove edge. Compared to the SR and NDVI, SAVI and EVI performed well in discrimination mangrove vegetation. Being optimized by the soil adjustment factor 'L', SAVI diminishes the impact of soil background reflectance on the overall classification, rendering a better and more distinct representation on the mangrove dominated areas (Figure 4.1d). Similarly, EVI output was also quite successful in discriminating the mangrove vegetation (Figure 4.1c). Though the only issue faced in the EVI output was the inaccurate classification of mangroves on the edges of saline blanks within the islands.

Apart from using the application of vegetation indices, we also attempted to classify and discriminate mangroves using three different target based classification indices, developed specifically for mangrove discrimination. Of the three selected mangrove classification indices, namely, MRI, CMRI and MVI, only the CMRI remains applied and tested over the Indian Sundarbans (Gupta et al., 2018). The MRI

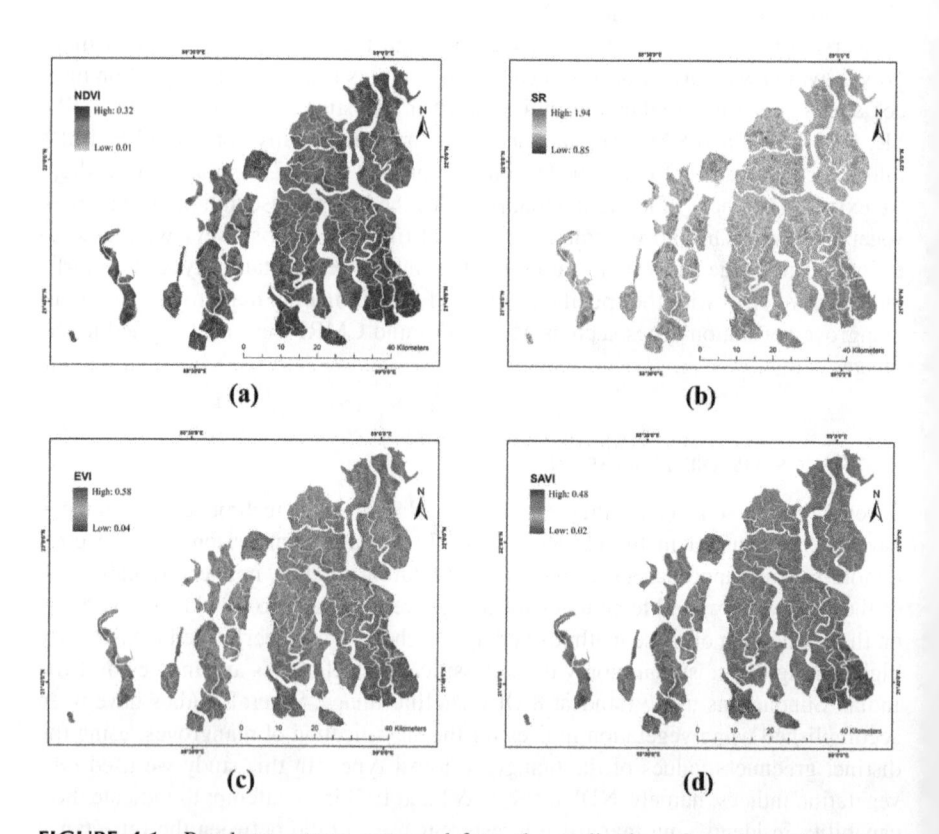

FIGURE 4.1 Resultant outputs generated from the application of vegetation indices on Landsat 8 OLI imagery over the Indian Sundarbans. (a) NDVI; (b) Simple Ratio; (c) Enhanced Vegetation Index; (d) SAVI.

uses the greenness value and results from the WI in low- and high-tide conditions to discriminate mangroves from non-mangrove vegetation. Compared to the MRI, the other two indices (CMRI and MVI) did not require any such prerequisite information of the environmental conditions, making them easier to execute. Though both the CMRI and MVI take into consideration the effect in moisture in the canopy spectra, unlike the MVI, the CMRI is a two-step process that involves the generation of NDVI and NDWI in the first step; in the next step the NDWI product is subtracted from the NDVI product to produce a distinct range of values. MVI on the other hand involved differential band ratios between the Green, NIR and SWIR 1 bands. Comparatively the execution of MVI is much simpler than the other two indices. Comparing the classification outputs, it was observed that the MRI captures the areas with dominant mangrove cover, but remains relatively unsuccessful in highlighting difference in the density of the canopy and identifies bald spots and saline blanks within the value range administered for mangrove dominated area. CMRI and MVI produced better discrimination of mangroves from the surrounding features, visually (Figure 4.2). The combined spectral signatures of greenness and moisture content allowed better distinction of mangrove dominated canopies as well as peripheral patches of sparsely distributed mangroves. The CMRI has also been tested in other mangrove forests adding up to the robustness of the index (Gupta et al., 2018), as for the MVI, being applied on both Landsat and Sentinel-2 satellite systems and rapid execution using an IDL-based MVI mapper (offline) and GEE interface (online), makes it a good choice for mapping and monitoring of mangroves (Baloloy et al., 2020). As technologies advance, the arena of remote sensing-based mapping and monitoring studies is also expected to advance, with better correction algorithms, new and improved indices answering the limitations of the existing ones.

4.6 CONCLUSION

Over the years several researchers have explored the extensive application of remote sensing as a tool to aid mangrove ecosystem based studies. Through this review, we provide a comprehensive overview of various types of remotely sensed data and techniques used in the mapping and monitoring of mangrove forests. Every year numerous studies are published focusing on various aspects of the mangrove ecosystem, which is an indication of the immense importance and growing interest in the field. Though several studies on this subject are published each year, a majority of these studies are focused on Asia, followed closely by Australia; North, Central and South America; and Africa. A bulk of these studies done in Asia are also focused on the Sundarbans (both Indian and Bangladesh).

For any study based on the mangrove ecosystem, knowledge of the spatial extent of the mangrove forests and its temporal change is an important objective. Remote sensing is till now the best available tool to aid this objective. As discussed in this chapter, mangrove mapping using remote sensing is a challenging task, considering the multitude of factors from sensor specifications to environmental parameters significantly influencing the studies. In remote sensing, the accuracy and applicability of data sets are majorly dependent on the return signals for the mangroves, which can be affected by a number of parameters. Different sensor mediums are sensitive

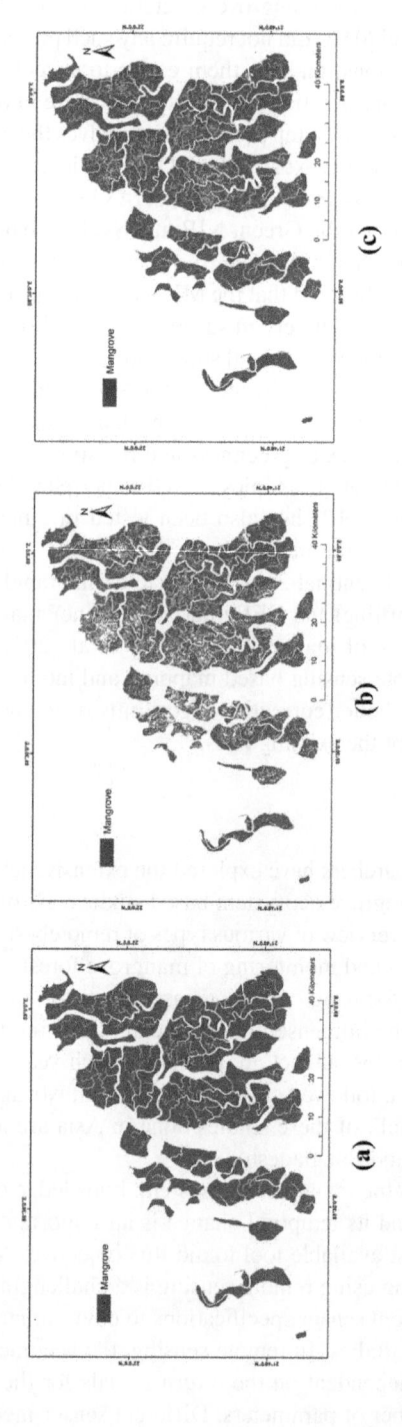

FIGURE 4.2 Resultant outputs generated from the application of mangrove classification indices on Landsat 8 OLI imagery over the Indian Sundarbans. ((a) Mangrove Recognition Index; (b) Combined Mangrove Recognition Index; (c) Mangrove Vegetation Index.)

to different factors, like optical sensors are majorly sensitive to spectral returns from leaves, branches, overall canopy cover and other parameters like background signals from underlying soil and water. Meanwhile, radar-based sensor mediums are sensitive to structural information and dielectric properties of the target. Geometry and orientation of the leaves, canopy structure, leaf moisture content, wetness of underlying soil and waterbodies extensively affect return signals from radar-based sensors. Another important factor that governs the accurate mapping and monitoring of these targets are the sensor specifications. For optical-based sensor mediums, bandwidth and spatial resolution are the two most important factors influencing its mapping accuracy. On the other hand, in radar-based sensor mediums, informations like incidence angle, polarization (HH, HV, VV, VH) and frequency of the signals are important factors to be considered for mapping.

In this chapter, we discussed about various data types, techniques and algorithms that have been developed and used for aiding mangrove ecosystem studies, though the selection of data sets and methods are specific to the desired goal of the studies which ranges from discrimination of mangroves from non-mangrove vegetation types; mangrove health; change detection studies; biomass retrieval; etc. Though the use of data sets and techniques varies according to the goals, classification of mangroves is a very basic step required for most studies. Over the years several vegetation indices like NDVI, EVI, SAVI, etc., have been used to classify/map mangrove forests, similarly, indices specifically designed for mangrove discrimination have also been developed for a better and more accurate mapping. A case study involving the application of some of the most widely used vegetation indices (NDVI, EVI, SAVI and SR) and relatively new mangrove classification indices (CMRI, MRI and MVI) were applied on Landsat 8OLI data sets, over the Indian Sundarbans discussed the use and performance of these methods for mangrove mapping. Though target-specific classification indices perform visually better in discriminating mangrove vegetation types, the challenge with such indices is their robustness and wide-scale application. It is expected that with the recent developments in sensors and sensor platforms, there will be a major leap in such studies, in terms of enhanced accuracy, less time consumption and independent of user accuracy.

REFERENCES

Aksomkoae, S., 1993. *Ecology and Management of Mangroves*. IUCN, Bangkok.

Alatorre, L.C., Sánchez-Andrés, R., Cirujano, S., Beguería, S. and Sánchez-Carrillo, S., 2011. Identification of mangrove areas by remote sensing: The ROC curve technique applied to the northwestern Mexico coastal zone using Landsat imagery. *Remote Sensing*, 3(8), pp. 1568–1583.

Alongi, D.M., 2008. Mangrove forests: Resilience, protection from tsunamis, and responses to global climate change. *Estuarine, Coastal and Shelf Science*, 76(1), pp. 1–13.

Baloloy, A.B., Blanco, A.C., Ana, R.R.C.S. and Nadaoka, K., 2020. Development and application of a new mangrove vegetation index (MVI) for rapid and accurate mangrove mapping. *ISPRS Journal of Photogrammetry and Remote Sensing*, 166, pp. 95–117.

Bandaranayake, W.M., 1998. Traditional and medicinal uses of mangroves. *Mangroves and Salt Marshes*, 2(3), pp. 133–148.

Barbier, E.B., 2006. Natural barriers to natural disasters: replanting mangroves after the tsunami. *Frontiers in Ecology and the Environment*, 4(3), pp. 124–131.

Barik, J. and Chowdhury, S., 2014. True mangrove species of Sundarbans delta, West Bengal, eastern India. *Check List*, 10(2), pp.329–334.

Bennett, E.L. and Reynolds, C.J., 1993. The value of a mangrove area in Sarawak. *Biodiversity & Conservation*, 2(4), pp. 359–375.

Birch, G., Nath, B. and Chaudhuri, P., 2015. Effectiveness of remediation of metal-contaminated mangrove sediments (Sydney estuary, Australia). *Environmental Science and Pollution Research*, 22(8), pp. 6185–6197.

Bouillon, S., Borges, A.V., Castañeda-Moya, E., Diele, K., Dittmar, T., Duke, N.C., Kristensen, E., Lee, S.Y., Marchand, C., Middelburg, J.J. and Rivera-Monroy, V.H., 2008. Mangrove production and carbon sinks: a revision of global budget estimates. *Global Biogeochemical Cycles*, 22(2).

Cannicci, S., Burrows, D., Fratini, S., Smith III, T.J., Offenberg, J. and Dahdouh-Guebas, F., 2008. Faunal impact on vegetation structure and ecosystem function in mangrove forests: a review. *Aquatic Botany*, 89(2), pp.186–200.

Carey, G., 1934. Further investigations on the embryology of viviparous seeds. In *Proc. Linn. Soc. NSW*. (Vol. 59, pp. 392–410).

Chakravortty, S., 2013. Analysis of end member detection and subpixel classification algorithms on hyperspectral imagery for tropical mangrove species discrimination in the Sunderbans Delta, India. *Journal of Applied Remote Sensing*, 7(1), p. 073523.

Chaffey, D.R., Miller, F.R. and Sandom, J.H., 1985. A forest inventory of the Sundarbans, Bangladesh.

Chaube, N.R., Lele, N., Misra, A., Murthy, T.V.R., Manna, S., Hazra, S., Panda, M. and Samal, R.N., 2019. Mangrove species discrimination and health assessment using AVIRIS-NG hyperspectral data. *Current Science*, 116, p. 1136.

Chaudhuri, P., Nath, B. and Birch, G., 2014. Accumulation of trace metals in grey mangrove Avicennia marina fine nutritive roots: The role of rhizosphere processes. *Marine Pollution Bulletin*, 79(1–2), pp. 284–292.

Cornejo, R.H., Koedam, N., Luna, A.R., Troell, M. and Dahdouh-Guebas, F., 2005. Remote sensing and ethnobotanical assessment of the mangrove forest changes in the Navachiste-San Ignacio-Macapule lagoon complex, Sinaloa, Mexico. *Ecology and Society*, 10(1): 16.

Cornforth, W.A., Fatoyinbo, T.E., Freemantle, T.P. and Pettorelli, N., 2013. Advanced land observing satellite phased array type L-Band SAR (ALOS PALSAR) to inform the conservation of mangroves: Sundarbans as a case study. *Remote Sensing*, 5(1), pp. 224–237.

Costanza, R., d'Arge, R., De Groot, R., Farber, S., Grasso, M., Hannon, B., Limburg, K., Naeem, S., O'neill, R.V., Paruelo, J. and Raskin, R.G., 1997. The value of the world's ecosystem services and natural capital. *Nature*, 387(6630), pp. 253–260.

Dahdouh-Guebas, F., 2002. The use of remote sensing and GIS in the sustainable management of tropical coastal ecosystems. *Environment, Development and Sustainability*, 4(2), pp. 93–112.

Dahdouh-Guebas, F., Jayatissa, L.P., Di Nitto, D., Bosire, J.O., Seen, D.L. and Koedam, N., 2005. How effective were mangroves as a defence against the recent tsunami? *Current Biology*, 15(12), pp. R443–R447.

Danielsen, F., Sørensen, M.K., Olwig, M.F., Selvam, V., Parish, F., Burgess, N.D., Hiraishi, T., Karunagaran, V.M., Rasmussen, M.S., Hansen, L.B. and Quarto, A., 2005. The Asian tsunami: a protective role for coastal vegetation. *Science*, 310(5748), pp. 643–643.

Díaz, B.M. and Blackburn, G.A., 2003. Remote sensing of mangrove biophysical properties: evidence from a laboratory simulation of the possible effects of background variation on spectral vegetation indices. *International Journal of Remote Sensing*, 24(1), pp. 53–73.

Dutta, D., Das, P.K., Paul, S., Sharma, J.R. and Dadhwal, V.K., 2015. Assessment of ecological disturbance in the mangrove forest of Sundarbans caused by cyclones using MODIS time-series data (2001–2011). *Natural Hazards*, 79(2), pp. 775–790.

Dwivedi, R.S., Rao, B.R.M. and Bhattacharya, S., 1999. Mapping wetlands of the Sundaban Delta and its environs using ERS-1 SAR data. *International Journal of Remote Sensing*, 20(11), pp. 2235–2247.

[FAO] Food and Agriculture Organization, 1994. Mangrove forest management guidelines

Gaodi, X., Lin, Z., Chunxia, L., Yu, X. and Wenhua, L., 2010. Applying value transfer method for eco-service valuation in China. *Journal of Resources and Ecology*, 1(1), pp. 51–59.

Ghosh, D. and Chakravortty, S., 2020. Change detection of tropical mangrove ecosystem with subpixel classification of time series hyperspectral imagery. In *Artificial Intelligence Techniques for Satellite Image Analysis* (pp. 189–211). Springer, Cham.

Giri, C., 2016. Observation and monitoring of mangrove forests using remote sensing: opportunities and challenges.

Giri, C., Pengra, B., Zhu, Z., Singh, A. and Tieszen, L.L., 2007. Monitoring mangrove forest dynamics of the Sundarbans in Bangladesh and India using multi-temporal satellite data from 1973 to 2000. *Estuarine, Coastal and Shelf Science*, 73(1–2), pp. 91–100.

Giri, C., Zhu, Z., Tieszen, L.L., Singh, A., Gillette, S. and Kelmelis, J.A., 2008. Mangrove forest distributions and dynamics (1975–2005) of the tsunami-affected region of Asia. *Journal of Biogeography*, 35(3), pp. 519–528.

Gopal, B. and Chauhan, M., 2006. Biodiversity and its conservation in the Sundarban mangrove ecosystem. *Aquatic Sciences*, 68(3), pp. 338–354.

Green, E.P., Clark, C.D., Mumby, P.J., Edwards, A.J. and Ellis, A.C., 1998. Remote sensing techniques for mangrove mapping. *International Journal of Remote Sensing*, 19(5), pp. 935–956.

Green, E.P., Mumby, P.J., Edwards, A.J. and Clark, C.D., 1996. A review of remote sensing for the assessment and management of tropical coastal resources. *Coastal Management*, 24(1), pp. 1–40.

Gupta, K., Mukhopadhyay, A., Giri, S., Chanda, A., Majumdar, S.D., Samanta, S., Mitra, D., Samal, R.N., Pattnaik, A.K. and Hazra, S., 2018. An index for discrimination of mangroves from non-mangroves using LANDSAT 8 OLI imagery. *MethodsX*, 5, pp. 1129–1139.

Hamilton, L.S. and Snedaker, S.C., 1984. *Handbook for Mangrove Area Management*. East-West Environment and Policy Institute, Honolulu.

Hati, J.P., Goswami, S., Samanta, S., Pramanick, N., Majumdar, S.D., Chaube, N.R., Misra, A. and Hazra, S., 2020. Estimation of vegetation stress in the mangrove forest using AVIRIS-NG airborne hyperspectral data. *Modeling Earth Systems and Environment*, pp. 1–13. doi: 10.1007/s40808-020-00916-5.

Huete, A., 1988. Huete, AR A soil-adjusted vegetation index (SAVI). Remote Sensing of Environment. *Remote Sensing of Environment*, 25, pp. 295–309.

Huete, A., Didan, K., Miura, T., Rodriguez, E.P., Gao, X. and Ferreira, L.G., 2002. Overview of the radiometric and biophysical performance of the MODIS vegetation indices. *Remote Sensing of Environment*, 83(1–2), pp. 195–213.

Iftekhar, M.S. and Islam, M.R., 2004. Managing mangroves in Bangladesh: A strategy analysis. *Journal of Coastal Conservation*, 10(1), pp. 139–146.

Jiang, Z., Huete, A.R., Didan, K. and Miura, T., 2008. Development of a two-band enhanced vegetation index without a blue band. *Remote sensing of Environment*, 112(10), pp. 3833–3845.

Juncosa, A.M., 1982. Developmental morphology of the embryo and seedling of Rhizophora mangle L. (Rhizophoraceae). *American Journal of Botany*, 69(10), pp. 1599–1611.

Karmaker, S., 2006. Study of Mangrove Biomass, Net Primary Production & Species Distribution using Optical & Microwave Remote Sensing Data (Doctoral dissertation, Indian Institute of Remote Sensing).

Kasischke, E.S., Melack, J.M. and Dobson, M.C., 1997. The use of imaging radars for ecological applications—A review. *Remote Sensing of Environment*, 59(2), pp. 141–156.

Kathiresan, K. and Bingham, B.L., 2001. Biology of mangroves and mangrove ecosystems. *Advances in Marine Biology*, 40, pp.84–254.

Kristensen, E., Bouillon, S., Dittmar, T. and Marchand, C., 2008. Organic carbon dynamics in mangrove ecosystems: A review. *Aquatic Botany*, 89(2), pp. 201–219.

Kuenzer, C., Bluemel, A., Gebhardt, S., Quoc, T.V. and Dech, S., 2011. Remote sensing of mangrove ecosystems: A review. *Remote Sensing*, 3(5), pp. 878–928.

Kumar, T. and Patnaik, C., 2013. Discrimination of mangrove forests and characterization of adjoining land cover classes using temporal C-band Synthetic Aperture Radar data: A case study of Sundarbans. *International Journal of Applied Earth Observation and Geoinformation*, 23, pp. 119–131.

Kumar, T., Mandal, A., Dutta, D., Nagaraja, R. and Dadhwal, V.K., 2019. Discrimination and classification of mangrove forests using EO-1 Hyperion data: A case study of Indian Sundarbans. *Geocarto International*, 34(4), pp.415–442.

Liu, H.Q. and Huete, A., 1995. A feedback based modification of the NDVI to minimize canopy background and atmospheric noise. *IEEE Transactions on Geoscience and Remote Sensing*, 33(2), pp. 457–465.

Lugo, A.E. and Snedaker, S.C., 1974. The ecology of mangroves. *Annual Review of Ecology and Systematics*, 5(1), pp. 39–64.

Lugo, A.E. and Snedaker, S.C., 1975. Properties of a mangrove forest in southern Florida. Inst. Food Agr. Sci., University of Florida.

Mandal, M.S.H., Kamruzzaman, M. and Hosaka, T., 2020. Elucidating the phenology of the Sundarbans mangrove forest using 18-year time series of MODIS vegetation indices. *Tropics*, 29(2), pp. 41–55.

Manjunath, K.R., Kumar, T., Kundu, N. and Panigrahy, S., 2013. Discrimination of mangrove species and mudflat classes using in situ hyperspectral data: A case study of Indian Sundarbans. *GIScience & Remote Sensing*, 50(4), pp. 400–417.

Manna, S., Nandy, S., Chanda, A., Akhand, A., Hazra, S. and Dadhwal, V.K., 2014. Estimating aboveground biomass in Avicennia marina plantation in Indian Sundarbans using high-resolution satellite data. *Journal of Applied Remote Sensing*, 8(1), p. 083638.

Manson, F.J., Loneragan, N.R., Skilleter, G.A. and Phinn, S.R., 2005. An evaluation of the evidence for linkages between mangroves and fisheries: A synthesis of the literature and identification of research directions. *Oceanography and Marine Biology*, 43, p. 483.

Mazda, Y., Magi, M., Kogo, M. and Hong, P.N., 1997. Mangroves as a coastal protection from waves in the Tong King delta, Vietnam. *Mangroves and Salt Marshes*, 1(2), pp. 127–135.

Mazda, Y., Magi, M., Nanao, H., Kogo, M., Miyagi, T., Kanazawa, N. and Kobashi, D., 2002. Coastal erosion due to long-term human impact on mangrove forests. *Wetlands Ecology and Management*, 10(1), pp. 1–9.

McLeod, E. and Salm, R.V., 2006. Managing mangroves for resilience to climate change. Gland: World Conservation Union (IUCN).

Mondal, B., Saha, A.K. and Roy, A., 2019. Mapping mangroves using LISS-IV and Hyperion data in part of the Indian Sundarban. *International Journal of Remote Sensing*, 40(24), pp. 9380–9400.

Mukhopadhyay, A., Mondal, P., Barik, J., Chowdhury, S.M., Ghosh, T. and Hazra, S., 2015. Changes in mangrove species assemblages and future prediction of the Bangladesh Sundarbans using Markov chain model and cellular automata. *Environmental Science: Processes & Impacts*, 17(6), pp. 1111–1117.

Mukhopadhyay, A., Wheeler, D., Dasgupta, S., Dey, A. and Sobhan, I., 2018. Aquatic salinization and mangrove species in a changing climate: Impact in the Indian Sundarbans. The World Bank.

Mumby, P.J., Edwards, A.J., Arias-González, J.E., Lindeman, K.C., Blackwell, P.G., Gall, A., Gorczynska, M.I., Harborne, A.R., Pescod, C.L., Renken, H. and Wabnitz, C.C., 2004. Mangroves enhance the biomass of coral reef fish communities in the Caribbean. *Nature*, 427(6974), pp. 533–536.

Mumby, P.J., Green, E.P., Edwards, A.J. and Clark, C.D., 1999. The cost-effectiveness of remote sensing for tropical coastal resources assessment and management. *Journal of Environmental Management*, 55(3), pp. 157–166.

Nagelkerken, I.S.J.M., Blaber, S.J.M., Bouillon, S., Green, P., Haywood, M., Kirton, L.G., Meynecke, J.O., Pawlik, J., Penrose, H.M., Sasekumar, A. and Somerfield, P.J., 2008. The habitat function of mangroves for terrestrial and marine fauna: A review. *Aquatic Botany*, 89(2), pp. 155–185.

Othman, M.A., 1994. Value of mangroves in coastal protection. *Hydrobiologia*, 285(1–3), pp. 277–282.

Parida, A.K., Das, A.B. and Mohanty, P., 2004. Defense potentials to NaCl in a mangrove, Bruguiera parviflora: differential changes of isoforms of some antioxidative enzymes. *Journal of Plant Physiology*, 161(5), pp. 531–542.

Pearson, R.L. and Miller, L.D., 1972. Remote mapping of standing crop biomass for estimation of the productivity of the shortgrass prairie. RSE, p. 1355.

Rahman, M., Ullah, R., Lan, M., Sri Sumantyo, J.T., Kuze, H. and Tateishi, R., 2013. Comparison of Landsat image classification methods for detecting mangrove forests in Sundarbans. *International Journal of Remote Sensing*, 34(4), pp. 1041–1056.

Ramsey III, E.W.R. and Jensen, J.R., 1996. Remote sensing of mangrove wetlands: Relating canopy spectra to site-specific data.

Ranjan, A.K. and Kanga, S., 2018. Dynamic changes in mangrove forest and Lu/Lc variation analysis over Indian Sundarban Delta in West Bengal (India) using multi-temporal satellite data. *i-Manager's Journal on Future Engineering and Technology*, 13(3), p. 9.

Reef, R. and Lovelock, C.E., 2015. Regulation of water balance in mangroves. *Annals of Botany*, 115(3), pp. 385–395.

Saenger, P., Hegerl, E.J. and Davie, J.D.S., 1983. Global status of mangrove ecosystems. Commission on Ecology Papers no. 3. World Conservation Union (IUCN), Gland, Switzerland.

Selvam, V., Ravichandran, K.K., Gnanappazham, L. and Navamuniyammal, M., 2003. Assessment of community-based restoration of Pichavaram mangrove wetland using remote sensing data. *Current Science*, 85, pp. 794–798.

Suárez, N. and Sobrado, M.A., 2000. Adjustments in leaf water relations of mangrove (Avicennia germinans) seedlings grown in a salinity gradient. *Tree Physiology*, 20(4), pp. 277–282.

Syed, M.A., Hussin, Y.A. and Weir, M., 2001, November. Detecting fragmented mangroves in the Sundarbans, Bangladesh using optical and radar satellite images. In *22nd Asian Conference on Remote Sensing* (Vol. 5, p. 9).

Tomlinson, P.B., 1986. *The Botany of Mangroves*. Cambridge University Press, London.

Turkar, V., Deo, R., Hariharan, S. and Rao, Y.S., 2011. Comparison of classification accuracy between fully polarimetric and dual-polarization SAR images. In *2011 IEEE International Geoscience and Remote Sensing Symposium* (pp. 440–443). IEEE.

UN Food and Agriculture Organization, 2007. The World's Mangroves 1980–2005: A thematic study prepared in the framework of the Global Forest Resources Assessment 2005.

Upadhyay, V.P. and Mishra, P.K., 2008. Population status of mangrove species in estuarine regions of Orissa coast, India. *Tropical Ecology*, 49(2), p. 183.

USGS LANDSAT, 2016. 8 (L8) Data Users Handbook Ver 2.0.

Vaiphasa, C., Ongsomwang, S., Vaiphasa, T. and Skidmore, A.K., 2005. Tropical mangrove species discrimination using hyperspectral data: A laboratory study. *Estuarine, Coastal and Shelf Science*, 65(1–2), pp. 371–379.

Walters, B.B., Rönnbäck, P., Kovacs, J.M., Crona, B., Hussain, S.A., Badola, R., Primavera, J.H., Barbier, E. and Dahdouh-Guebas, F., 2008. Ethnobiology, socio-economics and management of mangrove forests: A review. *Aquatic Botany*, 89(2), pp. 220–236.

Wang, L., Silván-Cárdenas, J.L. and Sousa, W.P., 2008. Neural network classification of mangrove species from multi-seasonal Ikonos imagery. *Photogrammetric Engineering & Remote Sensing*, 74(7), pp. 921–927.

Wells, S. and Ravilious, C., 2006. In the front line: Shoreline protection and other ecosystem services from mangroves and coral reefs (No. 24). UNEP/Earthprint.

Winarso, G., Purwanto, A. and Yuwono, D., 2014. New mangrove index as degradation health indicator using remote sensing data: Segara Anakan and Alas Purwo case study. In *12th Biennial Conference of Pan Ocean Remote Sensing Conference (PORSEC 2014)*.

Zhang, X. and Tian, Q., 2013. A mangrove recognition index for remote sensing of mangrove forest from space. *Current Science*, 105(8), pp. 1149–1154.

Zhang, X., Treitz, P.M., Chen, D., Quan, C., Shi, L. and Li, X., 2017. Mapping mangrove forests using multi-tidal remotely-sensed data and a decision-tree-based procedure. *International Journal of Applied Earth Observation and Geoinformation*, 62, pp. 201–214.

Section 2

Ecology of Sundarbans a Geo-Informatics Approach

5 Plant Diversity Assessment in Indian Sundarban Mangroves: A Geoinformatics Approach

Subrata Nandy, Muna Tamang, and S.P.S. Kushwaha
Indian Institute of Remote Sensing, Indian Space
Research Organisation, Dehradun, India

CONTENTS

5.1 MANGROVE ECOSYSTEMS

Biodiversity and climate change are inseparable for if we are to preserve the first one we must mark the second. Hence, to address climate change, we must protect our biodiversity. Various international organizations have set up goals to protect global biodiversity from climate change, though the leading effects of climate change on the distributions of species are yet to be fully recognized in conservation efforts. The coastal ecosystem is one of the most complex ecosystems in the world as it is the union of aquatic and terrestrial ecosystems. Mangroves are considered as the biologically diverse and most productive ecosystems on the Earth, delivering the major goods and services (Carugati et al., 2018). These are also known as one of the world's wealthiest warehouses of biological and genetic diversity (Sandilyan and Kathiresan, 2012). Yet, mangroves are declining rapidly due to natural and man-made activities (Giri et al., 2007). Notably, mangroves are vital nursery habitats and breeding spots for crustaceans, shellfishes, fishes, birds, reptiles and mammals; providing valuable ecological and economic resources, accretion ground for sediment, contaminants, nutrients, carbon and protection to coastal erosion. Mangrove plays a key role in the livelihoods and sustainability of humans as it is immensely used for food, fuel, timber and medicines (Alongi, 2002). Ecosystem services and coastal livelihood means worth US$ 1.6 billion are provided every year by the mangrove forests worldwide (Polidoro et al., 2010). Mangroves are one of the most carbon-rich forests in the tropics that maintain carbon balance (Donato et al., 2011) as well as provide an opportunity for tourism and recreation in coastal areas (Kuenzer et al. 2011).

5.2   MANGROVES OF THE WORLD

Mangroves are regarded as one of the most threatened ecosystems in the world. These forests include taxonomically diverse plants exhibiting common adaptive features. Mangroves are salt-tolerant evergreen forests that grow along lagoons, coastlines, estuaries and deltas in the tropics, subtropics and some temperate regions (FAO, 2020), primarily occurring between 30°N and 30°S latitudes (Giri et al., 2011). The latest FAO report, 'Global Forest Resource Assessment 2020' indicates a total of 113 countries having 14.79 Mha mangrove forests. The report further shows that Asia possesses the largest area under mangroves (5.55 Mha), followed by Africa (3.24 Mha), North and Central America (2.57 Mha), and South America (2.13 Mha). The smallest area of mangroves has been reported in Oceania (1.30 Mha). Indonesia (19%), Brazil (9%), Nigeria (7%) and Mexico (6%) together possess more than 40% of the global area under mangroves. Around 75% of the world's mangroves occur in 15 countries, of which merely 6.9% are protected under the existing protected areas network (Giri et al., 2011). Globally, India is one of the 15 mangrove prominence countries, while in South Asia the total mangrove cover is about 3% (FSI, 2019).

Mangroves are highly enriched with massive floral and faunal wealth. They have developed distinct modifications such as vivipary, strong interlocking, supporting and breathing root system, nutrient retention, and salt regulation (Kathiresan and Bingham, 2001). Mangroves are commonly divided into two categories 'true mangroves' and 'mangrove associates' (Ghosh, 2011). The tree structures of mangroves such as aerial roots and tree trunks support various epiphytic algae, lichens and vascular plants. About 80 tree species of true mangroves are present in the world (Lee et al., 2017). Although most mangrove species are extremely diversified, yet they have evolved certain common morphological, physiological, biological and ecological adaptability to flourish in the tidal habitat. The presence of pneumatophores (e.g. *Avicennia* spp. and *Sonneratia* spp.), stilt roots (e.g. *Rhizophora* spp. and *Bruguiera* spp.), viviparous water-dispersed propagules and salt excreting leaves are the most remarkable features they have developed.

5.3   MANGROVE BIODIVERSITY IN INDIA

The total mangrove cover in India is 4,975 km^2, which represents 0.15% of the country's total geographical area (FSI, 2019). In India, the largest area under mangroves occurs in West Bengal (42.45%), followed by Gujarat (23.66%) (FSI, 2019). In terms of mangrove diversity, India ranks third globally after Indonesia and Australia (FAO, 2020). The Indian coastline is nearly 12,700 km long with uneven distribution of mangroves. Around 80% of India's mangroves are found along the east coast and the largest areas are the Indian part of the Sundarbans and the Andaman and Nicobar Islands.

The Indian mangroves comprise 46 true mangrove species which belong to 14 families and 22 genera. The east coast of India has 40 mangrove species, while the west coast has 27 species. The Andaman and Nicobar Islands consist of 38 species of mangroves and are recognized as rich in terms of plant biodiversity (Ragavan et al., 2016). Some of the important mangrove species in India are *A. marina, R. mucronata, Morinda citrifolia, A. alba, S. alba, Heritiera littoralis, B. cylindrica, Excoecaria agallocha, Phoenix palludosa,* and *Ceriops tagal, A. marina,* with associates like *Sterculia apetala* and *B. cylindrica,* dominate the outer boundary of the estuaries whereas *R. mucronata,* with associates like *E. agallocha* and *B. cylindrica,* dominates the inner estuary (Banerjee, 2002).

5.4 MANGROVE BIODIVERSITY IN SUNDARBANS

Sundarbans, the world's greatest mangrove delta stretching over an area of about 10,000 km^2 of India and Bangladesh, accounts for approximately 2112.11 km^2 of India's mangroves (FSI, 2019). For its longstanding protection and management for livelihood measures, Sundarbans mangrove has been considered as UNESCO World Heritage site as well as Ramsar site. It provides habitat to many near-threatened and critically endangered species (Sreelekshmi et al., 2020). Recently, the Indian Sundarbans mangrove forest has been considered endangered under the Red List of Ecosystems (Sievers et al., 2020). About 75% of the forest area has a high biological richness and supports 100 plant species including 30 species of trees, 32 species of shrubs and the rest are herbs, grasses and ferns (Gopal and Chauhan, 2006). From selected islands of Indian Sundarbans, 27 true mangrove species were reported by Sreelekshmi et al. (2020). The dominant species were *A. alba* and *A. marina* followed by *E. agallocha, S. apetala* and *B. gymnorhiza.* Many species like *A. alba, C. tagal, R. apiculata, S. griffithii, Xylocarpus mekongensis* and *Lumnitzera racemosa* were found in the mixed mangrove forest of Sundarbans (Giri et al., 2014). *Avicennia, Phoenix,* mixed mangroves and mangrove scrub cover 23.21% of the total geographical area of Sundarban Biosphere Reserve (SBR) (Nandy and Kushwaha, 2011).

5.5 ROLE OF REMOTE SENSING IN PLANT DIVERSITY ASSESSMENT OF WORLD'S MANGROVES

Remote sensing (RS) is a vital tool for mapping and monitoring natural resources (Nandy et al., 2007, 2011; Bhatt et al., 2013a; Kumar et al., 2014; Bagaria et al., 2017; Ghosh et al., 2017; Kushwaha et al., 2018; Navalgund et al., 2019; Nandy et al., 2020; Srinet et al., 2020). As the field survey is challenging and laborious in mangrove swamps, RS can play a significant role in the mapping and monitoring of the mangroves (Giri et al., 2014; Heenkenda et al., 2014). It provides synoptic coverage and serves as a crucial tool for providing up-to-date information that helps in mangrove conservation and planning. Although the benefits of using RS data include repetitive, synoptic coverage, and availability of free or low-cost satellite data, yet RS data do not entirely replace the field surveys (Dwivedi et al. 1999; Nandy et al., 2019). RS techniques have exhibited great potential to identify, map and monitor changes in the mangrove conditions (Kuenzer et al., 2011). With the advancement of high spectral resolution satellite data and progress in digital image classification algorithms, there is the potentiality to digitally classify mangroves at the species level (Heenkenda et al., 2014). Landsat TM, SPOT XS and IRS LISS-III are some of the common sensors which have been used to map mangroves (Nandy and Kushwaha, 2011). It can be observed that most of the earlier studies have used parametric classifiers to map mangroves, and among these, the visual image interpretation technique was found to be the most prevalent method for mangrove mapping (Nandy and Kushwaha, 2011). However, in recent decades, most of the studies (Table 5.1) have used non-parametric machine learning algorithms, like Random Forest (RF), Support Vector Machine (SVM) and neural networks for effectively mapping mangroves worldwide. Table 5.1 shows the various satellite data and classification techniques used for mangrove mapping worldwide in recent decades. Mangrove species are difficult to identify in coarser resolution images (Wang et al. 2004) whereas the utilization of high-resolution multi-spectral satellite imagery was found to be beneficial (Satyanarayana et al., 2011). Spectral information played a more important role in classifying mangrove species (Wang et al. 2004). Huang et al. (2009) mapped *A. germinans, Laguncularia racemosa* and *R. mangle* by integrating the spectral and textural features derived from the IKONOS image. The classification was done using SVM in which the overall accuracy was found to be 92.1%. Heenkenda et al. (2014) differentiated mangroves at the species level in Rapid Creek coastal mangrove forest of Australia using high spatial resolution RS data. They combined WorldView-2 images with images captured by an UltraCamD camera for mangrove classification and classified the mangroves at species level using a SVM algorithm with best-fit parameters. *A. marina, B. exaristata, C. tagal, R. stylosa* and *L. racemosa* were the five mangrove species mapped by them. The overall classification accuracy attained was 89%. It was also found that the SVM algorithm was highly effective in mangrove species mapping with WorldView-2 data than with aerial images. Valderrama-Landeros et al. (2018), using Sentinel-2, Landsat-8, SPOT-5, and WorldView-2 data, assessed the accuracy of mangrove forest classification for the pacific coast of Mexico and reported the highest overall accuracy of 93% with WorldView-2 data and 64% with Landsat-8 data. There was over-estimation of the extent of *L. racemosa* and under-estimation of the extent of *R. mangle* by other sensors compared to WorldView-2. The study revealed that by applying the same classification approach for mapping of mangrove forests, the spatial resolution has great importance. Wang et al. (2018a) mapped mangroves of the Nansha Wetland Park in Guangzhou city of China using three machine learning algorithms, such as SVM, RF and Decision Tree (DT) to evaluate the pixel-based and object-based image analysis methods. They noted overall higher performance by object-based method compared to pixel-based one. The highest overall accuracy of 79.63% was achieved with SVM for pixel-based image analysis whereas RF achieved the highest overall accuracy (82.40%) for object-based image analysis. *S. apetala*, dominant mangrove species, as well as species, viz., *Hibiscus tiliaceus*, which occurs partially mixed with other species, could be well-classified. Wan et al. (2020) used GF-5 hyperspectral data for mapping six mangrove species. They compared RF and SVM for classification where RF performed better.

TABLE 5.1

A Summary of the Various Approaches Adopted for Mangrove Mapping in Recent Decades

Sl. No.	Study Area	Satellite Data	Time of the Satellite Data Acquisition	Classification Techniques	Accuracy	Species Mapped	References
1.	Punta Galeta, Caribbean coast of Panama,	IKONOS and QuickBird	13 June 2000 and 28 July 2002	Maximum likelihood (ML)	Overall- 72.2%–75.5%	*Avicennia germinans, Laguncularia racemosa* and *Rhizophora mangle*	Wang et al. (2004)
2.	Sungai Belungkor, Johor, Malaysia	IKONOS	16 March 2001	ML, Minimum Distance to Mean and Contextual Logical Channel	Overall – 68%, 64% and 82%	*R. apiculata* and *R. mucronata*	Kanniah et al. (2007)
3.	Punta Galeta, Caribbean coast of Panama	IKONOS	13 June 2000	SVM	Overall – 92.1%	*A. germinans, L. racemosa* and *R. mangle*	Huang et al. (2009)
4.	Sundarbans Biosphere Reserve, West Bengal, India	IRS LISS-III	2002	Visual Interpretation (VI)	Overall – 91.67%	*Avicennia* sp. and *Phoenix* sp.	Nandy and Kushwaha (2011)
5.	Southeast Queensland, Australia	CASI-2, hyperspectral data	29 July 2004	Spectral angle mapper, Linear spectral unmixing and object-based image analysis	Overall – 69%, 56% and 76%,	*Avicennia* sp., *Ceriops* sp., and *Rhizophora* sp.	Kamal and Phinn (2011)
6.	Tumpat, Malaysia	QuickBird	29 April 2006	ML	Overall – 54%	*A. alba, Nypa fruticans* and *Sonneratia caseolaris*	Satyanarayana et al. (2011)
7.	Sundarbans Biosphere Reserve, West Bengal, India	Landsat TM/ETM	1999 and 2010	ML	Overall – 80% and 85.71%	*Avicennia* sp., *Excoecaria* sp., *Phoenix* sp., *Bruguiera* sp. and *Ceriops* sp.	Giri et al. (2014)

(Continued)

TABLE 5.1 *(Continued)*

A Summary of the Various Approaches Adopted for Mangrove Mapping in Recent Decades

SI. No.	Study Area	Satellite Data	Time of the Satellite Data Acquisition	Classification Techniques	Accuracy	Species Mapped	References
8.	Mai Po Nature Reserve, Hong Kong	Hyperion and SPOT-5	21 November 2008	Nearest Neighbour (NN)	Overall- 88%	*Acanthus ilicifolius, Aegiceras corniculatum, A. marina,* and *Kandelia obovata*	Jia et al. (2014)
9.	Rapid Creek, Australia	WorldView-2	5 June 2010	SVM	Overall – 89%	*A. marina, C. tagal, Bruguiera exaristata, Lumnitzera racemosa,* and *R. stylosa.*	Heenkenda et al. (2014)
10.	Sundarbans mangrove forest, India and Bangladesh	Landsat MSS/ TM/ETM+/OLI	1977–2015	ML	Overall – 72%–89%	*Heritiera fomes, Excoecaria aglocha, C. decandra, S. apetala* and *Xylocarpus mekongensis*	Ghosh et al. (2016)
11.	Nusa Lembongan, Bali, Indonesia	IKONOS, Geoeye and Worldview-2	2001 and 2014	ML	Overall – 65%–80%	*A. marina, B. gymnorhiza, R. apiculata, S. alba, R. stylosa, X. granatum*	Viennois et al. (2016)
12.	Agua Brava coastal lagoon, Pacific coast of Mexico	Landsat OLI, SPOT-5, Sentinel-2 and WorldView-2	2014–2016	DT	Overall – 64%–93%	*R. mangle* and *L. racemosa*	Valderrama-Landeros et al. (2018)
13.	Nansha Wetland, China	Pléiades-1	22 September 2012	DT, SVM, and RF	Overall- 65.74%–82.40%	*S. apetala* and *Hibiscus tiliaceus*	Wang et al. (2018a)

(Continued)

TABLE 5.1 (*Continued*)

A Summary of the Various Approaches Adopted for Mangrove Mapping in Recent Decades

Sl. No.	Study Area	Satellite Data	Time of the Satellite Data Acquisition	Classification Techniques	Accuracy	Species Mapped	References
14.	Qi'ao Island, Guangdong Province, China	UAV hyperspectral image	15 October 2016	k-nearest neighbour and SVM	Overall- 76.12% and 82.39%	*K. candel, S. apetala, Acrostichum aureum, A. corniculatum, A. ilicifolius, H. littoralis and Thespesia populnea*	Cao et al. (2018)
15.	DeepBay, Hong Kong	WorldView-2	14 November 2010	Convolution Neural Network	Overall- 98.81%	*A. ilicifolius, S. caseolaris, A. corniculatum, K. obovate, A. marina and S. apetala*	Wan et al. (2019)
16.	Mai Po Nature Reserve, Hong Kong	Worldview-3 and Radarsat-2	1 January 2015 and 8 November 2014	Rotation Forest, RF, and SVM	Overall- 85.23%, 84.09% and 79.55%	*K. obovata, A. marina, A. ilicifolius* and *A. corniculatum*	Zhang et al. (2018)
17.	Dongzhaigang National Nature Reserve, Hainan Island, China	Sentinel-2, Landsat 8 and Pléiades-1	9 December, 14 February 2016, and 4 February 2014	RF	Overall- 70.95%, 68.57% and 78.57%	*B. sexangula, C. tagal, R. stylosa, K. candel, A. marina and L. racemosa*	Wang et al. (2018b)
18.	Mai Po Nature Reserve, Hong Kong	WorldView-3 and Airborne LiDAR	20 December 2017 and March 21 2018	RF and SVM	Overall- 87% and 88%	*A. corniculatum, A. ilicifolius, A. marina, K. obovata* and *S. apetala*	Li et al. (2019)
19.	Mai Po Nature Reserve, Hong Kong	Gaofen-5 (GF-5)	5 October 2018	RF and SVM	Overall-87.12%	*A. ilicifolius, A. corniculatum, A. marina, and K. obovata*	Wan et al. (2020)
20.	Futian Mangrove Nature Reserve, Shenzhen, China	Sentinel-2 and Google Earth imagery	25 Januay 2017	SVM	Overall-53.78%–73.10%	*K. obovata, A. marina, A. corniculatum, A. ilicifolius, S. apetala and S. caseolaris*	Li et al. (2020)

The mapping of mangrove distribution is vital for biodiversity assessment, conservation as well as to better estimate their ecological service value (Li et al., 2020). For mangrove mapping and biodiversity characterization, the inclusion of ideal spectral and spatial resolution should be considered while selecting data (Heenkenda et al., 2014; Valderrama-Landeros et al., 2018). In India, several studies have been carried out using RS to assess the biological richness (Roy and Tomar, 2000; Kushwaha et al., 2005; Nandy and Kushwaha, 2010; Bhatt et al., 2013b).

5.6   RS-BASED PLANT DIVERSITY ASSESSMENT OF SUNDARBAN MANGROVES

The name Sundarban comes either from 'sundari' (*Heritiera fomes*), the dominant tree species in Sundarban, or from the sheer beauty of the forest. This deltaic ecosystem is the single largest continuous habitat in the world for the Royal Bengal Tiger. Nandy and Kushwaha (2011) used IRS 1D LISS-III satellite imagery (23.5 m spatial resolution with four spectral bands, viz., green, red, near infrared and shortwave infrared) of January and February 2002 for mangrove mapping in SBR of India. Different approaches, such as the on-screen visual interpretation, supervised and unsupervised classifications were used for mapping. Four mangroves classes, viz., *Avicennia*, *Phoenix*, mixed mangroves, mangrove scrub, and eight non-mangrove classes were mapped using the above-mentioned approaches. The four mangrove classes fall under the *Littoral and Swamp Forest (4B)* described by Champion and Seth (1968). The highest classification accuracy was obtained using on-screen visual interpretation (91.67%) compared to supervised (79.90%) and unsupervised (71.08%) classifications. The study reported that 23.21% of the total geographical area of SBR was covered by mangroves, of which 18.31% area was under the mixed mangroves. The species composition of the mapped mangrove classes was different. *A. alba, A. marina, E. agallocha, S. apetala, Aegialitis rotundifolia, Aechmanthera gossypina* and *B. gymnorhiza* were the dominant species found in *Avicennia* forest and species such as *P. paludosa, C. tagal* and *E. agallocha* dominated the *Phoenix* forest. The species composition of mixed mangrove forest was *E. agallocha, A. alba, S. apetala, A. gossypina, X. moluccensis, A. marina* and *C. tagal*, whereas mangrove scrub consisted of *S. apetala, B. gymnorhiza, A. rotundifolia, X. moluccensis, A. gossypina, E. agallocha, C. tagal* and *C. decandra*.

The biological richness of SBR was assessed by Nandy and Kushwaha (2010). They used SPatial LAndscape Modelling (SPLAM) software, developed for this purpose, for landscape analysis and biological richness modelling; wherein vegetation type/land use map was used as an input to the SPLAM. The satellite image (IRS 1D LISS-III) was used for vegetation/land use stratification. Different landscape parameters such as fragmentation (Forman and Godron, 1986), patchiness (Romme, 1982), porosity (Forman and Godron, 1986), interspersion (Lyon, 1983) and juxtaposition (Lyon, 1983) were generated and combined with biotic interference (roads and settlements) buffers to map the disturbance index. It was observed that most of the vegetation types were undisturbed in SBR. The low disturbed area occupied 57.55% of the total forest area, which indicated that most of the forest area is intact. Highly disturbed forests occupied 10.41% of the forest area. The cause of the disturbance was found to be both natural and anthropogenic; the conversion of mangrove forest to pisciculture was observed to be the main reason for disturbance (Nandy, 2009). The natural transformation of mangroves to waterbody was due to erosion and saline water ingression.

The biological richness levels were mapped by using SPLAM in which information on disturbance index, species diversity (Shannon and Weaver, 1949), information on rare, endangered, endemic, threatened species and species importance value (Belal and Springuel, 1996) were integrated. Species richness was calculated from the field inventory data. The uniqueness of the mangrove classes was determined based on their importance and critical habitat value in the landscape. Based on the total importance value of each species, the biodiversity value of each class was determined. Thus, biological richness at the landscape was determined as a function of ecosystem uniqueness, species richness, biodiversity value and disturbance index. The multi-criteria analysis procedure was followed for biological richness assessment. The biological richness map reflected the plant diversity status of the SBR (Figure 5.1). It was found that 17.56% of the SBR area, possessing 75.63% of the mangroves, was biologically highly rich. The mixed mangrove class showed the highest richness compared to other mangrove classes.

Giri et al. (2014) used Landsat TM and ETM satellite data of 1999 and 2010, respectively, to identify the mangrove species present in the Indian Sundarbans. The spectral profile of mostly four species such as *Avicennia, Phoenix, Excoecaria* and *Ceriops* was generated using the hyperspectral image of 2011. The digital classification using the maximum likelihood classifier yielded the overall accuracy for the 1999 and 2010 periods as 80% and 85.71%, respectively. The most dominant species were found to be *Avicennia* followed by *Phoenix* and *Excoecaria*. However, they observed that the aerial distribution of *Avicennia, Ceriops*, and *Bruguiera* sp. has shrunk over time. Ghosh et al. (2016) mapped the mangrove species of Sundarbans using Landsat MSS, TM, ETM and OLI, and detected changes over 38 years. The maximum likelihood classification was used to classify and map the mangrove species. Both *E. agallocha* and *H. fomes* showed a decrease of 9.9%, while *C. decandra, X. mekongensis* and *S. apetala* increased by 12.9%, 380.40%, and 57.3%, respectively.

5.7　CONCLUSIONS

Mangroves are regarded as the most threatened ecosystems in the world. The area of mangroves has shrunk by 1.04 Mha since 1990 (FAO, 2020). The area under mangrove forest in SBR was 55.01% in 1999, which has shrunk to 50.63% by 2010 (Giri et al., 2014). As wetlands, mangroves are facing rapid decline. The depletion and degradation have affected 81% of inland wetland species and 36% of the marine and coastal species since 1970 (Gopal and Chauhan, 2006). Species like *H. fomes, P. paludosa* and *Nypa fruticans* are rapidly declining (Gopal and Chauhan, 2006). Over the past decades, large areas of SBR mangroves have been converted to pisciculture. An ever-increasing human population, and the resultant pressure on mangroves, continues to threaten biodiversity. A high level of interference has also affected the freshwater inflows negatively and exerted pressure on biological resources. The

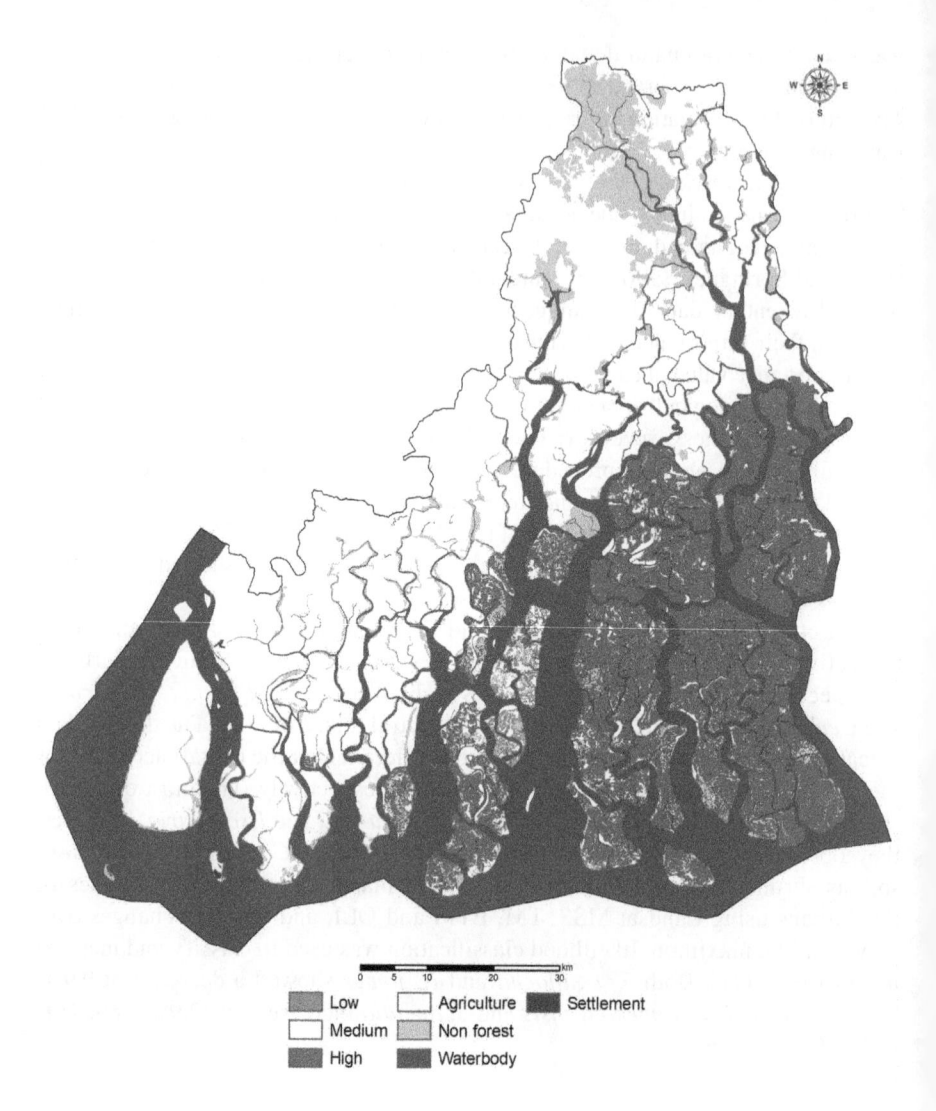

FIGURE 5.1 Biological richness map of Sundarban Biosphere Reserve. (From Nandy and Kushwaha, 2010.)

increase in salinity has also been observed in mangrove areas in absence of flushing by freshwater from upstream areas.

It is now imperative to initiate efficient conservation and management strategies for saving and restoring mangrove forests to meet the needs of the local people. The role of mangrove forests as a 'bio-shield' that preserved vulnerable coastal communities in the Indian Ocean tsunami of December 2004 and other natural disasters have been already highlighted. The species, which are endemic to mangrove habitats, are the most endangered due to mangrove loss (Lee et al., 2017). The global targets set by the UN Sustainable Development Goals, the Aichi Biodiversity Targets, the Paris

Agreement on Climate Change, and Land Degradation Neutrality can be achieved only by healthy functioning of wetlands. Various protection and conservation measures such as the Ramsar Convention on Wetlands and the Kyoto Protocol are being implemented by international programmes to prevent further loss of the mangroves. With the development of improved spectral, spatial, temporal and radiometric resolutions, RS data can play a significant role not only in mapping and monitoring of mangroves but also in rapid biodiversity assessment, which will be vital for the management and conservation of this important natural resource.

REFERENCES

Alongi, D.M., 2002. Present state and future of the world's mangrove forests. *Environmental Conservation*, 29(3), pp. 331–349.

Bagaria, P., Nandy, S., Mitra, D., Lal, P. and Sivakumar, K., 2017. Application of different satellite image classification techniques for mapping land use and land cover of east Godavari river estuarine landscape, Andhra Pradesh, India. *International Journal of Advancement in Remote Sensing, GIS and Geography*, 5(1), pp. 1–16.

Banerjee, L.K., 2002. Sundarbans biosphere reserve. In: Singh, N.P., and Singh K.P. (eds.) *Floristic Diversity and Conservation Strategies in India*, vol. V, Botanical Survey of India. Ministry of Environment and Forests, New Delhi, pp. 2801–2829.

Belal, A.E. and Springuel, I., 1996. Economic value of plant diversity in arid environments. *Nature and Resources*, 32(1), pp. 33–39.

Bhatt, G.D., Kushwaha, S.P.S., Nandy, S. and Bargali, K., 2013a. Vegetation types and land uses mapping in south Gujarat using remote sensing and geographic information system. *International Journal of Advancement in Remote Sensing, GIS and Geography*, 1(1), pp.20–31.

Bhatt, G.D., Kushwaha, S.P.S., Nandy, S., Bargali, K., Tadvi, D., Nagar, P.S. and Daniel, M., 2013b. Plant richness modelling in south Gujarat using remote sensing and geographic information system. *Indian Forester*, 139(9), pp. 757–768.

Cao, J., Leng, W., Liu, K., Liu, L., He, Z. and Zhu, Y., 2018. Object-based mangrove species classification using unmanned aerial vehicle hyperspectral images and digital surface models. *Remote Sensing*, 10(1), p. 89.

Carugati, L., Gatto, B., Rastelli, E., Martire, M.L., Coral, C., Greco, S. and Danovaro, R., 2018. Impact of mangrove forests degradation on biodiversity and ecosystem functioning. *Scientific Reports*, 8(1), pp. 1–11.

Champion, H.G. and Seth, S.K., 1968. *A Revised Survey of the Forest Types of India*. Manager of Publications, Delhi.

Donato, D.C., Kauffman, J.B., Murdiyarso, D., Kurnianto, S., Stidham, M. and Kanninen, M., 2011. Mangroves among the most carbon-rich forests in the tropics. *Nature Geoscience*, 4(5), pp. 293–297.

Dwivedi, R. S., Rao, B. R. M., Kushwaha, S. P. S. and Bhattacharya, S. (1999). Mapping wetlands of the Sundarban Delta and it's environs using ERS-1 SAR data. *International Journal of Remote Sensing*, 20(11), pp. 2235–2247.

FAO 2020. Global Forest Resources Assessment 2020 Main Report. Food and Agriculture Organization of the United Nations, Rome.

Forman, R.T.T. and Gordon, M., 1986. *Landscape Ecology*. John Wiley, New York, p. 619.

FSI 2019. Mangrove resources of the country. India State of Forest Report 2019. Forest Survey of India, Ministry of Environment, Forest and Climate Change, Government of India, Dehradun.

Ghosh, D., 2011. Mangroves. *Resonance*, 16(1), pp. 47–60.

Ghosh, M.K., Kumar, L. and Roy, C., 2016. Mapping long-term changes in mangrove species composition and distribution in the Sundarbans. *Forests*, 7(12), p. 305.

Ghosh, S., Nandy, S., Patra, S., Kushwaha, S.P.S., Kumar, A.S. and Dadhwal, V.K., 2017. Land cover classification using ICESat/GLAS full waveform data. *Journal of the Indian Society of Remote Sensing*, 45(2), pp. 327–335.

Giri, C., Ochieng, E., Tieszen, L.L., Zhu, Z., Singh, A., Loveland, T., Masek, J. and Duke, N., 2011. Status and distribution of mangrove forests of the world using earth observation satellite data. *Global Ecology and Biogeography*, 20(1), pp. 154–159.

Giri, C., Pengra, B., Zhu, Z., Singh, A. and Tieszen, L.L., 2007. Monitoring mangrove forest dynamics of the Sundarbans in Bangladesh and India using multi-temporal satellite data from 1973 to 2000. *Estuarine, Coastal and Shelf Science*, 73(1–2), pp. 91–100.

Giri, S., Mukhopadhyay, A., Hazra, S., Mukherjee, S., Roy, D., Ghosh, S., Ghosh, T. and Mitra, D., 2014. A study on abundance and distribution of mangrove species in Indian Sundarban using remote sensing technique. *Journal of Coastal Conservation*, 18(4), pp. 359–367.

Gopal, B. and Chauhan, M., 2006. Biodiversity and its conservation in the Sundarban mangrove ecosystem. *Aquatic Sciences*, 68(3), pp. 338–354.

Heenkenda, M.K., Joyce, K.E., Maier, S.W. and Bartolo, R., 2014. Mangrove species identification: Comparing WorldView-2 with aerial photographs. *Remote Sensing*, 6(7), pp. 6064–6088.

Huang, X., Zhang, L. and Wang, L., 2009. Evaluation of morphological texture features for mangrove forest mapping and species discrimination using multispectral IKONOS imagery. *IEEE Geoscience and Remote Sensing Letters*, 6(3), pp. 393–397.

Jia, M., Zhang, Y., Wang, Z., Song, K. and Ren, C., 2014. Mapping the distribution of mangrove species in the Core Zone of Mai Po Marshes Nature Reserve, Hong Kong, using hyperspectral data and high-resolution data. *International Journal of Applied Earth Observation and Geoinformation*, 33, pp. 226–231.

Kamal, M. and Phinn, S., 2011. Hyperspectral data for mangrove species mapping: A comparison of pixel-based and object-based approach. *Remote Sensing*, 3(10), pp. 2222–2242.

Kanniah, K.D., Wai, N.S., Shin, A. and Rasib, A.W., 2007. Per-pixel and sub-pixel classifications of high-resolution satellite data for mangrove species mapping. *Applied GIS*, 3(8), pp. 1–22.

Kathiresan, K. and Bingham, B.L., 2001. Biology of mangroves and mangrove ecosystems. *Advances in Marine Biology*, 40, pp. 84–254.

Kuenzer, C., Bluemel, A., Gebhardt, S., Quoc, T.V. and Dech, S., 2011. Remote sensing of mangrove ecosystems: A review. *Remote Sensing*, 3(5), pp. 878–928.

Kumar, R., Nandy, S., Agarwal, R. and Kushwaha, S.P.S., 2014. Forest cover dynamics analysis and prediction modeling using logistic regression model. *Ecological Indicators*, 45, pp. 444–455.

Kushwaha, S.P.S., Nandy, S., Shah, M.A., Agarwal, R. and Mukhopadhyay, S., 2018. Forest cover monitoring and prediction in a Lesser Himalayan elephant landscape. *Current Science*, 115(3), pp. 510–516.

Kushwaha, S.P.S., Padmanaban, P., Kumar, D. and Roy, P.S., 2005. Geospatial modeling of plant richness in Barsey Rhododendron Sanctuary in Sikkim Himalayas. *Geocarto International*, 20(2), pp. 63–68.

Lee, S.Y., Jones, E.B.G., Diele, K., Castellanos-Galindo, G.A. and Nordhaus, I., 2017. Biodiversity. In: Rivera-Monroy, V.H., Lee, S., Kristensen, Y.E., and Twilley, R.R. (eds.) *Mangrove Ecosystems: A Global Biogeographic Perspective*. Springer, Cham, pp. 55–86.

Li, Q., Wong, F.K.K. and Fung, T., 2019. Classification of mangrove species using combined WordView-3 and LiDAR data in Mai Po Nature Reserve, Hong Kong. *Remote Sensing*, 11(18), p. 2114.

Li, H., Han, Y. and Chen, J., 2020. Combination of Google Earth imagery and Sentinel-2 data for mangrove species mapping. *Journal of Applied Remote Sensing*, 14(1), p. 010501.

Lyon, J.G., 1983. Landsat-derived land-cover classifications for locating potential kestrel nesting habitat. *Photogrammetric Engineering and Remote Sensing*, 49(2), pp. 245–250.

Nandy, S. and Kushwaha, S.P.S., 2010. Geospatial modelling of biological richness in Sunderbans. *Journal of the Indian Society of Remote Sensing*, 38(3), pp. 431–440.

Nandy, S. and Kushwaha, S.P.S., 2011. Study on the utility of IRS 1D LISS-III data and the classification techniques for mapping of Sunderban mangroves. *Journal of Coastal Conservation*, 15(1), pp. 123–137.

Nandy, S., 2009. Geospatial modelling of biological richness for conservation prioritization of Sunderban mangroves (Doctoral dissertation, Ph. D Thesis, Forest Research Institute University, Dehradun).

Nandy, S., Ghosh, S., Kushwaha, S.P.S. and Kumar, A.S., 2019. Remote sensing-based forest biomass assessment in northwest Himalayan landscape. In *Remote Sensing of Northwest Himalayan Ecosystems* (pp. 285–311). Springer, Singapore.

Nandy, S., Kushwaha, S.P.S. and Dadhwal, V.K., 2011. Forest degradation assessment in the upper catchment of the river Tons using remote sensing and GIS. *Ecological Indicators*, 11(2), pp. 509–513.

Nandy, S., Kushwaha, S.P.S. and Mukhopadhyay, S., 2007. Monitoring the Chilla–Motichur wildlife corridor using geospatial tools. *Journal for Nature Conservation*, 15(4), pp. 237–244.

Nandy, S., Lakshmi, M.N. and Kushwaha, S.P.S., 2020. Habitat suitability analysis of Himalayan Musk Deer (Moschus leucogaster) in part of Western Himalaya, India. *Journal of the Indian Society of Remote Sensing*, 48, pp. 1–11.

Navalgund, R.R., Kumar, A.S. and Nandy, S., 2019. *Remote Sensing of Northwest Himalayan Ecosystems* (Vol. 1). Springer, Singapore.

Polidoro, B.A., Carpenter, K.E., Collins, L., Duke, N.C., Ellison, A.M., Ellison, J.C., Farnsworth, E.J., Fernando, E.S., Kathiresan, K., Koedam, N.E. and Livingstone, S.R., 2010. The loss of species: Mangrove extinction risk and geographic areas of global concern. *PLoS One*, 5(4), p. e10095.

Ragavan, P., Saxena, A., Jayaraj, R.S.C., Mohan, P.M., Ravichandran, K., Saravanan, S. and Vijayaraghavan, A., 2016. A review of the mangrove floristics of India. *Taiwania*, 61(3), pp. 224–242.

Romme, W.H., 1982. Fire and landscape diversity in subalpine forests of Yellowstone National Park. *Ecological Monographs*, 52(2), pp. 199–221.

Roy, P.S. and Tomar, S., 2000. Biodiversity characterization at landscape level using geospatial modelling technique. *Biological Conservation*, 95(1), pp. 95–109.

Sandilyan, S. and Kathiresan, K., 2012. Mangrove conservation: A global perspective. *Biodiversity and Conservation*, 21(14), pp. 3523–3542.

Satyanarayana, B., Mohamad, K.A., Idris, I.F., Husain, M.L. and Dahdouh-Guebas, F., 2011. Assessment of mangrove vegetation based on remote sensing and ground-truth measurements at Tumpat, Kelantan Delta, East Coast of Peninsular Malaysia. *International Journal of Remote Sensing*, 32(6), pp. 1635–1650.

Shannon, C.E. and Weaver, W., 1949. The Mathematical Theory of Communication, by CE Shannon (and Recent Contributions to the Mathematical Theory of Communication), W. Weaver. University of illinois Press.

Sievers, M., Chowdhury, M.R., Adame, M.F., Bhadury, P., Bhargava, R., Buelow, C., Friess, D.A., Ghosh, A., Hayes, M.A., McClure, E.C. and Pearson, R.M., 2020. Indian Sundarbans mangrove forest considered endangered under Red List of Ecosystems, but there is cause for optimism. *Biological Conservation*, 251, p. 108751.

Sreelekshmi, S., Nandan, S.B., Kaimal, S.V., Radhakrishnan, C.K. and Suresh, V.R., 2020. Mangrove species diversity, stand structure and zonation pattern in relation to environmental factors—A case study at Sundarban delta, east coast of India. *Regional Studies in Marine Science*, 35, p. 101111.

Srinet, R., Nandy, S., Padalia, H., Ghosh, S., Watham, T., Patel, N.R. and Chauhan, P., 2020. Mapping plant functional types in Northwest Himalayan foothills of India using random forest algorithm in Google Earth Engine. *International Journal of Remote Sensing*, 41(18), pp. 7296–7309.

Valderrama-Landeros, L., Flores-de-Santiago, F., Kovacs, J.M. and Flores-Verdugo, F., 2018. An assessment of commonly employed satellite-based remote sensors for mapping mangrove species in Mexico using an NDVI-based classification scheme. *Environmental Monitoring and Assessment*, 190(1), p. 23.

Viennois, G., Proisy, C., Feret, J.B., Prosperi, J., Sidik, F., Rahmania, R., Longépé, N., Germain, O. and Gaspar, P., 2016. Multitemporal analysis of high-spatial-resolution optical satellite imagery for mangrove species mapping in Bali, Indonesia. *IEEE Journal of Selected Topics in Applied Earth Observations and Remote Sensing*, 9(8), pp. 3680–3686.

Wan, L., Lin, Y., Zhang, H., Wang, F., Liu, M. and Lin, H., 2020. GF-5 hyperspectral data for species mapping of mangrove in mai po, Hong Kong. *Remote Sensing*, 12(4), p. 656.

Wan, L., Zhang, H., Lin, G. and Lin, H., 2019. A small-patched convolutional neural network for mangrove mapping at species level using high-resolution remote-sensing image. *Annals of GIS*, 25(1), pp. 45–55.

Wang, D., Wan, B., Qiu, P., Su, Y., Guo, Q. and Wu, X., 2018a. Artificial mangrove species mapping using Pléiades-1: An evaluation of pixel-based and object-based classifications with selected machine learning algorithms. *Remote Sensing*, 10(2), p. 294.

Wang, D., Wan, B., Qiu, P., Su, Y., Guo, Q., Wang, R., Sun, F. and Wu, X., 2018b. Evaluating the performance of Sentinel-2, Landsat 8 and Pléiades-1 in mapping mangrove extent and species. *Remote Sensing*, 10(9), p. 1468.

Wang, L., Sousa, W.P., Gong, P. and Biging, G.S., 2004. Comparison of IKONOS and QuickBird images for mapping mangrove species on the Caribbean coast of Panama. *Remote Sensing of Environment*, 91(3–4), pp. 432–440.

Zhang, H., Wang, T., Liu, M., Jia, M., Lin, H., Chu, L.M. and Devlin, A.T., 2018. Potential of combining optical and dual polarimetric SAR data for improving mangrove species discrimination using rotation forest. *Remote Sensing*, 10(3), p. 467.

6 Spatio-Temporal Distribution of Microforams in the Sundarbans Mangrove Forest, Bangladesh: Response to Ecological Imbalance

Tumpa Saha
Bangladesh Atomic Energy Commission

Mahmud Al Noor Tushar
University of Dhaka

Premanondo Debnath
Center for Environmental and Geographic
Information Services (CEGIS)

Subrota Kumar Saha
University of Dhaka

CONTENTS

6.1   INTRODUCTION

The Sundarbans, the world's largest mangrove wetland and contiguous ecosystem (Chaudhuri and Naithani, 1985), is located at the southern extremity of the Ganges River delta, that is, the plain bordering the northern margin of the Bay of Bengal. Late Holocene deposits of the GBM river system formed the riverine delta and are spread across coastal parts of India and Bangladesh (Stanley and Hait, 2000). About two centuries ago, the forests extended north beyond the city of Calcutta (Banerjee, 1964) and the Khulna and Bakerganj districts as far-east of the Meghna River in Bangladesh (Siddiqi, 2001). Islam (2001) stated that about 9,000–8,000 years BP the mangroves ecosystem may have existed in most parts of Jessore and Narail, parts of Jhenaidah and Magura. He also showed that from ca. 5,000–3,000 years BP, the pollen record implies that a dense mangrove forest belt existed along the south eastern belt including Satkhira, Khulna, Pirojpur and Patuakhali districts in Bangladesh. This mangrove ecosystem is the region of transition between the fresh water of the Ganges distributaries system and the saline water of the Bay of Bengal, spanning about 350 km in width in the southern Bangladesh and the state of West Bengal in India. Within the geographical area, forests cover nearly 10,000 km^2, of which about 6,000 km^2 in Bangladesh. It has a total land area of about 4,143 km^2 and water area about 1,874 km^2 (Danda, 2010). UNESCO has declared Sundarbans as the World Heritage Site since 1987 and it is the only mangrove tiger habitat (Manna et al. 2010).

In the modern marine environment, foraminifera constitute an abundant diverse group of shelled microorganisms. They are the most ancient and plentiful fossils and also the most efficient of rock builders that's why they are geologically significant (Flint, 1899). Different abiotic factors such as coastal geomorphology, temperature, salinity, tidal amplitude and duration, dissolved oxygen and nutrients govern the environmental setting of mangrove ecosystem. Competition, space and food supply are also included in the biotic factors of this ecosystem. In time and space, such kind of macro-level environmental factors of mangroves undergo changes (Ghosh et al., 2014). These changes are well reflected in the distribution of foraminifera. Reconstruction of paleo-environmental condition and ecology may be established by understanding the development of mangrove communities in recent times. This ecosystem consists of various species of fauna and flora. Royal Bengal tiger is the most significant and endangered species. The flora of the Sundarbans presents a natural buffer and acts as a rampart against coastal erosion and saltwater intrusion into one of the most over-populated regions of the world. Many invertebrates, phytoplankton, fungi, bacteria, zooplankton, benthic invertebrates and mollusks are common here. A large number of distributaries of Ganges and Brahmaputra River flow through this ecosystem. A complex network of estuaries, beaches, tidal creeks, tidal inlets, mudflats, sandflats and mangrove swamps exists in this tide-dominated wetland (Mitra et al., 2009). It is a classic example of tide-dominated wetland (Selvam, 2003). With a high tidal range, high rainfall and high annual sediment load of several rivers debouching sediments into the region, Sundarbans, is environmentally dynamic. Moreover, the constant threat of cyclones, storm surges, sea level rise and reduced flow of freshwater into the mangrove system are common in this ecoregion. As foraminifera are so much sensitive to environment, these incidents highly affect them.

The Sundarbans region is highly inaccessible because of its geographic location. Moreover, this forest is a waterlogged jungle where tigers and other wild animals thrive and intersected by numerous river channels and creeks. So, biodiversity of modern foraminifera and their distribution in the estuaries of the western coastal region of Bangladesh were largely overlooked. Embroidery of tidal creeks, encompassing the islands formed of Alfisols (older alluvial soil) and Ardisols (coastal saline soil), characterizes Sundarbans and offshore linear tidal shoals aligned perpendicular to the shore line and separated by swales (Figure 6.2).

This cumbersome study is a combination of field work, laboratory analysis and interpretation. A number of foraminiferal parameters are generated from several statistical analyses. Then, the effects of the ecology of the given area on these parameters symbolize how the environment reacts with different variables that are calculated from analysis.

6.2 METHODOLOGY

However, in this study, three steps research methodology (Figure 6.1) was used: (i) conceptual development; (ii) field work and (iii) laboratory work. The following flow diagram is a complete representation of the methodology applied to the study.

The study area had been chosen for its uniqueness and then to get something new because the micropaleontology of that area was completely unexplored and unknown. A total of 32 samples were collected from ten locations (Figure 6.2) of the Sundarbans (Table 6.1).

At first, the samples were washed carefully over a 63 μm sieve to remove the unwanted fractions of mud particles. Then, it was dried in an oven at 100°C temperature for 30–40 minutes. Every specimen was observed under Leica EZ4E stereomicroscope (Leica Corporation) at 35× magnification and bright-field images were captured using cellSens software (Leica).

6.3 RESULTS AND DISCUSSIONS

A total 17 species of foraminifera representing nine families were identified in this study based on the plates of foraminifera introduced by Loeblich and Tappan, 1987. According to their origin, the identified species are benthic and the most significant and abundant genera are *Nonionina, Nonionella, Ammonia, Elphidium, Rosalina, Haplophragmoides, Jadammina, Trochammina, Asterorotalia, Quinqueloculina, Miliammina*, etc. (Figure 6.3). The percentage of recorded species based on sampling points has been shown by Table 6.1.

Based on the count of all specimens present in representative aliquots or in the whole sample, microfaunal quantitative analysis was conducted. Frequency distribution shows different species abundant in different locations (Figure 6.4). *Noninella miocenica* shows highest frequency in Bangabandhu Island where *Ammonia beccarii* is abundant in Dublar char. Overall, *Nonionella, Nonionina*, displays a coarser mode than the other genera.

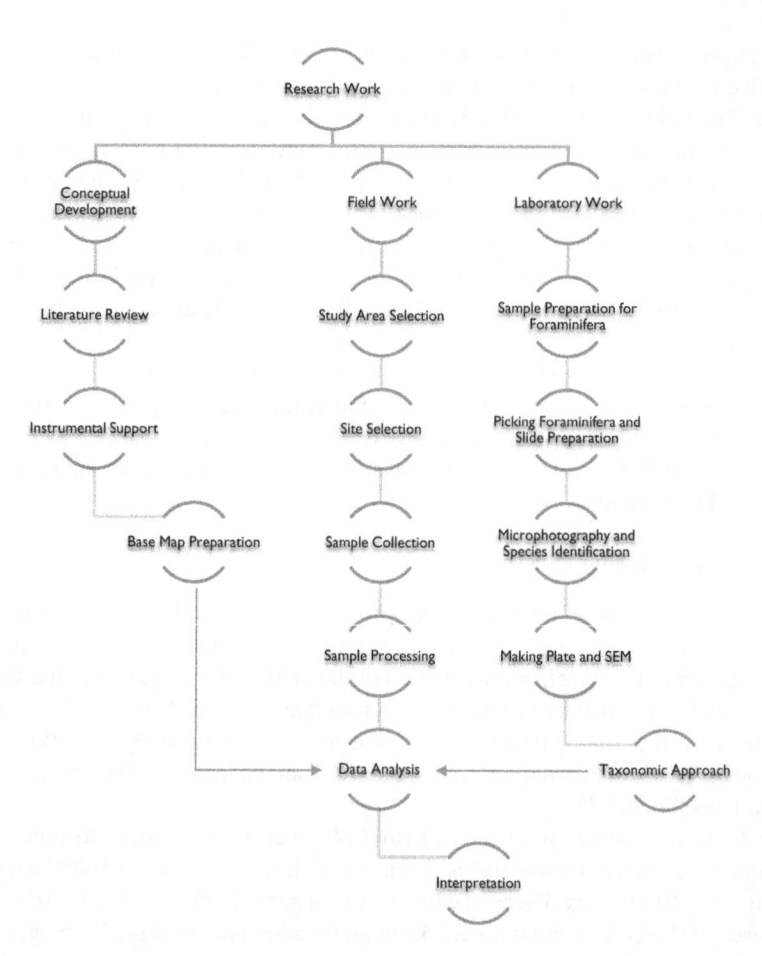

FIGURE 6.1   Flow diagram of methodology.

TABLE 6.1
List of Samples with Geographic Coordinates

Location Name	District	Latitude	Longitude	Number of Samples
Dimer Char (Dc)	Bagerhat	21°50′ 56.580″ N	21°50′ 56.580″ N	3
Kochikhali (Ko)	Bagerhat	21°51′ 32.638″ N	21°51′ 32.638″ N	3
Jamtola (Jm)	Bagerhat	21°50′ 46.689″ N	21°50′ 46.689″ N	3
Katka (Kt)	Bagerhat	21°51′ 14.974″ N	21°51′ 14.974″ N	4
Chandpai Range (Cr)	Bagerhat	21°45′ 37.436″ N	21°45′ 37.436″ N	3
Dublar Char (Db)	Bagerhat	21°42′ 59.757″ N	21°42′ 59.757″ N	4
Bangabandhu Island (Bi)	Khulna	21°42′ 36.308″ N	21°42′ 36.308″ N	3
Nalian Range (Nr)	Khulna	21°43′ 58.820″ N	21°43′ 58.820″ N	2
Putney Island (Pi)	Khulna	21°41′ 21.206″ N	21°41′ 21.206″ N	3
Madarbaria (Mb)	Satkhira	21°40′ 29.087″ N	21°40′ 29.087″ N	4

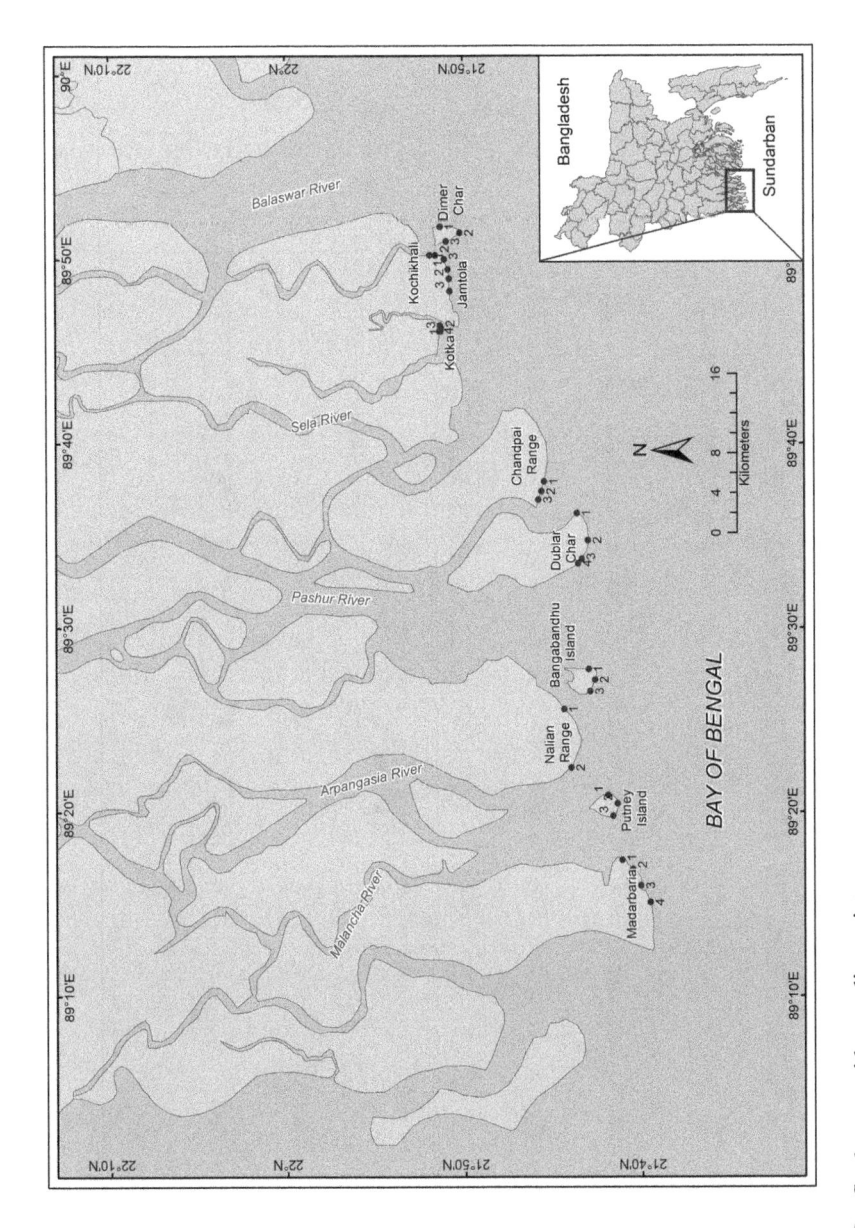

FIGURE 6.2 Study area with sampling points.

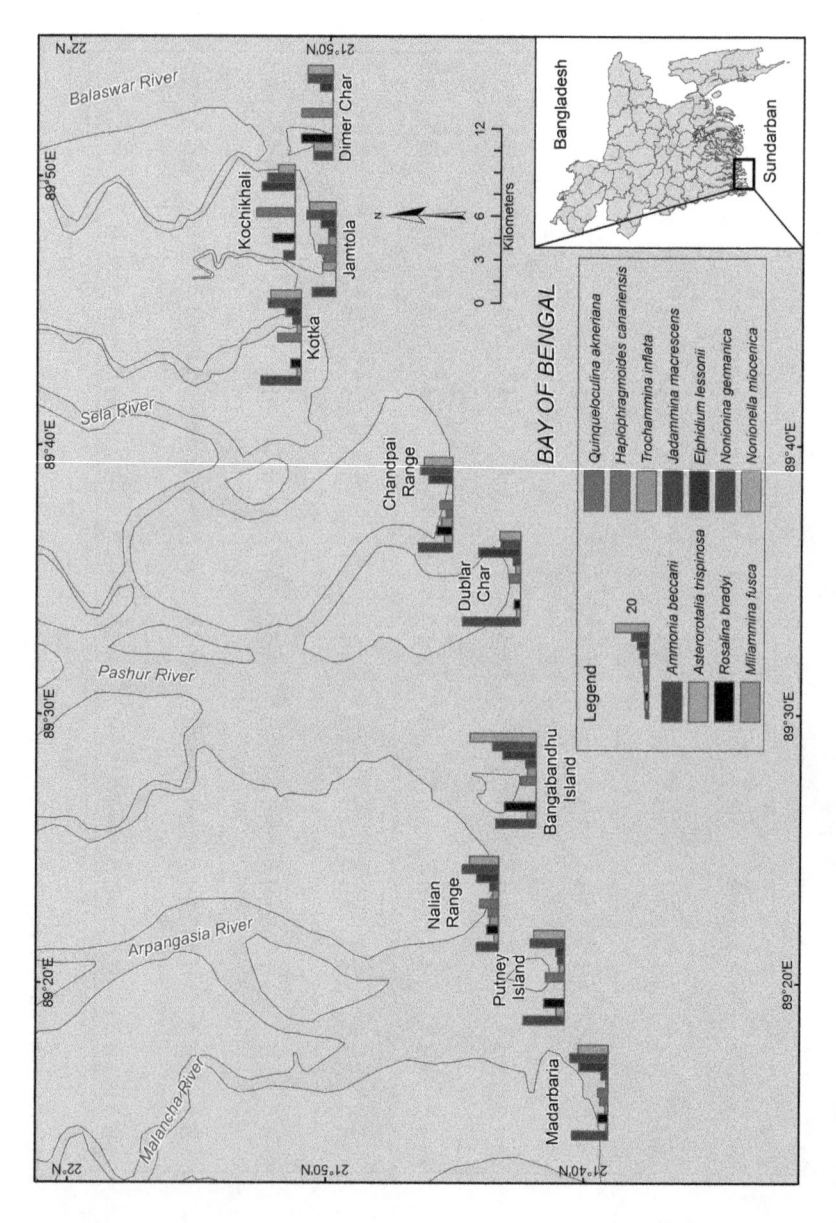

FIGURE 6.3 Spatial distribution of foraminiferal species based on relative abundance.

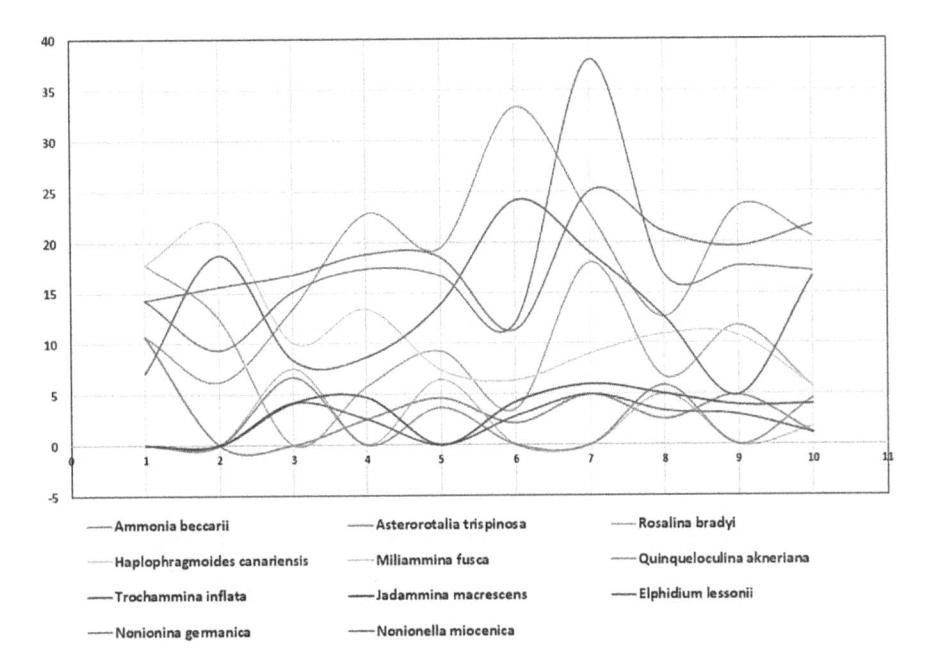

Legend:
— Ammonia beccarii   — Asterorotalia trispinosa   — Rosalina bradyi
— Haplophragmoides canariensis   — Miliammina fusca   — Quinqueloculina akneriana
— Trochammina inflata   — Jadammina macrescens   — Elphidium lessonii
— Nonionina germanica   — Nonionella miocenica

FIGURE 6.4 Frequency distribution of abundant species in different locations.

Hierarchical cluster analysis (HCA) was carried out in order to recognize groups of samples with homogeneous foraminiferal content using sum of specimens to create a matrix. The relative abundance of commonly occurring species was included where number of specimens ranging 50–280 for each of 10 locations. The Euclidean distance coefficient was applied to compare samples and the ward's method of minimum variance to assemble clusters for the HCA (Van Hengstum and Scott, 2011).

Here, HCA may be divided into three major parts (Figure 6.5). Cluster A shows similarity between locations Dimer char and Kochikhali. These samples carry a high dominance of *H. canariensis* (18%–22%) and *R. bradyi* (13%–18%) (Table 6.2). They display the lowest species diversity of the study area (H: 1.4–1.6) and show higher foraminiferal abnormality index (FAI) value (7–8) (Table 6.3).

Part C includes several locations: Bangabandhu Island (Bi), Katka (Kt), Dublar Char (Db), Nalian Range (Nr), Putney Island (Pi), Chandpai Range (Cr) and Madarbaria (Mb). It shows minimum dissimilarities and all locations are below the standard line. Among them, Cr and Mb are similar and they are characterized by high dominance of *A. beccarii* (19%–21%) and *N. germanica* (19%–22%) with *N. miocenica* (17%) (Table 6.2). Mb shows lower FAI (3) when Kt has the higher FAI (9) and Nr shows higher species diversity (H) than other 6 locations (H: 3.9) with lowest FAI value (2) (Table 6.3).

Cluster B relates Jamtala (Jm) with part C and Jm displays the maximum dissimilarities. Jm shows the highest FAI value (11) (Table 6.3) in the study area.

Species richness can be termed as the number of species in a sample or study area. It is also known as species diversity or biodiversity and diversity index (H) relates to the number of individuals sampled in an area. Using the statistical package

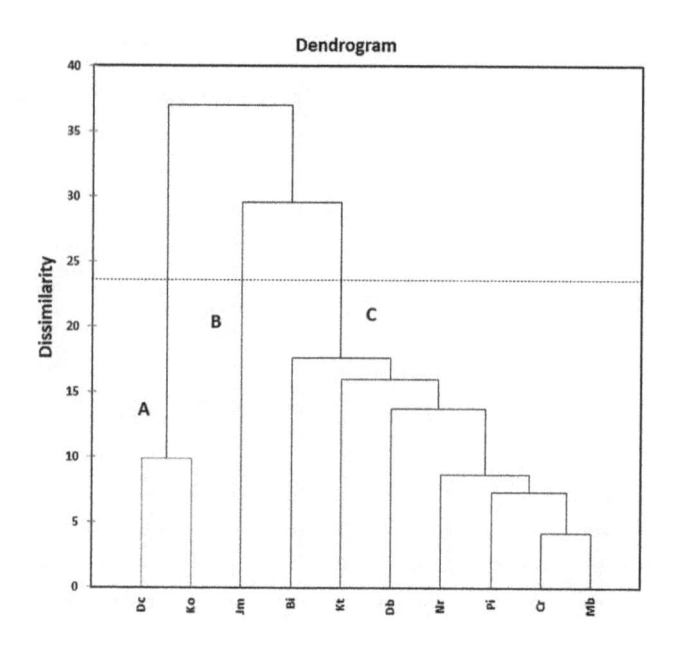

FIGURE 6.5   HCA of microforams based on relative abundance.

TABLE 6.2
Observed Foraminiferal Species (%) Across Study Locations

Sample	Ammonia beccarii	Ammonia aoteana	Ammonia tepida	Asterorotalia trispinosa	Rosalina sp	Rosalina bradyi	Miliammina fusca	Quinqueloculina akneriana	Haplophragmoides canariensis	Nonionellina sp	Trochammina inflata	Jadammina macrescens	Elphidium lessonii	Cribroelphidium vadescens	Nonionina germanica	Nonionella miocenica
Dc	11	0	0	11	0	18	0	0	18	0	0	0	7.1	7.1	14	14
Ko	6.3	0	0	0	0	13	0	0	22	0	0	0	19	16	16	9.4
Jm	13	0	1.7	0	4.2	0	7.6	6.7	10	1.7	4.2	4.2	8.4	5.9	17	15
Kt	23	1.8	1.4	2.5	0	5.8	0	0	13	0	2.5	4.7	8.7	0	19	17
Cr	19	0	0	4.6	0	9.3	6.5	3.7	7.4	0	0	0	14	0	19	17
Db	33	0	0	2.1	0	3.5	0	0	6.4	0	2.8	4.3	24	0	11	12
Bi	23	0	0	5	0	18	0	0	9	0	5	6	19	0	25	38
Nr	13	0	0	2.5	0	6.7	5	5.9	11	1.7	3.4	5	13	0	21	17
Pi	24	0	0	4.9	0	12	0	0	11	0	2.9	3.9	4.9	0	20	18
Mb	21	0	0	1.1	0	5.7	1.7	4.6	5.7	0	1.1	4	17	0	22	17

TABLE 6.3
Faunal Parameter of Foraminiferal Assemblages:
Diversity Index (H), FAI

Location	Diversity Index, H	FAI
1. Dimer Char (Dc)	1.6	8
2. Kochikhali (Ko)	1.4	7
3. Jamtala (Jm)	2.1	11
4. Katka (Kt)	1.8	9
5. Chandpai Range (Cr)	2.1	3
6. Dublar Char (Db)	1.6	6
7. Bangabandhu Island (BI)	2.6	3
8. Nalian Range (Nr)	3.9	2
9. Putney Island (Pi)	1.9	2
10. Madarbariea (Mb)	3.4	3

PAleontological STatistics (Hammer et al., 2001; Hammer and Harper, 2006), the species richness was calculated, given by the Shannon index, H (Shannon, 1948). Then, FAI index, that is, percentage of deformed specimens (Coccioni et al., 2005) was determined, also referred as the relative abundance of deformed shell (Table 6.3).

In the diversity case, a log series distribution is used for Fisher alpha index (Fisher et al., 1943) to predict the species number. The range of Fisher's alpha diversity values can compare the locations of study area (Murray, 2006).

The diversity index, H, shows number of species versus number of individuals in the study locations. Species diversity was considered as potential indicators of environmental stress and reliable ecological proxy. Overall the H-index shows low diversity of foraminiferal assemblages in the Sundarbans region. All samples belong between index value ($H = 1$–4) where maximum satisfies the value (2–3) (Table 6.3 and Figure 6.7).

The Sundarbans is a complex network of estuaries, beaches, tidal creeks, tidal inlets, mudflats, sandflats and mangrove swamps (Mitra et al. 2009). That is the reason most of all locations are not easily accessible there. But compared to other locations, Dimer char (Dc), Kochikhali (Ko), Jamtala (Jm) and Katka (Kt) are easily accessible and has been become one of the must going tourist destinations. During the peak season, a large number of ships and steamboats go there. Every year about 100,000 domestic and international tourists visit this very part of Sundarbans (Biswas et al., 2011). Again, Mongla port is located at northern part of the Sundarban and the second busiest seaport of Bangladesh with a harboring capacity of over 70 oceangoing ships at a time (Aziz & Paul, 2015). Numerous international shipping companies have used Mongla port as an alternative because of accretive congestion in Chattogram port. Due to its location at the confluence of the Passur River that drains about 85% of the water passes through this region (Aziz & Paul, 2015), Dublar char is affected mostly. Simultaneously, Dublar char is one of the most popular tourist spots in Sundarbans and well-known for dry fish processing.

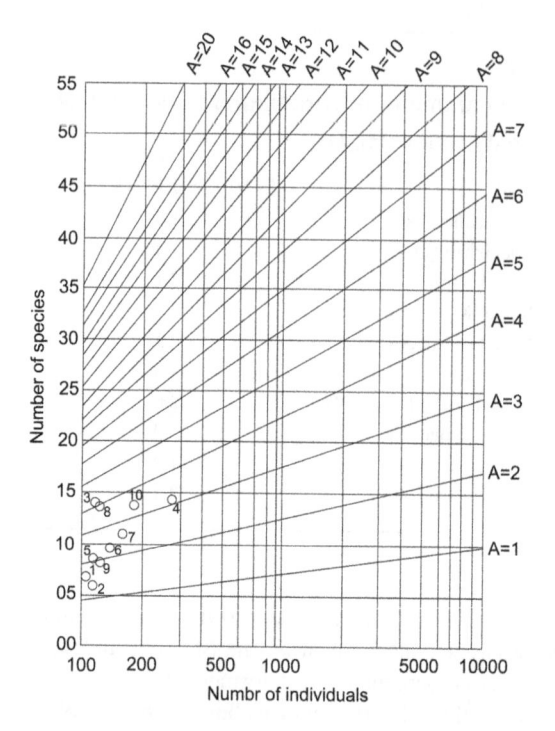

FIGURE 6.6   Diversity index (H) of foraminiferal assemblage of the Sundarban. (Modified from Fisher et al., 1943.)

The above-mentioned areas show lower H with higher FAI (index of abnormality) values that implies samples from these very areas yield minimum number of species but most of them show morphological abnormality. On the other hand, the situation is more or less reverse in the areas like Bangabandhu Island (Bi), Nalian Range (Nr), Putney Island (Pi), Chandpai Range (Cr) and Madarbaria (Mb). These areas suffer less human intervention because of their geographic location and lack of easy access, that is, only accessible with great difficulty through steamboats and ships (Table 6.3 and Figure 6.6). Considerable reduction of some characteristic flora and fauna as a response to human intervention like increasing metal concentrations spilling away from waterways into major draining rivers like Passur, Sela, Sibsa, etc., waste from the Mongla port and coal-bearing power plant has also been reported (Aziz & Paul, 2015). Here the reduction of species diversity and abundance as well as increased development of abnormalities indicate anthropogenically impacted ecological imbalance that justifies an overwhelming benthic foraminiferal response to environmental stress. This chapter highlights the development of some morphological abnormalities (Plate 6.1) like aberrant chamber size and shape, protuberance on one or more chambers (Alve, 1991), incomplete development of the last whorl (Yanko et al., 1998), etc.

As a natural level for unstressed population, 1% abnormal tests can be considered (Alve, 1991), but FAI index ranges between 2 and 11 (Table 6.3) which exceeds the natural threshold in all locations of study area. As the abnormalities also present

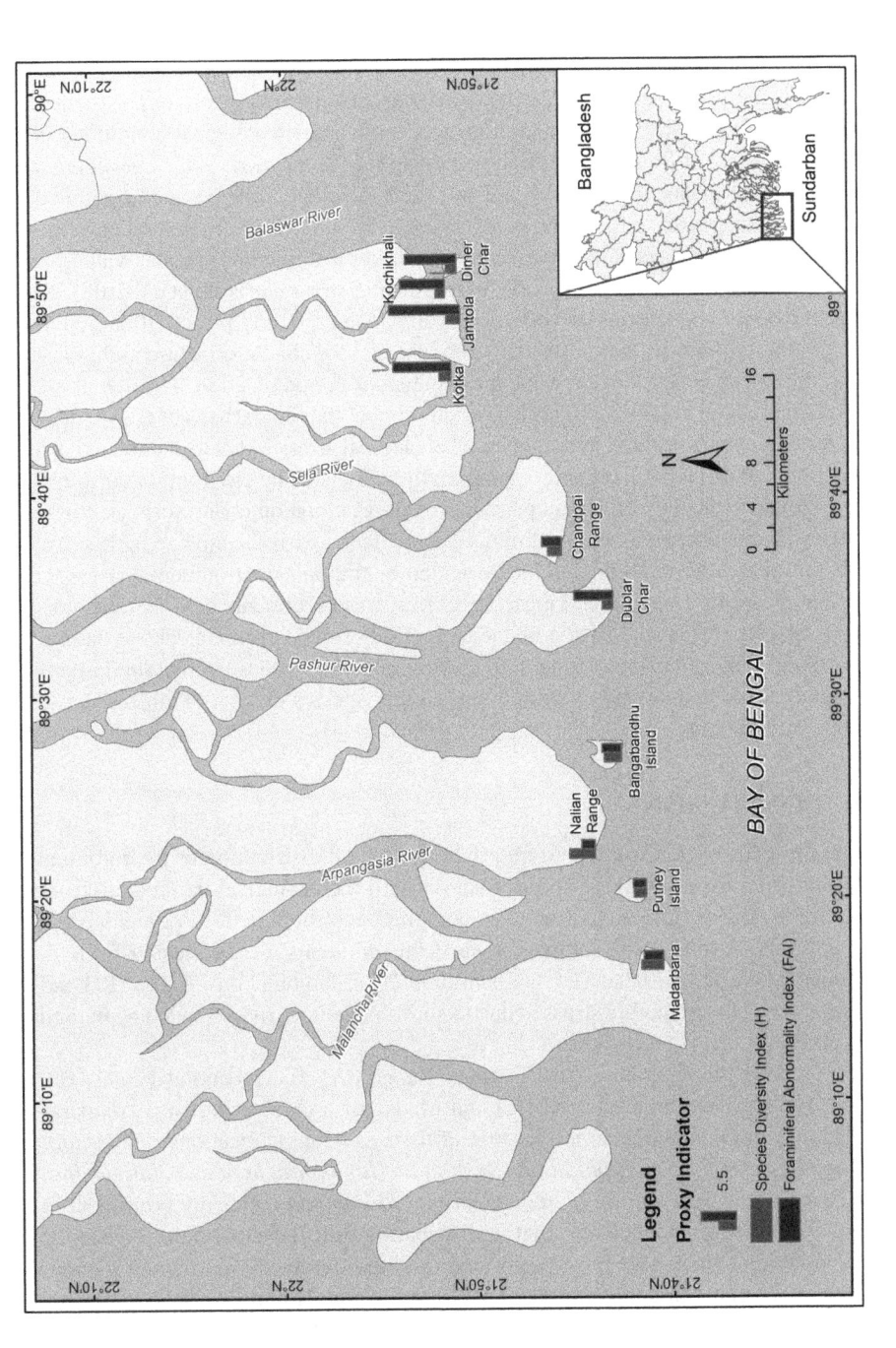

FIGURE 6.7 Spatial distribution of diversity index (H) and FAI.

in the agglutinated tests (*Jadammina macrescens*), the heavy metal ions may not acquire during the test building. The effects of heavy metal on cytological activities (i.e. organic linings) of the cell could link the development of the morphological abnormalities (Bergamin et al., 2019).

Low diversity, low density and high percentages of deformed specimens character-ize foraminiferal assemblage, which highlight unfavorable ecological conditions in the western coast of Bangladesh. High percentages of *A. beccarii* and *N. germanica* are recorded in all locations which pointed their tolerance to heavy metal polluted environments (according to Yanko et al., 1999). It is not typical of stressed environ-ment as indifferent species in apparent contradiction with the low H-index values of the assemblage, indicating significant environmental stress (Jorissen et al., 2018).

High density and species diversity, short life span and good preservation poten-tial of benthic foraminifera make them an ideal tool for characterization and moni-toring of marine ecosystems as ecological indicator (Gooday, 2003; Jorissen et al., 2007; Schönfeld et al., 2012). Benthic microforams of the Sundarbans of Bangladesh were studied, characterized and applied as ecological stress indicators in this chap-ter. Among all of faunal parameters, potentially influenced by environmental stress, species diversity is one of the major stress proxies. For biomonitoring surveys, a pre-liminary characterization of microforams is suitable to assess the most reliable stress index for an area never studied before, affected by multiple environmental stressors of different origin. Heavy metal enrichment of marine sediments of stressed condi-tions is also characterized by the abundance of deformed specimens above natural background. Future biomonitoring activities for ecological status in the Sundarbans may apply a stress index based on the lowering of species diversity with respect to reference conditions.

6.4 CONCLUSIONS

The main focus of this study is to highlight the spatial distribution of microforams in terms of foram assemblage and to relate it with the ecological parameters from Sundarbans, Bangladesh. This study is a compilation of both field work and labora-tory analysis. A total of 32 samples of surficial sediments were collected from 10 locations where Dimer Char (DC), Kochikhali (Ko), Jamtala (Jm), Katka (Kt) and Dublar Char (Db) are highly stressed due to anthropogenic activities aided by human interventions.

The rest of the locations like Chandpai Range (Cr), Bangabandhu Island (Bi), Nalian Range (Nr), Putney Island (Pi) and Madarbaria (Mb) experience compara-tively lower environmental stress because of their geographic locations. *Nonionina, Nonionella, Ammonia, Elphidium, Rosalina, Haplophragmoides, Jadammina, Trochammina, Asterorotalia, Quinqueloculina, Miliammina*, etc., are prominent in terms of relative abundance and spatial distribution over the study area. Frequency distribution and HCA have been performed to figure out the similarities of homog-enous microforams in different locations. In this study, H and FAI have also been computed to investigate the environmental status. HCA finds three different clusters namely the cluster A, cluster B and cluster C. Cluster A represents the similarity between the locations Dimer char and Kochikhali with a species diversity of 1.4–1.6

and a FAI ranges from 7 to 8. Cluster B relates Jamtala (Jm) with cluster C and Jm displays the maximum dissimilarities. Part C includes several locations, namely, Bangabandhu Island, Katka, Dublar Char, Nalian Range, Putney Island, Chandpai Range and Madarbaria. It shows minimum dissimilarities and all locations are below the standard line. *H. canariensis*, *Q. akneriana* and *N. germanica* are the dominant species for cluster A, B and C, respectively. In terms of species diversity (H), samples from the location Nalian Range (Nr) rise on to the top with an index value of 3.9. Based on overall H it can be concluded that the foram assemblages show a low diversity in the Sundarbans region with high FAI. For naturally unstressed foram population group the FAI value is 1; beyond this the FAI indicates environmental stress. In this study samples from each and every location exceeds the natural threshold FAI value. Low species diversity and FAI value > 1 in each and every location and high percentages of pollution tolerant species *A. beccarii* and *N. germanica* directly indicate an environmentally stressed condition with significant ecological imbalance. Increased metal concentrations spilling away into different river like Passur, Baleswar, Sela, Sibsa from waterways including the Mongla port, coal-bearing power plant waste and hydrodynamic features associated to human-induced activities point out the effects of the terrestrial contribution to the mangrove and marine environments and its effect on biota which cause several morphological abnormalities. This study may be the first attempt of investigation of ecological imbalance based on microforams and so it will be able to provide baseline information for further microfossils-based environmental, climatic and ecological investigations.

APPENDIX: SEM MICROGRAPHS OF RECORDED FORAMS

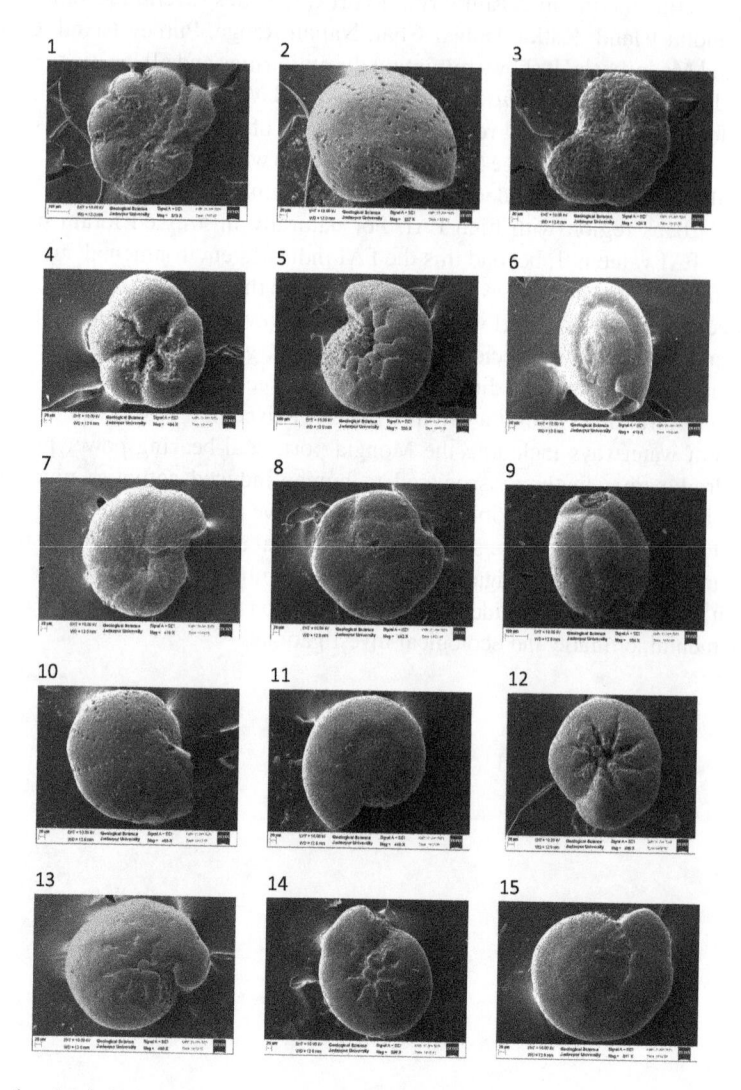

PLATE 1: 1. *Asterorotalia trispinosa*. 2. *Cribroelphidium vadescens*. 3. *Jadammina macrescens*. 4. *Trochammina inflata*. 5. *Rosalina bradyi*. 6. *Quinqueloculina akneriana*. 7. *Nonionellina* sp. 8. *Trochammina* sp. 9. *Miliammina fusca*. 10. *Elphidium lessonii*. 11. *Rosalina* sp. 12. *Nonionina germanica*. 13. *Haplophragmoides canariensis*. 14. *Nonionella miocenica*. 15. *Ammonia beccarii*.

REFERENCES

Alve, E., 1991. Benthic foraminifera in sediment cores reflecting heavy metal pollution in Sorfjord, western Norway. *The Journal of Foraminiferal Research*, *21*(1), pp. 1–19.

Aziz, A. and Paul, A.R., 2015. Bangladesh Sundarbans: present status of the environment and biota. *Diversity*, *7*(3), pp. 242–269.

Banerjee, A.K., 1964. Forests of Sundarbans. *Forest Department, Calcutta, West Bengal, Centenary Volume*, pp. 166–185.

Bergamin, L., Di Bella, L., Ferraro, L., Frezza, V., Pierfranceschi, G. and Romano, E., 2019. Benthic foraminifera in a coastal marine area of the eastern Ligurian Sea (Italy): Response to environmental stress. *Ecological Indicators*, *96*, pp. 16–31.

Biswas, M.H.A., Ara, M., Haque, M.N. and Rahman, M.A., 2011. Application of control theory in the efficient and sustainable forest management. *International Journal of Scientific and Engineering Research*, *2*(3), pp. 26–33.

Chaudhuri, A.B. and Naithani, H.B., 1985. A comprehensive survey of tropical mangrove forests of Sundarbans and Andamans (pt. 1).-A comprehensive survey of epiphytes (pt. 2).-Lianas, climber and shrubby climbers (pt. 3).-A comprehensive ecological-botanical survey of the grasses (Gramineae) and sedges (Cyperaceae) (pt. 4).-A comprehensive eco-botanical survey of monocotyledons (pt. 5).

Coccioni, R., Frontalini, F., Marsili, A. and Troiani, F., 2005. Benthic foraminifera and trace metals: environmental implications. *The evolutionary dynamics of the coastal strip between the mouths of the Foglia and Metauro rivers: Towards the integrated management of a coast of high environmental value. Quaderni del Centro di Geobiologia, Univ Urbino*, *3*, pp. 57–92.

Danda, A., 2010. Sundarbans: Future imperfect climate adaptation report. *World Wide Fund for Nature, New Delhi*.

Fisher, R.A., Corbet, A.S. and Williams, C.B., 1943. The relation between the number of species and the number of individuals in a random sample of an animal population. *The Journal of Animal Ecology*, *12*, pp. 42–58.

Flint, J.M., 1899. *Recent Foraminifera: A descriptive catalogue of specimens dredged by the US Fish Commission Steamer Albatross*. Smithsonian Institution. United States National Museum.

Ghosh, A., Biswas, S. and Barman, P., 2014. Marsh foraminiferal assemblages in relation to vegetation in Sunderban, India. *Journal of the Geological Society of India*, *84*(6), pp. 657–667.

Gooday, A.J., 2003. Benthic foraminifera (Protista) as tools in deep-water palaeoceanography: environmental influences on faunal characteristics.

Hammer, Ø., Harper, D.A.T. and Ryan, P.D., 2001. PAST: Paleontological Statistic Software package for education and data analysis (reference manual). *Oslo: Natural History Museum, University of Oslo*.

Hammer, Ø. and Harper, D.A.T. 2006. *Paleontological Data Analysis*. Blackwell Publishing, Oxford.

Islam, M.S., 2001. *Sea-Level Changes in Bangladesh: The Last Ten Thousand Years*. Asiatic Society of Bangladesh, Bangladesh.

Jorissen, F., Nardelli, M.P., Almogi-Labin, A., Barras, C., Bergamin, L., Bicchi, E., El Kateb, A., Ferraro, L., McGann, M., Morigi, C. and Romano, E., 2018. Developing Foram-AMBI for biomonitoring in the Mediterranean: Species assignments to ecological categories. *Marine Micropaleontology*, *140*, pp. 33–45.

Jorissen, F.J., Fontanier, C. and Thomas, E., 2007. Proxies in Late Cenozoic Paleoceanography (Pt. 2).

Manna, S., Chaudhuri, K., Bhattacharyya, S. and Bhattacharyya, M., 2010. Dynamics of Sundarban estuarine ecosystem: Eutrophication induced threat to mangroves. *Saline Systems*, 6(1), p. 8.

Mitra, A., Gangopadhyay, A., Dube, A., Schmidt, A.C. and Banerjee, K., 2009. Observed changes in water mass properties in the Indian Sundarbans (northwestern Bay of Bengal) during 1980–2007. *Current Science*, 97, pp. 1445–1452.

Murray, J.W., 2006. *Ecology and Applications of Benthic Foraminifera*. Cambridge University Press, Cambridge.

Schönfeld, J., Alve, E., Geslin, E., Jorissen, F., Korsun, S. and Spezzaferri, S., 2012. The FOBIMO (FOraminiferal BIo-MOnitoring) initiative-Towards a standardised protocol for soft-bottom benthic foraminiferal monitoring studies. *Marine Micropaleontology*, 94, pp. 1–13.

Selvam, V., 2003. Environmental classification of mangrove wetlands of India. *Current Science*, 84(6), pp. 757–765.

Shannon, C.E., 1948. A mathematical theory of communication. *The Bell System Technical Journal*, 27(3), pp. 379–423.

Siddiqi, N.A., 2001. Mangrove forestry in Bangladesh. Institute of Forestry and Environmental Sciences, University of Chittagong.

Stanley, D.J. and Hait, A.K., 2000. Holocene depositional patterns, neotectonics and Sundarban mangroves in the western Ganges-Brahmaputra delta. *Journal of Coastal Research*, 16, pp. 26–39.

van Hengstum, P.J. and Scott, D.B., 2011. Ecology of foraminifera and habitat variability in an underwater cave: Distinguishing anchialine versus submarine cave environments. *The Journal of Foraminiferal Research*, 41(3), pp. 201–229.

Yanko, V., Arnold, A.J. and Parker, W.C., 1999. Effects of marine pollution on benthic foraminifera. In *Modern Foraminifera* (pp. 217–235). Springer, Dordrecht.

Yanko, V., Ahmad, M. and Kaminski, M., 1998. Morphological deformities of benthic foraminiferal tests in response to pollution by heavy metals; implications for pollution monitoring. *The Journal of Foraminiferal Research*, 28(3), pp. 177–200.

7 Change Analysis of Biophysical Parameters of Mangrove Forests over Indian Sundarbans using Geospatial Techniques: A Special Emphasis on Leaf Area Index and Percentage Canopy Cover

Apratim Biswas and Chalantika Laha Salui
Indian Institute of Engineering Sciences and Technology

CONTENTS

7.1 INTRODUCTION

By definition, mangroves are trees or shrubs that grow in chiefly tropical coastal swamps that are flooded at high tide. Commonly these have a large number of inter-tangled roots above the ground which form a dense network. Mangroves work as a shield for the coastal areas (Hochard et al., 2019). The areas which ought to be devastated by the Tsunami and cyclones are damaged much lesser because of these shrubs. They provide safety by both canopy and their roots. They have other importance as well which we will deal with later. It grows in the coastal waters which get flooded twice a day by salty sea water. Thus, it can be safely assumed that mangroves are very much immune to the harshest and toughest conditions possible for fauna and still provide so many advantageous traits. These species can live where other species would never survive and they have devised some very unique traits to help them achieve this feat. The most famous trait that mangroves are known for is their roots. Mangroves live in a kind of environment where they are submerged by sea water twice daily. Most tree species are not capable of surviving to flood even by freshwaters and mangroves survive salt-water flooding. Technically they can tolerate water 100 times saltier than most species.

There are near about 70–80 species of different kinds of mangroves which can be mainly divided into three categories (Golley et al., 1962). First are the Fringing (or seaward) zone which is the closest to the ocean. The trees have their roots sticking into the ocean water. These are also the most popular kinds of mangroves because of their numbers, size and easily recognizable roots. These are called 'Prop roots'; these roots are used to hold the tree in place as these trees are the ones that get most disturbed by the currents and winds during cyclones. These although look pretty much like normal trees from the above green and bushy but they have reddish tinted roots that are fully submerged in water during tides and because these infringement zones are the most soil erosion-prone zone which these roots help keep in place.

Mangroves play an important or probably the most prolific role in storm attenuation and surge damage reduction as well as sediment trapping and wave erosion reduction (Danielsen et al., 2005; Kovacs et al., 2011). Apart from this the coastal shrubs have a list of other secondary importance. The list extends from fisheries such as crabs, prawns, reef fish and larvae to firewood and building material (Odum & McIvor, 1990; Naylor et al., 2000; Kathiresan & Bingham, 2001; Manson et al., 2005; Walters et al., 2008; Ewel et al., 1998). Along these, mangroves are an important source of tourism and raw material resource for the industries which deal with chemical goods (Basak et al., 1996). These mangroves are very good sources of organic nutrients as well (Odum & Heald, 1972; Hutching et al., 1987; Ewel et al., 1998; Duke et al., 2007). Regardless of all these commercial uses these mangroves are a big part of life for the local people as well as the local wildlife habitat. These mangroves in several parts of the world provide coral reefs and seagrass beds from effluent and siltation as well (Mumby et al., 2004).

According to Polidoro et al. (2010) these mangroves worldwide provide near US\$ 1.6 billion. Without all the ecological and physical benefits mangroves are a huge resource of carbon which is known as 'blue carbon stock' as well, which vastly supports in the mitigation against climate changes (Alongi, 2012; Chmura et al., 2003; Donato et al., 2011; Murdiyarso et al., 2013). In reports by Donato et al. (2011) and Fujimoto et al. (1999), the carbon stocks of mangroves in only the Indo-Pacific region are in the range of nearly 1,000 mgC ha^{-1}. According to the report of Donato et al. (2011), these carbon stocks are several times higher than the carbon stocks of tropical forests of dry land (per unit area).

These mangrove habitats are due to various unfortunate reasons degrading rapidly. Kovacs et al. (2001) used Landsat data and reported that the mangrove system is undergoing massive degradation. Valiela et al. (2001) and Wilkie and Fortuna (2003) stated that these forests despite all these advantages are among the most threatened where over 50% of the global mangrove population is lost since 1900 and 35% since 1980. Between 1980 and 1990 1.5% annual global mangrove loss is seen worldwide which by the year 2000 may reach as low as 15 million ha worldwide (Wilkie and Fortuna et al., 2003). In the tropical parts of Southeast Asia, almost 10 thousand ha of mangrove land were deforested in the span of the last 15 years for aquaculture and agriculture (Richards and Friess, 2016). According to Duke et al. (2007), mangrove forests may as well be extinct totally within 100 years if these degradation rates are kept on. With the degradation in mangrove population, the rise of sea level shifts in meteorological patterns which may modify the salinity of water and increased tropical storm activity (Wanless, 1998; Gilman et al., 2007).

Thus, it emerges as a very important matter at hand to protect and monitor these coastal forests against land-use change and degradation (Castillo et al., 2017). First, we need to identify the reasons due to which these forests are losing their mass. Anthropogenic activities stand at the start of the line such as aquaculture agriculture, urbanization, pollution and tourism (Barbier and Cox, 2004). Tidal variations due to global climatic change have a big part to play in it as well (Xue et al., 2004; Allen et al., 2011). Some other climate change-born factors such as sediment rate alteration, nutrient and/or freshwater input rate changes along with inundation patterns all of these affect the crisis which is brought down upon the mangrove forests (Hogarth, 1999; De Lacerda and Linneweber, 2002; Walters et al., 2008). Coastal development and freshwater diversion are a heavy reason for this degradation of mangrove forests as well (Ellison and Farnsworth, 1996; Farnsworth and Ellison, 1997). Many of these tropical mangrove forests are located in countries that are developing and because of that short-term economic gain is a popular culture that leads to exploitation of mangrove habitats through shrimp pond conversion and timbering (Alongi, 2002). These have such effects that throughout various studies the rate of loss of mangrove is very impactful, to say the least (Wilkie and Fortuna et al., 2003; FAO, 2007; Giri et al., 2011).

Another huge aspect of mangrove degradation is natural causes. One of the primary reasons is the change in salinity regimes which calls for such a huge global mangrove structure change (Ewel et al., 1998; McDonald et al., 2003; Mitsch et al., 2009). Tsunami and cyclones do a massive degradation of these mangroves (Primavera, 2000). These threats to the mangrove in turn pose threats to the marine ecosystem and wildlife (Farnsworth and Ellison, 1997; Verheyden et al., 2002; Krauss et al., 2008; Bosire et al., 2008;). As a consequence, natural tidal systems get blocked, sedimentation decreases, the antibiotic impact from aquaculture like toxic pollution happen along with land subsidence, exposure to wave surges (Alongi, 2002; Tong et al., 2004; Danielsen et al., 2005; Primavera and Agbayani, 1997; Gilman et al., 2008; Thu and Populus, 2007).

The need for mapping and monitoring of mangrove ecosystems is thus very important. This can be carried out in the traditional way as well as the geospatial way. Many studies throughout the years have been carried out traditionally and with ground measurements like Ball (1998), Ewel et al. (1998) and Cole et al. (1999). The downside is as these studies are purely traditional and ground measurement dependent these are very hard to undertake, perform and logistically very much expensive as the forested wetlands are very remotely accessible and movement is very restricted. Thus, these studies are too difficult to pursue at a national or global level (Kovacs et al., 2011). The temporal aspect of monitoring is as important for forest assessment as the spatial aspect which is very difficult for the cost constraints of the traditional methods. This is where the demand of geospatial method grows.

Sundarbans is one of the largest deltas and mangrove forest in the world. It is partially situated in India and partially in Bangladesh. The Sundarbans is famous for its varied range of vegetation and wildlife. This includes 260 species of birds, the Bengal tiger, and other compromised species, such as the estuarine crocodile and the Indian python. The site holds exceptional biodiversity in its earthly, sea-going and marine habitats; going from micro to macro wildlife and vegetation. The Sundarbans is of much significance worldwide for its jeopardized species including the Royal Bengal Tiger, Ganges and Irawadi dolphins, estuarine crocodiles, and the basically endangered river terrapin (Batagur Baska). It is the only mangrove habitat on the planet containing Panthera Tigris species. Mangroves face various kinds of threats starting form natural to anthropogenic causes.

The damage to the swamps is very critical as it directly affects the livelihood of coastal communities. The enormous damage to the vital ecosystems, especially the mangroves, affects the fishery resource in the area. In simple words the local ecosystem as well as the human habitat. We for the scope of this paper are going to work on only the Indian portion of the vast mangrove forest known as the Sundarbans which is demarcated in Figure 7.1.

The main objective of our study is to map out the extents of the Sundarbans mangrove ecosystems in order to assess the damage. We have to gather temporal imagery for different timelines separated by a specific time gap. We look forward to work with the multi-spectral imagery data. These imageries will be temporal and will let us derive different vegetation indices to calculate our next two parameters from.

Next, our objective is to calculate our first parameter Leaf Area Index (LAI) from the normalized difference vegetation index (NDVI) derived from the multi-spectral data according to past studies for derivation of LAI in mangrove canopies. Our focus will be to successfully derive the LAI layers for all three temporal images and measure whether there was any change in the LAI for the study area.

The last of our parameters is the % Canopy Cover which according to past studies has strong correlation with the Simple Ratio (SR) and Modified Simple Ratio (MSR) vegetation indices. We need to successfully extract the temporal layers of the SR and MSR for the study area and we have to generate percent canopy cover for that area. Our objective then will be to evaluate whether there was any change in the % canopy cover for the area which we are studying.

The main aspect of this study is to assess the damage done to the Sundarbans mangroves due to various Anthropogenic and natural causes. We in this study look forward to assess the damage of a certain mangrove forest and for that we need to evaluate the changes in the two parameters, namely, LAI and % canopy cover as past studies have concluded these two while in a decreasing spree can imply mangrove degradation.

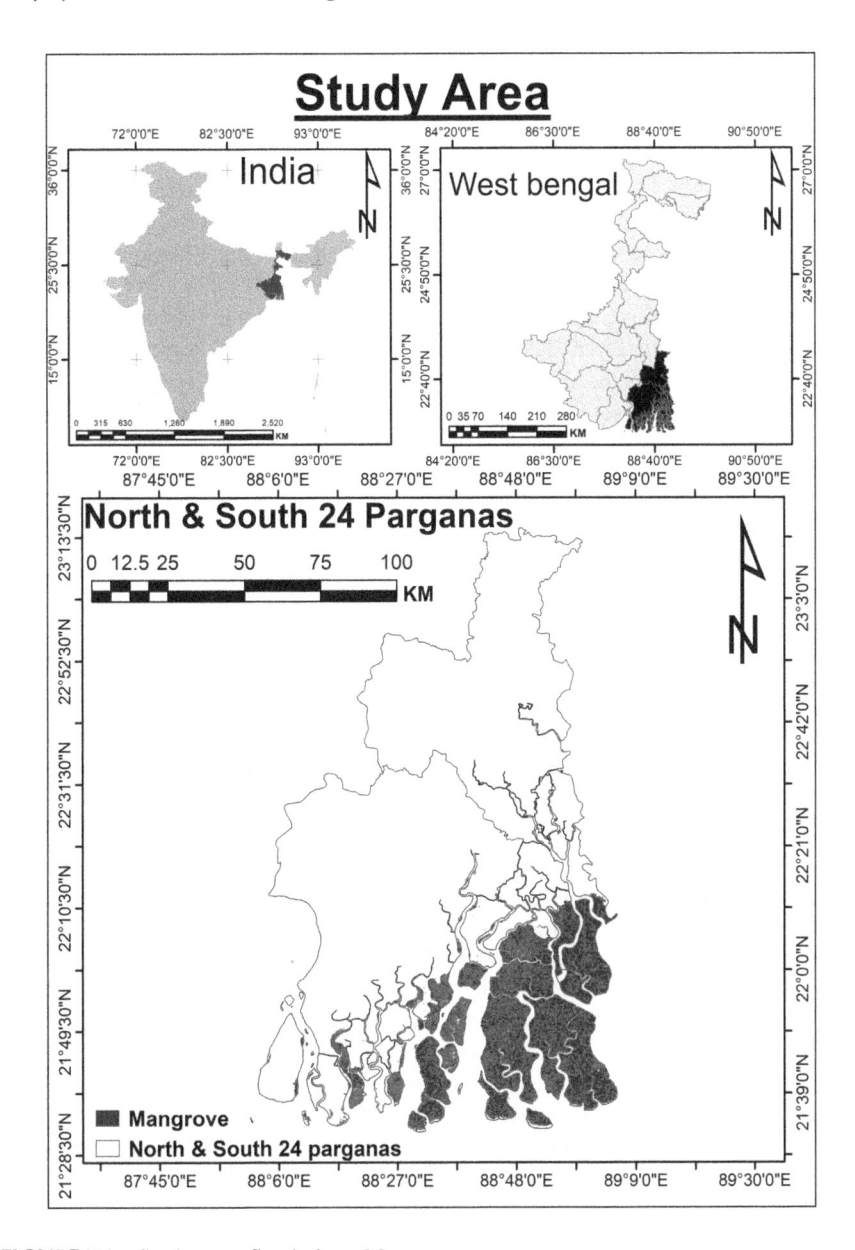

FIGURE 7.1   Study area: Sundarbans Mangroves.

7.2   DATA USED AND METHODOLOGY

Multispectral data sets were used for the acquisition of the vegetation indices required to produce the parametric layers. Two different sources of data had to be used due to constraint in temporal availability. As Sentinel-2 data were not available prior to 2015, Landsat data sets were used for the 1990 and 2005 imagery in spite of the lacking spatial resolution. Mangrove delineation shapefiles were used from the Global Mangrove Watch for three different time periods.

Landsat-5 Thematic Mapper images were used for 1990 and 2005 with a spatial resolution of 30 m in the concerned bands. The data representing 1990 were captured on January 14, 1990. The path-row combination was 138-045. The image was a collection 1, tier-1 category precision and terrain corrected data. The 2005 data were of the same path-row combination of 138-045, collection 1, tier-1 category precision and terrain corrected data captured on the January 7 of 2005. The third, fourth and fifth bands, namely, Red, Near Infrared and Short-Wave Infrared bands are used to acquire the parametric layers. L1TP or the precision and terrain corrected data were used which nullified the need of any correctional pre-processing.

The Sentinel-2 data set had a slightly different path and thus we had to acquire three different tiles of Sentinel-2B data set, namely, T45QXD, T45QXE and T45QYE captured on 15th March, 15th March and 17th January of 2020, respectively, to avoid cloud covers and acquire clean imagery. The 4th, 8th and 11th bands of the sentinel MSI sensor were used to generate the aforementioned layers. The 11th band has a spatial resolution of 20 m and thus had to be resampled to match the other two bands. L1C or the Level-1C data were used which needed no radiometric or geometric corrections before processing.

All the multispectral data were acquired from the USGS Global Visualization Viewer (GloVis) data portal (glovis.usgs.gov).

Sentinel-2 multispectral data have one of the best spatial resolutions among the non-commercial data sources with only 10 m spatial resolution on most of its bands. This is the reason this data are favoured by many researchers working on fairly recent multispectral satellite data. However, the oldest sentinel data available are of 2015, so for the older observations of 1990 and 2005, we had to resort to Landsat-5 TM data regardless of its much weaker band spatial resolution of 30 m. The 138-045 path row combination of the Landsat-5 covers the whole Indian Sundarbans forest, which saved us of downloading multiple imageries for the same timeline. For Sentinel-2, three separate images with the tile numbers 45QXD, 45QXE and 45QYE had to be mosaicked using the data management tools. For any timeline we had two parameters which in turn called for in total three individual multispectral bands, namely, Red (0.63–0.69 μm), Near Infra-Red (0.77–0.90 μm) and Short Wave Infra-Red (1.55–1.75 μm). After mosaicking the SWIR or eleventh band in the sentinel-2 imagery had to be resampled to 10 m. For the calculation of LAI using the method of Green et al. (1997), NDVI was the necessary vegetation index using the Red and Near Infra-Red bands

$$NDVI = (NIR - Red) / (NIR + Red)$$

where NDVI = normalized difference vegetation index, NIR = Near Infra-Red Band values and Red = visible red band values. This entire process is depicted in Figure 7.2.

The new NDVI layer of the three timelines was then converted to LAI using

$$LAI = 12.74 \times NDVI + 1.34$$

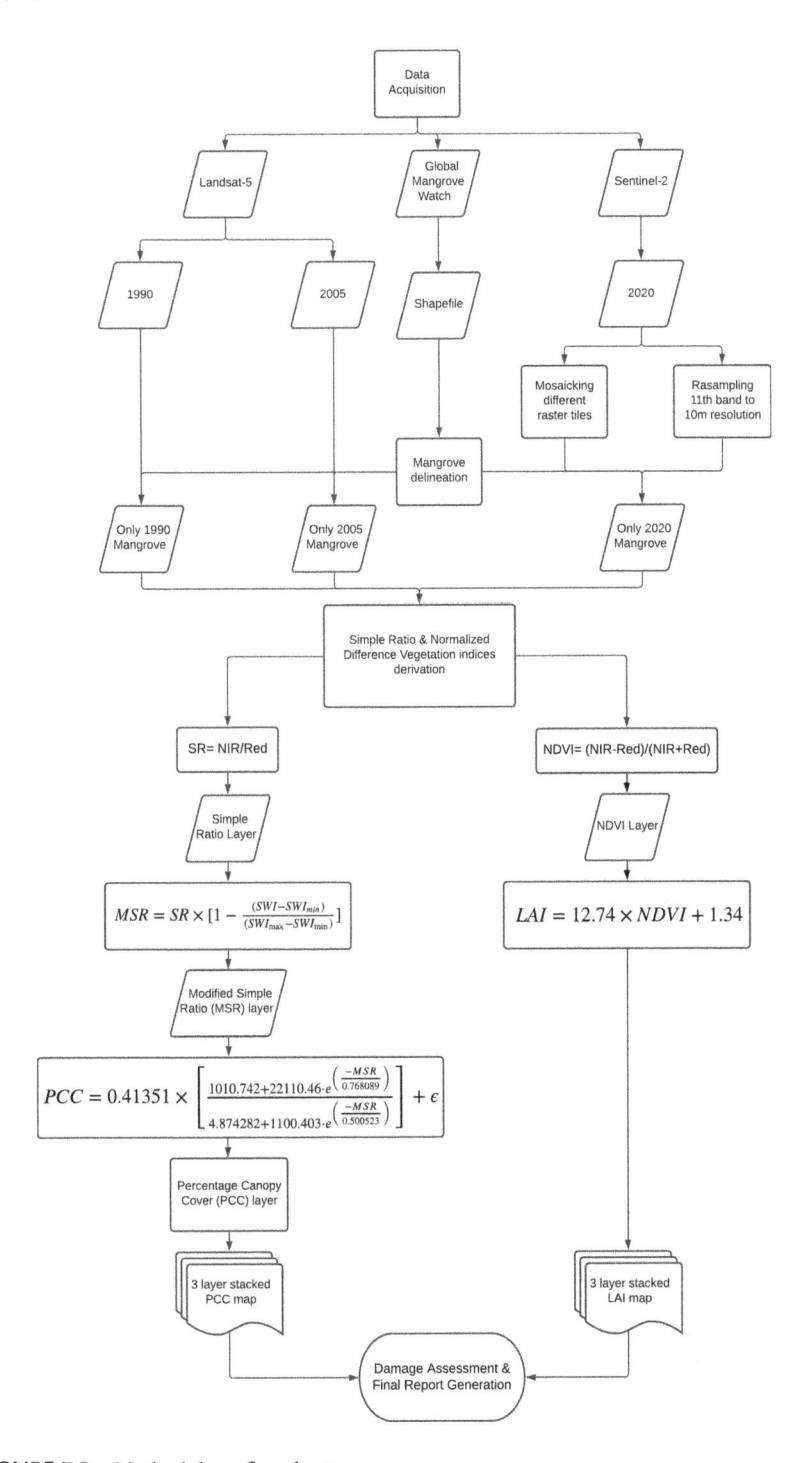

FIGURE 7.2   Methodology flowchart.

where NDVI=normalized difference vegetation index layer values and LAI=leaf area index values.

For the next parameter of Percentage Canopy Cover (PCC) two of the necessary vegetation indices were the SR and the MSR vegetation indices. The Red and Near Infra-Red bands were required to calculate SRVI.

$$SR = NIR / Red$$

where NIR = Near Infra-Red Band values and Red = visible red band values.

$$MSR = SR \times \left[1 - \left\{ (SWI - SWI_{min}) / (SWI_{max} - SWI_{min}) \right\} \right]$$

where SWI=short-wave infrared band values, MSR=Modified Simple Ratio, and SR=Simple Ratio.

The MSR layers were then used to predict the PCC of the area from a formula devised by Staben and Evans (2008)

Canopy Cover% = 0.41351*

$$\left(1010.742 + 22110.46e^{((-MSR)/0.768089)} \right) / \left(4.874282 + 1100.403e^{((-MSR)/0.500523)} \right)$$

where MSR = Modified Simple Ratio values.

All of these calculations were processed using the Raster Calculator from the spatial analyst toolbox. We had to then delineate the mangrove forest areas to process the raster layers of different timelines. Rather than digitizing manually we resorted to a secondary data source, namely, Global Mangrove Watch (www.globalmangrovewatch.org). Three different global shapefiles were acquired from the aforementioned source consecutively for three different timelines. The Indian Sundarbans area was masked out manually from the shapefiles using a common raster mask and applied then to precisely delineate the mangrove forest pixels. A total of six layers (3 LAI, 3 PCC) were generated and made analysis ready. The layers finally were analyzed using the raster calculator to get the relative changes between years in the chosen parameters.

7.3   RESULTS AND DISCUSSION

Two different parameters were under analysis in this study, namely, LAI and PCC. LAI is essentially used to evaluate the health of trees on a large scale as in a forest. And the PCC is used to measure the extent of closure the leaves have.

7.3.1   LEAF AREA INDEX

The LAI layer was calculated for three different timeline satellite images of 1990, 2005 and 2020 with interval of 15 years. Due to the calculation being empirical some of the layers have illustrated negative LAIs which is practically impossible as both the numerator and the denominator are units of area.

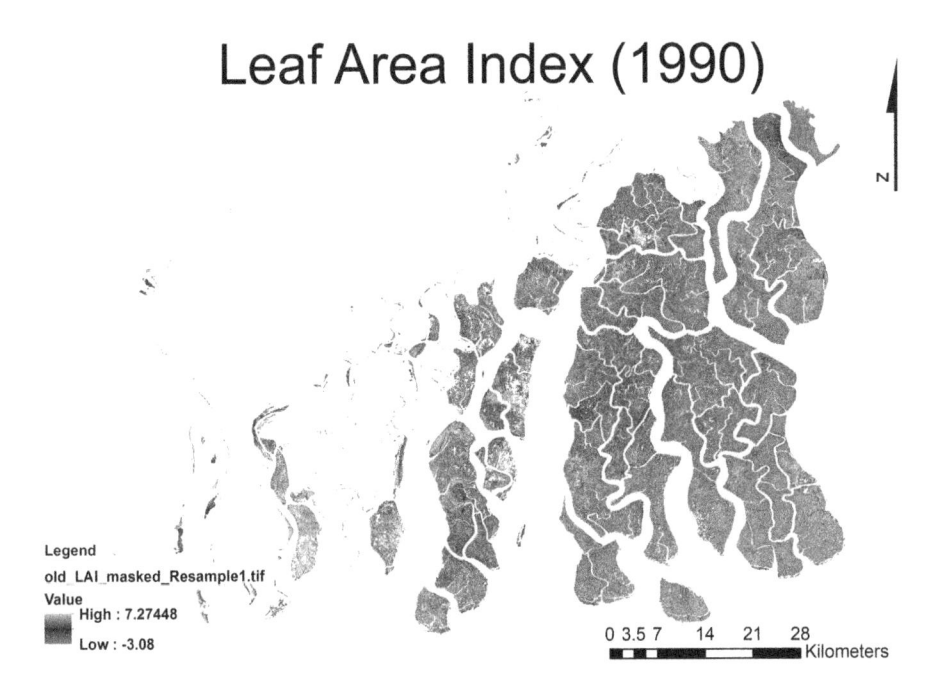

FIGURE 7.3   Leaf area index (1990).

The LAI layer and the resultant information are shown in Figure 7.3. This observation is from the satellite data of year 1990. Maximum LAI is 7.27 and the mean of the individual LAI values is 4.27. The light blue signifies higher value of LAI, which in turn means healthier or mature trees. As we can see dominance of light blue towards the southern side of the forest, we can assume healthier mature trees to be found in the fringing regions of the near coastal regions. A miniscule standard deviation shows that most of the values are closer to the mean and lower or higher values can mainly be classified as exceptions.

Figure 7.4 shows LAI layer and consequent information from the year 2005. Maximum LAI was 8.49 with a mean of 6.03. The mangroves in the southern region growing up to be more mature can be implied from even brighter shades of blue in the near coastal regions. Still a very small standard deviation which means most of the pixels have LAI values near the mean value of 6.03.

Figure 7.5 shows the LAI layer and statistical information from the year 2020. Maximum and mean LAI is 10.93 and 7.52, respectively. Brighter shades of blue are visible not only towards the southern region but also towards the inland. As LAI can be a signifier of plant health as well as maturity, we can imply that the trees can be seen growing throughout these years of study. The younger trees towards inland have also come of age as the years have passed and some narrow red band like shapes can be seen towards the extreme coastal zones which may signify lesser LAI or damaged forest stands of mangrove due to various reasons but as the images for 2020 were sensed in the month of march these degradations could be caused by anything but the super cyclone Amphan. The observations of the Year-wise mean Leaf Area Index is depicted in Figure 7.6.

Leaf Area Index (2005)

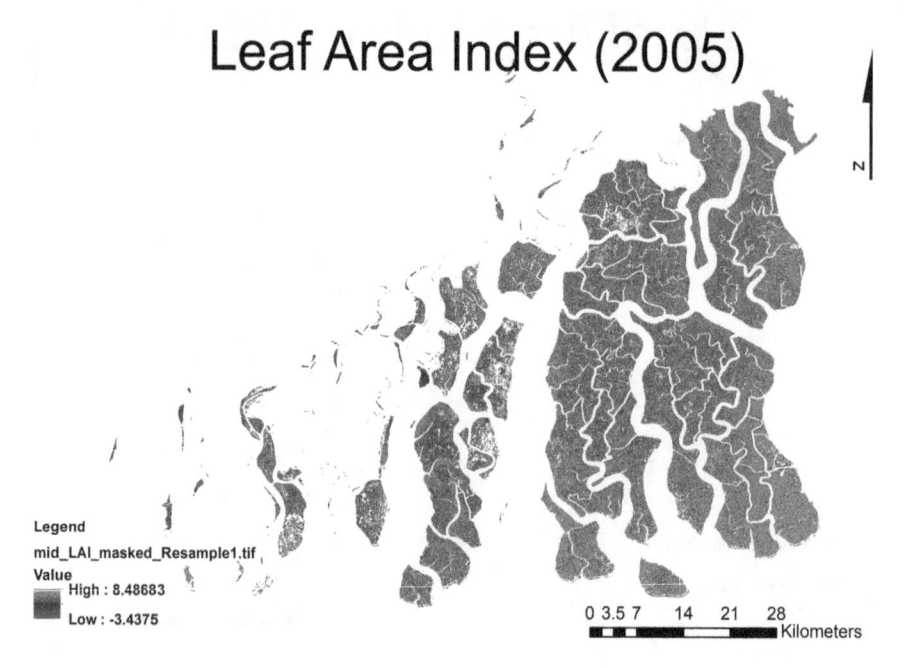

FIGURE 7.4 Leaf area index (2005).

Leaf Area Index (2020)

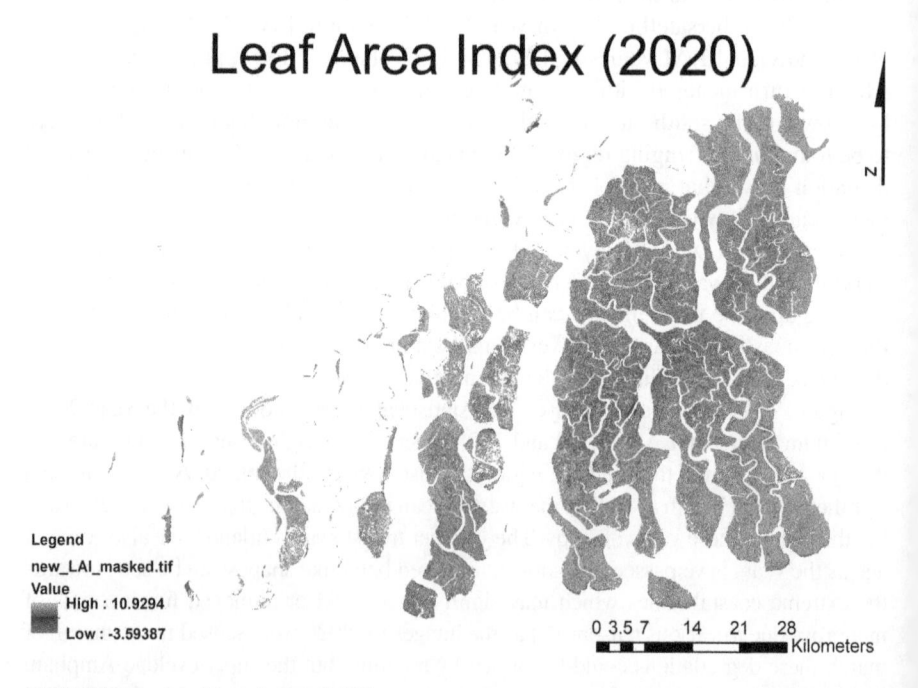

FIGURE 7.5 Leaf area index (2020).

Leaf Area Index change

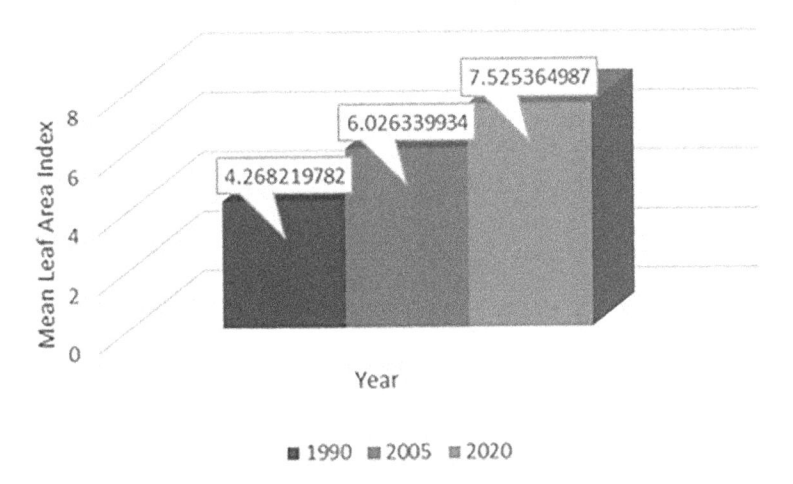

FIGURE 7.6   Year-wise graph of mean Leaf Area Index.

7.3.2 PERCENT CANOPY COVER

The PCC was calculated for three different years similar to the LAI. Some of the findings similar to LAI can be observed here which in turn support the theories we have to back up the result.

Figure 7.7 portrays the PCC layer for 1990 with the statistics like 45% being the maximum canopy cover percentage illustrated by dark red and the mean % canopy cover for the year 1990 of 23%. Similar to the LAI, PCC can also be an illustrator of a tree stand's health and maturity. We can see comparatively bushy trees with more canopy cover towards the coastal areas and the most closure near the marked inlands.

Figure 7.8 illustrates the % canopy cover of the year 2005. The maximum is 77% and mean canopy cover for 2005 is 35%. We can see increase in canopy cover at the coastal areas even more with brighter red spots. The marked areas with the maximum canopy cover in the year 1990 have been going through degradation and are portraying significant blue shades. From the statistics we can claim there is a significant amount of increase in canopy cover from 1990 to 2005.

Figure 7.9 shows the % canopy cover layer for the year 2020. From the statistics table we get the maximum % cover at 89% and the mean for the year 2020 at 59%. The coastal area canopies have increased in cover as well as the marked areas which were degraded in the 2005 imagery. Those areas have once again grown to be among the most canopy covered areas. The coastal areas have brighter and more spread red spotted areas than the previous years, further propagating the theory that these tree stands in the coastal areas have grown in age and maturity over all the years of the study. We can even see build-up of canopy cover along the inland forests as well. But, since the 2020 imagery is sensed before the Cyclone Amphan in May, the blue banded areas near the far coastal regions are degradations propagated by some other cause than the aforementioned. The observations of the Year-wise mean Percentage Canopy Cover is depicted in Figure 7.10.

% Canopy Cover (1990)

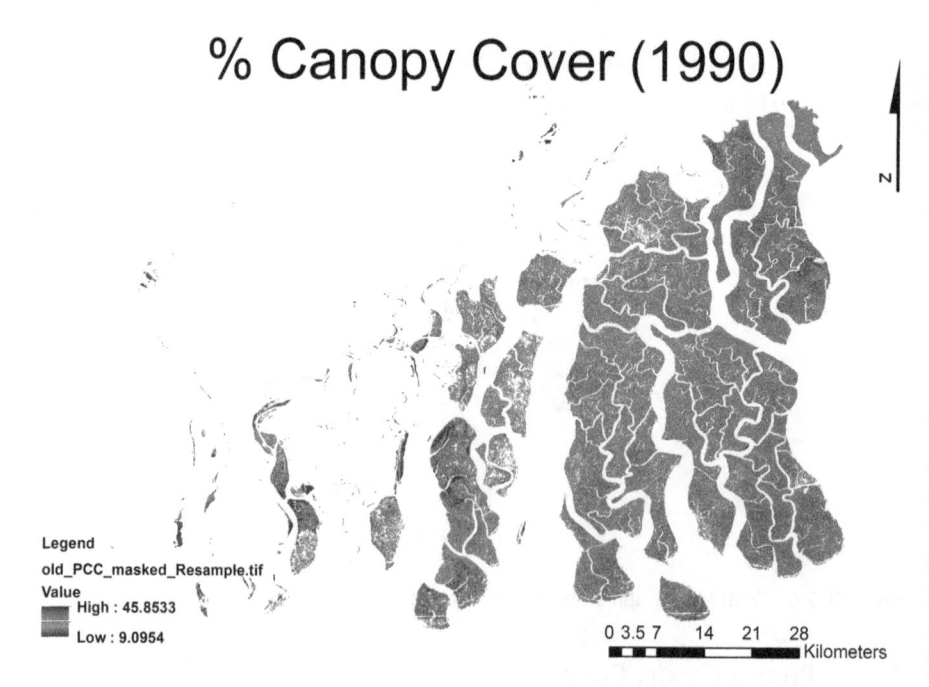

Legend
old_PCC_masked_Resample.tif
Value
High : 45.8533
Low : 9.0954

0 3.5 7   14   21   28
Kilometers

FIGURE 7.7 % Canopy cover (1990).

% Canopy Cover (2005)

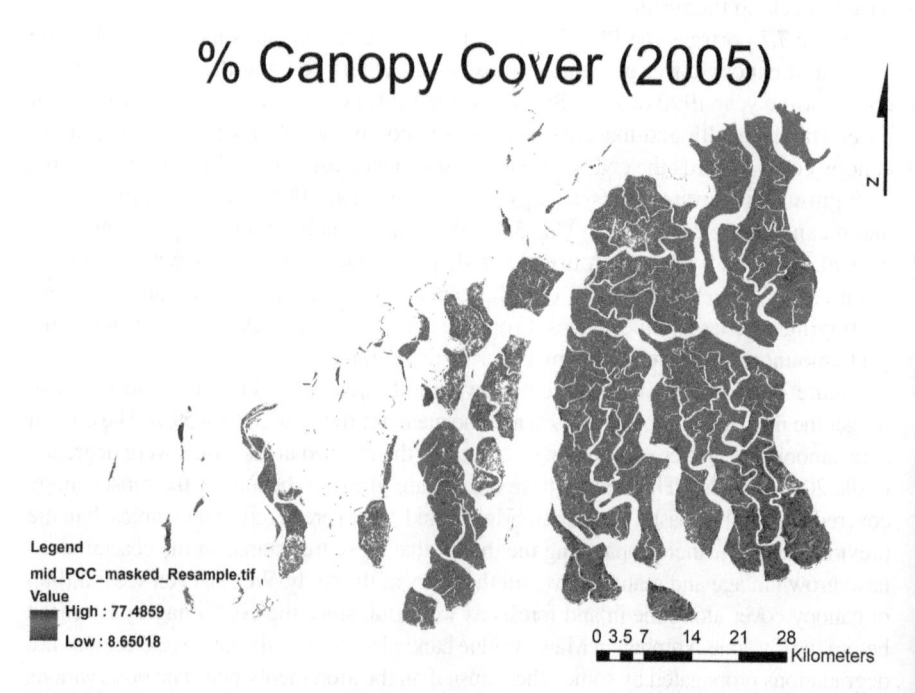

Legend
mid_PCC_masked_Resample.tif
Value
High : 77.4859
Low : 8.65018

0 3.5 7   14   21   28
Kilometers

FIGURE 7.8 % Canopy cover (2005).

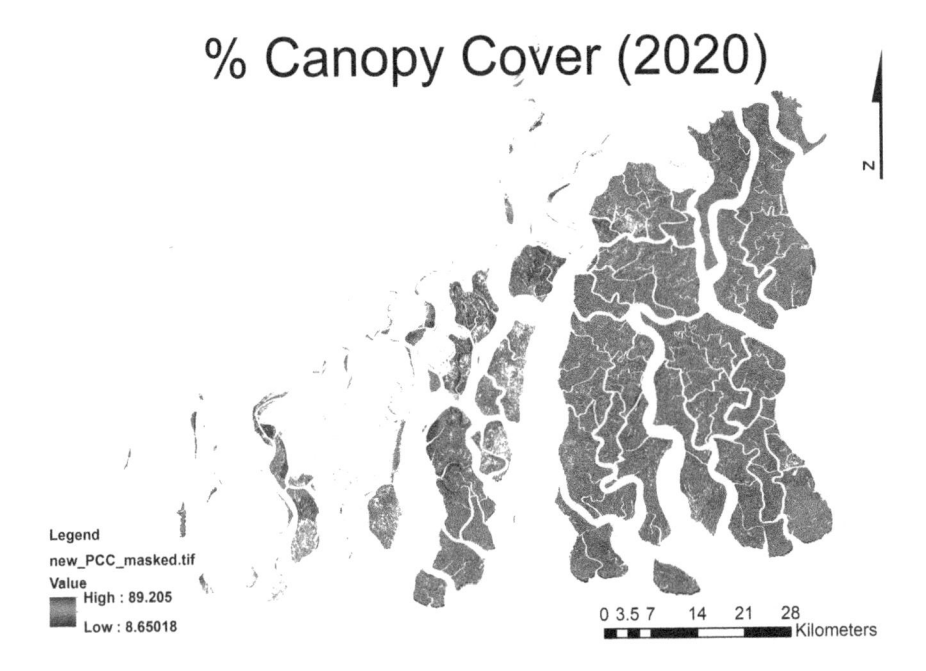

FIGURE 7.9   % Canopy cover (2020).

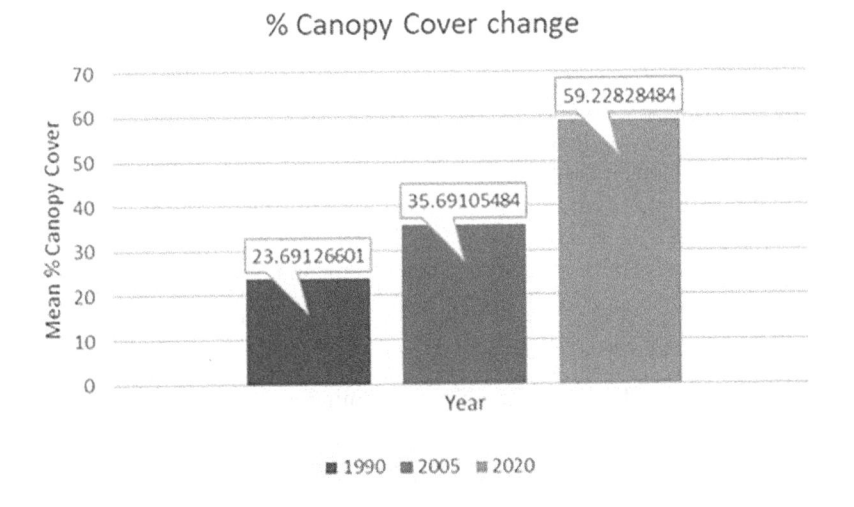

FIGURE 7.10   Year-wise graph of mean Percentage Canopy Cover.

7.3.3 AREAL EXTENT

Other than these two pre-planned parameters another observation was made from the layers generated for LAI and PCC analysis. This discrepancy was generated from the shapefiles we used to delineate the mangrove forests and in turn the area covered

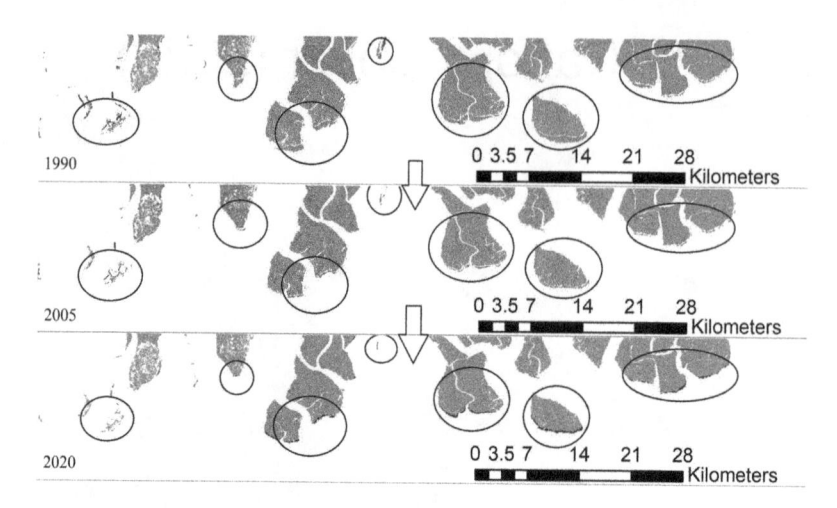

FIGURE 7.11 Areal extent change.

TABLE 7.1
Areal Extent

Timeline	Area (km²)	% Reduction from Origin Year	% Reduction from Previous Year
1990	1978.04	-	-
2005	1861.7	5.88	5.88
2020	1851.44	6.4	0.55

by them. The landmasses at the far-coastal regions can be seen to creep inland which may imply land submergence.

We can notice in all of the marked areas in Figure 7.11 that those areas are decreasing in a very gentle pace. From the shapefile metadata we have acquired the areal extents and tabulated the data analysing the decrement in Table 7.1.

7.4 CONCLUSION

We can imply from the steady growth, and coming to maturity of the mangroves in Sundarbans that this is a result of the mangrove conservation strategies undertaken by the authorities. Cyclonic storm and disturbances are one of the main natural reasons of mangrove degradation. During the period of our study massive cyclones from the 1999 Odisha cyclone to Aila in 2009 and Hudhud in 2014 have struck the Sundarbans. Thus, a steady increase in the health parameters can only point towards conscious and efficient conservation and reforestation strategies. This theory is supported by a study from Hazra and Samanta (2016).

Sundarbans is a very dynamic estuary (Raha, 2014), which goes through various changes but the change noticed in the shapefiles or the mangrove extents seems steady and caused by something more severe than some common tidal movements of mangrove swamps. As the levels rise the coastal areas submerge and the coastline regresses slowly and steadily. According to a study of Hazra et al. (2002), the rate of decrement of coastal areas is correlated with the rise of sea levels. This coastal erosion and flooding propagate increase in the rate of homeless environmental refugees by the thousands (Bera, 2013).

Due to the limitations of cloud cover in multispectral imagery we had to acquire data with zero to least cloud cover. This was not problem for the 1990 and 2005 imagery as a whole year's data were available to acquire from. However, in May 2020 the cyclone Amphan struck the coastal lines of West Bengal doing massive damage. This was a good opportunity to feed of from for a study on mangrove damage assessment. However, due to aforementioned restrictions of cloud cover and in turn suitable data unavailability it was not possible to take Amphan in the scope of this study. This leaves an opportunity for future researchers to work on by incorporating the effects of Amphan in a similar study of mangrove damage assessment.

REFERENCES

Allen, D., Dalal, R.C., Rennenberg, H. and Schmidt, S., 2011. Seasonal variation in nitrous oxide and methane emissions from subtropical estuary and coastal mangrove sediments, Australia. *Plant Biology*, 13(1), pp. 126–133.

Alongi, D.M., 2002. Present state and future of the world's mangrove forests. *Environmental Conservation*, 29(3), pp. 331–349.

Alongi, D.M., 2012. Carbon sequestration in mangrove forests. *Carbon Manage*, 3 (3), pp. 313–322.

Ball, M., 1998. Mangrove species richness in relation to salinity and waterlogging: a case study along the Adelaide River floodplain, northern Australia. *Global Ecology & Biogeography Letters*, 7(1), pp. 73–82.

Barbier, E.B. and Cox, M., 2004. An economic analysis of shrimp farm expansion and mangrove conversion in Thailand. *Land Economics*, 80(3), pp. 389–407.

Basak, U.C., Das, A.B. and Das, P., 1996. Chlorophylls, carotenoids, proteins and secondary metabolites in leaves of 14 species of mangrove. *Bulletin of Marine Science*, 58(3), pp. 654–659.

Bera, M.K., 2013. Environmental refugee: A study of involuntary migrants of Sundarban islands. In *Proceedings of the 7th International Conference on Asian and Pacific Coasts (APAC 2013)* Bali, Indonesia (pp. 916–925).

Bosire, J.O., Dahdouh-Guebas, F., Walton, M., Crona, B.I., Lewis Iii, R.R., Field, C., Kairo, J.G. and Koedam, N., 2008. Functionality of restored mangroves: A review. *Aquatic Botany*, 89(2), pp. 251–259.

Castillo, J.A.A., Apan, A.A., Maraseni, T.N. and Salmo III, S.G., 2017. Estimation and mapping of above-ground biomass of mangrove forests and their replacement land uses in the Philippines using Sentinel imagery. *ISPRS Journal of Photogrammetry and Remote Sensing*, 134, pp. 70–85.

Chmura, G.L., Anisfeld, S.C., Cahoon, D.R. and Lynch, J.C., 2003. Global carbon sequestration in tidal, saline wetland soils. *Global Biogeochemical Cycles*, 17(4), pp. 1–12.

Cole, T.G., Ewel, K.C. and Devoe, N.N., 1999. Structure of mangrove trees and forests in Micronesia. *Forest Ecology and Management*, 117(1–3), pp. 95–109.

Danielsen, F., Sørensen, M.K., Olwig, M.F., Selvam, V., Parish, F., Burgess, N.D., Hiraishi, T., Karunagaran, V.M., Rasmussen, M.S., Hansen, L.B. and Quarto, A., 2005. The Asian tsunami: a protective role for coastal vegetation. *Science*, 310(5748), pp. 643–643.

De Lacerda, L.D. and Linneweber, V., 2002. *Mangrove Ecosystems: Function and Management*. Springer Science & Business Media, Berlin.

Donato, D.C., Kauffman, J.B., Murdiyarso, D., Kurnianto, S., Stidham, M. and Kanninen, M., 2011. Mangroves among the most carbon-rich forests in the tropics. *Nature Geoscience*, 4(5), pp. 293–297.

Duke, N.C., Meynecke, J.O., Dittmann, S., Ellison, A.M., Anger, K., Berger, U., Cannicci, S., Diele, K., Ewel, K.C., Field, C.D. and Koedam, N., 2007. A world without mangroves? *Science*, 317(5834), pp. 41–42.

Ellison, A.M. and Farnsworth, E.J., 1996. Anthropogenic disturbance of Caribbean mangrove ecosystems: Past impacts, present trends, and future predictions. *Biotropica*, 28, pp. 549–565.

Ewel, K., Twilley, R. and Ong, J.I.N., 1998. Different kinds of mangrove forests provide different goods and services. *Global Ecology & Biogeography Letters*, 7(1), pp. 83–94.

Joint FAO/WHO Expert Committee on Food Additives. Meeting and World Health Organization, 2007. Evaluation of certain food additives and contaminants: Sixty-eighth report of the Joint FAO/WHO Expert Committee on Food Additives (Vol. 68). World Health Organization.

Farnsworth, E.J. and Ellison, A.M., 1997. The global conservation status of mangroves. *Ambio* (Sweden).

Fujimoto, K., Imaya, A., Tabuchi, R., Kuramoto, S., Utsugi, H. and Murofushi, T., 1999. Belowground carbon storage of Micronesian mangrove forests. *Ecological Research*, 14(4), pp. 409–413.

Gilman, E., Ellison, J. and Coleman, R., 2007. Assessment of mangrove response to projected relative sea-level rise and recent historical reconstruction of shoreline position. *Environmental Monitoring and Assessment*, 124(1–3), pp. 105–130.

Gilman, E.L., Ellison, J., Duke, N.C. and Field, C., 2008. Threats to mangroves from climate change and adaptation options: A review. *Aquatic Botany*, 89(2), pp. 237–250.

Giri, C., Long, J. and Tieszen, L., 2011. Mapping and monitoring Louisiana's mangroves in the aftermath of the 2010 Gulf of Mexico oil spill. *Journal of Coastal Research*, 27(6), pp. 1059–1064.

Golley, F., Odum, H.T. and Wilson, R.F., 1962. The structure and metabolism of a Puerto Rican red mangrove forest in May. *Ecology*, 43, pp. 9–19.

Green, E.P., Mumby, P.J., Edwards, A.J., Clark, C.D. and Ellis, A.C., 1997. Estimating leaf area index of mangroves from satellite data. *Aquatic Botany*, 58(1), pp. 11–19.

Hazra, S., Ghosh, T., DasGupta, R. and Sen, G., 2002. Sea level and associated changes in the Sundarbans. Science and Culture, 68(9/12), pp.309–321.

Hazra, S. and Samanta, K., 2016. Temporal Change Detection (2001–2008): Study of Sundarban (No. id: 10526).

Hochard, J.P., Hamilton, S. and Barbier, E.B., 2019. Mangroves shelter coastal economic activity from cyclones. *Proceedings of the National Academy of Sciences*, 116(25), pp. 12232–12237.

Hogarth, P.J., 1999. *The Biology of Mangroves*. Oxford University Press (OUP), Oxford.

Hutching, G., Potton, C. and Forest, R., 1987. *Forests, Fiords & Glaciers New Zealand's World Heritage the Case for a South-West New Zealand World Heritage Site*. Hodder & Stoughto, Wellington.

Kathiresan, K. and Bingham, B.L., 2001. Biology of mangroves and mangrove ecosystems. *Advances in Marine Biology*, 40, pp. 84–254.

Kovacs, J.M., Wang, J. and Blanco-Correa, M., 2001. Mapping disturbances in a mangrove forest using multi-date Landsat TM imagery. *Environmental Management*, 27(5), pp. 763–776.

Kovacs, J.M., Liu, Y., Zhang, C., Flores-Verdugo, F. and de Santiago, F.F., 2011. A field based statistical approach for validating a remotely sensed mangrove forest classification scheme. *Wetlands Ecology and Management*, 19(5), p. 409.

Krauss, K.W., Lovelock, C.E., McKee, K.L., López-Hoffman, L., Ewe, S.M. and Sousa, W.P., 2008. Environmental drivers in mangrove establishment and early development: A review. *Aquatic Botany*, 89(2), pp. 105–127.

Manson, F.J., Loneragan, N.R., Harch, B.D., Skilleter, G.A. and Williams, L., 2005. A broad-scale analysis of links between coastal fisheries production and mangrove extent: A case-study for northeastern Australia. *Fisheries Research*, 74(1–3), pp. 69–85.

McDonald, K.O., Webber, D.F. and Webber, M.K., 2003. Mangrove forest structure under varying envrionmental conditions. *Bulletin of Marine Science*, 73(2), pp. 491–505.

Mitsch, W.J., Gosselink, J.G., Zhang, L. and Anderson, C.J., 2009. *Wetland Ecosystems*. John Wiley & Sons, New York.

Mumby, P.J., Edwards, A.J., Arias-González, J.E., Lindeman, K.C., Blackwell, P.G., Gall, A., Gorczynska, M.I., Harborne, A.R., Pescod, C.L., Renken, H. and Wabnitz, C.C., 2004. Mangroves enhance the biomass of coral reef fish communities in the Caribbean. *Nature*, 427(6974), pp. 533–536.

Murdiyarso, D., Kauffman, J.B. and Verchot, L.V., 2013. Climate change mitigation strategies should include tropical wetlands. *Carbon Management*, 4(5), pp. 491–499.

Naylor, R.L., Goldburg, R.J., Primavera, J.H., Kautsky, N., Beveridge, M.C., Clay, J., Folke, C., Lubchenco, J., Mooney, H. and Troell, M., 2000. Effect of aquaculture on world fish supplies. *Nature*, 405(6790), pp. 1017–1024.

Odum, W.E. and Heald, E.J., 1972. Trophic analyses of an estuarine mangrove community. *Bulletin of Marine Science*, 22(3), pp. 671–738.

Odum, W.E. and McIvor, C.C., 1990. Mangroves. Ecosystems of Florida, pp. 517–548.

Polidoro, B.A., Carpenter, K.E., Collins, L., Duke, N.C., Ellison, A.M., Ellison, J.C., Farnsworth, E.J., Fernando, E.S., Kathiresan, K., Koedam, N.E. and Livingstone, S.R., 2010. The loss of species: mangrove extinction risk and geographic areas of global concern. *PloS One*, 5(4), p. e10095.

Primavera, J.H., and Agbayani, R.F., 1997. Comparative strategies in community-based mangrove rehabilitation programs in the Philippines. In Community Participation in Conservation, Sustainable Use and Rehabilitation of Mangroves in Southeast Asia. Proceedings of the ECOTONE V, 8–12 January 1996, Ho Chi Minh City, Vietnam, pp. 229–243. United Nations Educational Scientific and Cultural Organisation.

Primavera, J.H., 2000. Development and conservation of Philippine mangroves: Institutional issues. *Ecological Economics*, 35(1), pp. 91–106.

Raha, A.K., 2014. Sea level rise and submergence of Sundarban Islands: A time series study of estuarine dynamics. *Journal of Ecology and Environmental Sciences*, 5, pp. 114–123. ISSN: 0976-9900.

Richards, D.R. and Friess, D.A., 2016. Rates and drivers of mangrove deforestation in Southeast Asia, 2000–2012. *Proceedings of the National Academy of Sciences*, 113(2), pp. 344–349.

Staben, G.W. and Evans, K.G., 2008. Estimates of tree canopy loss as a result of Cyclone Monica, in the Magela Creek catchment northern Australia. *Austral Ecology*, 33(4), pp. 562–569.

Thu, P.M. and Populus, J., 2007. Status and changes of mangrove forest in Mekong Delta: Case study in Tra Vinh, Vietnam. *Estuarine, Coastal and Shelf Science*, 71(1–2), pp. 98–109.

Tong, P.H.S., Auda, Y., Populus, J., Aizpuru, M., Habshi, A.A. and Blasco, F., 2004. Assessment from space of mangroves evolution in the Mekong Delta, in relation to extensive shrimp farming. *International Journal of Remote Sensing*, 25(21), pp. 4795–4812.

Valiela, I., Bowen, J.L. and York, J.K., 2001. Mangrove Forests: One of the World's Threatened Major Tropical Environments: At least 35% of the area of mangrove forests has been lost in the past two decades, losses that exceed those for tropical rain forests and coral reefs, two other well-known threatened environments. *Bioscience*, 51(10), pp. 807–815.

Verheyden, A., Dahdouh-Guebas, F., Thomaes, K., De Genst, W., Hettiarachchi, S. and Koedam, N., 2002. High-resolution vegetation data for mangrove research as obtained from aerial photography. *Environment, Development and Sustainability*, 4(2), pp. 113–133.

Walters, B.B., Rönnbäck, P., Kovacs, J.M., Crona, B., Hussain, S.A., Badola, R., Primavera, J.H., Barbier, E. and Dahdouh-Guebas, F., 2008. Ethnobiology, socio-economics and management of mangrove forests: A review. *Aquatic Botany*, 89(2), pp. 220–236.

Wanless, H.R., 1998. Mangroves, hurricanes, and sea level rise. South Florida Study Group, The Conservancy, Naples, Florida.

Wilkie, M.L. and Fortuna, S., 2003. Status and trends in mangrove area extent worldwide. Forest Resources Assessment Programme. Working Paper (FAO).

Xue, X., Hong, H. and Charles, A.T., 2004. Cumulative environmental impacts and integrated coastal management: The case of Xiamen, China. *Journal of Environmental Management*, 71(3), pp. 271–283.

Section 3

Ecosystem Services
Analysis using GIS

8 Economic Valuation of Ecosystem Services of Sundarbans Natural Reserve Region, India

Srikanta Sannigrahi, Francesco Pilla,
Bidroha Basu, and Arunima Sarkar Basu
University College Dublin

P.S. Roy
World Recourses Institute India

P.K. Joshi
Jawaharlal Nehru University (JNU)

CONTENTS

8.1 INTRODUCTION

The human impact on natural earth resources is inevitable nowadays, and it is responsible for changing the function, structure, composition, and configuration of an ecosystem substantially (Vitousek et al., 1997). Among all driving factors, land-use degradation is considered to be the most crucial aspect of ecosystem degradation. However, monitoring and evaluating the effects of land-use changes on ESs are challenging (Richmond et al., 2007; Li et al., 2014; Brouwer et al., 2013; Eigenbrod et al., 2010; Zhao et al., 2004; Sannigrahi, Zhang, Joshi, et al., 2020; Sannigrahi, Zhang, Pilla, et al., 2020). With the growing economy and enforcement of different land use land cover (LULC) policies in developing and also developed nations, there has been a drastic change in LULC around the world. In order to improve overall human well-being and to limit the deterioration of ecology and the concerned environment, its economic analysis is highly required.

In the era of artificial intelligence and soft computing, many model building approaches have been evolved to build a conceptual framework for estimating ESV's from a local to a global scale. Such models include ARtificial Intelligence for Ecosystem Service (ARIES) for spatially explicit mapping of natural resources (http://aries.integratedmodelling.org/), Toolkit for Ecosystem Service Site-Based Assessment (TESSA) (http://tessa.tools/) and Integrated Valuation of Ecosystem Services and Tradeoffs (InVEST) for mapping and valuation of 18 ESs designated to the terrestrial, coastal and marine ecosystem (https://www.naturalcapitalproject.org/invest/). The other relevant models are Multi-Scale Integrated Model of Ecosystem Services (MIMES), Social Values for Ecosystem Services (SolVES), EcoServ-GIS, Ecosystem Services Identification and Inventory (ESII), Ecosystem Services Review for Impact Assessment (ESR), Marine Integrated Decision Analysis System (MIDAS), ecosystem management decision support (EMDS), Envisions, InFOREST, Water Erosion Prediction Project (WEPP) Soil and Water Assessment Tool (SWAT), Global Unified Metamodel of the Biosphere (GUMBO) to name a few (Bagstad et al., 2013; Snell, 2016; Grêt-Regamey et al., 2016; Vigerstol and Aukema, 2011; Farrell and Marion, 2002; Duarte et al., 2016).

As the uses of ESs have multiple roles nowadays, like raising awareness about nature and green services, analyzing the trade-offs between socio-economic development and environmental service losses, and its profound application in sustainable urban and regional planning, a careful evaluation of the process, methods, and approaches involved in ES valuation are necessary to increase accountability and applicability in decision-making and policy implementation (Burkhard et al., 2009; Costanza et al., 2014; Comberti et al., 2015). According to different classification systems including the Common International Classification of Ecosystem Services (CIESS) ((Potschin and Haines-Young, 2013; Haines-Young and Potschin, 2018), National Ecosystem Service Classification system (NESCS), US EPA Classification System for Final Ecosystem Goods and Services (USEPA), MEA (2005) and TEEB (2010), the ESs have been classified into four major groups which include *provisioning services* (like food production, raw material production, freshwater supply, medicinal resources), *regulating services* (like local climate and air quality regulation, carbon sequestration and oxygen release, storm and flood management, wastewater treatment, erosion control and soil formation, sediment retention and nutrient recycling, pollination, biological control), *supporting services* (like habitat provision, genetic diversity), and *cultural services* (like recreation, tourism, aesthetic, spiritual experience) (TEEB, 2010) (Figure 8.1).

This generalized methodology adopted in this study could be replicated for any ecosystem evaluation. For a complex eco-region like Sundarbans, a combination of biophysical models is required for the quantification of ESV. The sensitivity of

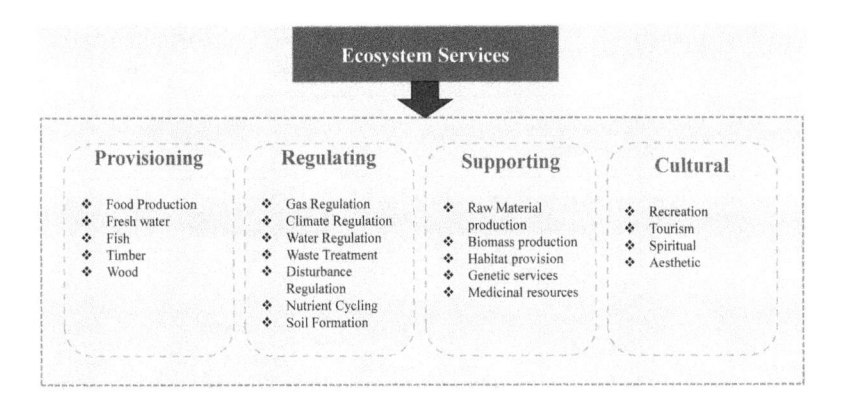

FIGURE 8.1 Typology of ESs defined by Millennium Ecosystem Assessment, 2005.

the machine learning (ML) models was tested, but the effects of erroneous model specification on ES valuation have not been substantially explored. This seeks special attention and further research. Multiple similarity and dissimilarity assessment techniques were employed to evaluate the performances of ML models. This approach could be followed to address the issues that exist in land use optimization and modelling.

8.1.1 Valuation and Quantification of Natural Capital and Ecosystem Services

Several approaches and methods have been developed in the last few decades for accurate quantification of ESs and ESVs. These include cost-benefit trade-off analysis, benefits transfer valuation, market price-based valuation, biophysical surrogate market approach, contingent valuation, willingness to pay valuation, travel cost valuation, avoided cost approach, expert opinion value transfer approach, statistical value transfer approach and spatially explicit value transfer, etc. (Costanza et al., 1997, 2014; Braat and de Groot, 2012, Scolozzi et al., 2014, Daily et al., 2009, Nelson et al., 2009, Fisher, Turner and Morling, 2009). Among these, the benefits transfer valuation approach has been used widely for regional to global ES valuation due to its simplicity and feasibility. Costanza et al. (1997) have popularized the simple benefit transfer approach through his work global ES valuation for 16 major eco-regions and 17 ES functions in 1994. However, this method is severely criticized for being insensitive in explaining the spatial explicit variation of ESVs, and it does not cover the spatiotemporal differences and uncertainties while estimating ESVs in local to global scale. Using the same approach, Costanza et al. (2014) have quantified the biome-specific global ESVs for 1997 and 2011 land area, considering both 1997- and 2011-unit values based on 2007 US$ to demonstrate the uncertainties and discrepancies of the simple benefit transfer method. Using the same approach, Liu et al. (2010) had evaluated the ESVs of 12 ESs of 11 eco-regions in New Jersey. This study had observed that the benefits transfer approach might produce an overestimated or underestimated measurement in a complex landscape, as it entirely depends on the

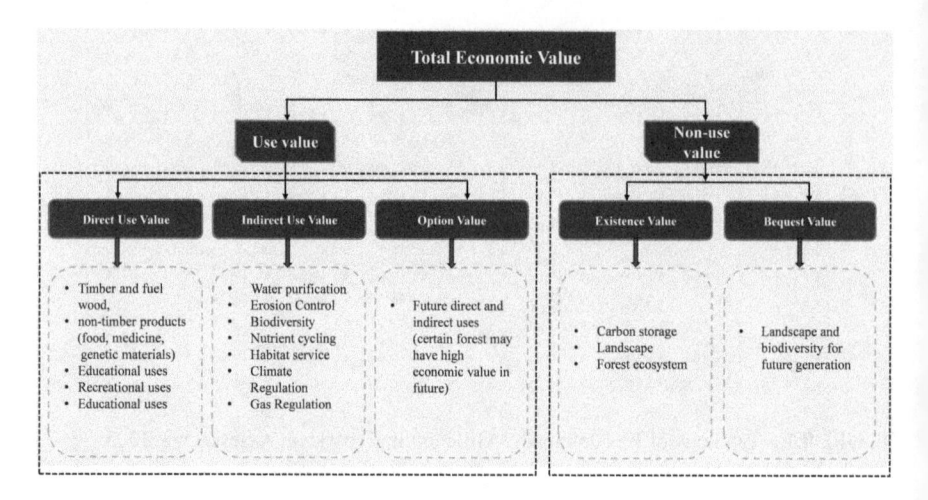

FIGURE 8.2   Types of valuation methods and approaches.

spatial homogeneity and does not explain the inter-dependencies and spatiotemporal heterogeneity in the valuation process.

Figure 8.2 shows the types of valuation approaches available for the estimation of ESs. The total economic value is the sum of use-value and non-use value of an ES. The use-value consists of three approaches including the *direct use value* (timber and fuel-wood, non-timber products, recreation, tourism, educational, medicinal, fishing, cultural uses, human habitat), *indirect use value* (water regulation, water purification, biodiversity maintenance, erosion control, local climate and air quality regulation, gas regulation, nutrient cycling) and *option value* (future potential uses of any services). However, the non-use value comprises *existence value* (carbon sequestration, biodiversity value) and *bequest value* (landscape, biodiversity for future generation). According to TEEB (2010), a single valuation method is not enough for addressing the variability that exists in the market valuation of natural capital. Therefore, a broad and multi-functional valuation framework is required, which could be useful for calibrating economic variabilities in estimating ES valuation. To establish such a framework; three steps need to be followed.

- **Step 1:** Identify and assess the individual and group of ESs affected by several indigenous and passive alternatives and would be beneficial or detrimental to society.
- **Step 2:** Emphasizing on estimating and demonstrating ESs using relevant and identical valuation methods and approaches. It includes scale- and time-based cost-benefit analysis, trade-offs, and synergy, etc.
- **Step 3:** Quantifying and estimating ESVs, evaluating the uncertainties in valuation that causes over/underestimation of service values, identification and developing appropriate proxies and surrogate markets, payment for ESs, strengthening and enforcing biodiversity protection and conservation for climate and societal problem alleviation (TEEB, 2010; Pandeya et al., 2016).

8.1.2 Social Cost of Carbon and Carbon Taxation

Climate change is real, and multiple shreds of evidence are showing that climate change is happening. Outlining and computing economic damages caused by climate change is challenging, which induced decisive market failure. In order to address market failure, one needs to fix the economic values and social cost of CO_2 emissions. The social cost of carbon (SCC) is defined by the amount of economic cost connected with the climate change benefits or damages that caused from the release of one tonne of additional carbon dioxide (tCO_2) (Ricke et al., 2018). The SCC also measures the economic damages (in US Dollar) from any climate change impacts from producing one ton of extra carbon dioxide into the atmosphere. Although the SCC is one of the most robust and trustworthy carbon cost estimates available, it has been severely criticised as it does not include all the recognised and well-known methodical and monetary effects of climate change that mainly due to the emission of carbon and added into the atmosphere. Therefore, SCC delivers a monetary estimate of the marginal impacts of climate change. According to Ricke et al. (2018), the SCC value ranges from about US\$10 $tCO_2{}^{-1}$ to nearly US\$1,000 $tCO_2{}^{-1}$ across the countries.

8.1.3 Application of Remote Sensing and Machine Learning Approaches in Ecosystem Service Valuation

The advancement relevant geospatial approaches and its wide application in ecosystem studies have been thoroughly researched. Likewise, the network and locational choice-based analysis and public participatory geographic information system are quintessential, especially for sustainable cultural and recreational ES valuation. Gravity model volunteered geographic information and citizen science (to name a few) have often been used in social and cultural ES valuation across the world. However, in several cases, the accurate quantification of indirect regulating and supporting services found problematic due to (i) unavailability of real valuation market; (ii) improper selection of proxies and surrogate market; (iii) the lack of concrete information about the production function or the trade-off and synergies among ESs; and (iv) the lack of a valid baseline, benchmark or threshold for evaluation (Polasky et al., 2011; Schmidt, Moore and Alber, 2014).

The Sundarbans Biosphere Region in India is an immensely important ecosystem, for its biodiversity and ecological value. Despite having a great significance in producing several key ESs, including provisioning (honey, crab, fish etc.), regulating (storm and flood protection, coastal erosion, climate regulation, nutrient retention etc.), cultural (tourism, recreation), limited efforts and data exist to quantify the biophysical and economic values of these ESs. Therefore, a careful evaluation is required to establish a feasible valuation framework for estimating the spatially explicit ESVs and thereby identify the principal driving factors responsible for ESs degradation.

8.2 MATERIALS AND METHODS

8.2.1 LULC CLASSIFICATION

Moderate-resolution remote sensing data such as Landsat Multi-Spectral Scanner for 1973, Thematic Mapper for 1988, 2003, and Operational Land Imager for 2013 and 2018 with path 138, and rows 044 and 045 were utilized for LULC classification of the regions for different years. Figure 8.3 shows the methodological approaches followed in this study. Costanza et al. (1997, 2014) equivalent biomes were approximated from the LULC classification. For the classification of LULC, the processed Landsat satellite images were collected from Earth Explorer (https://earthexplorer.usgs.gov/). Satellite images were collected during December/January months so as to have least concentration of cloud and hazes of the wintertime over the Sundarbans region. The entire area was classified into six major eco-regions, that is, cropland (CL), mixed/natural vegetation (MV), mangrove (MA), waterbody (WB), sandy coast (SC), and urban (UB) using ten advanced supervised ML algorithms [i.e., Artificial Neural Network (ANN), Decision Tree (DT), Bayes, Gradient Boosted Tree (GBT), Linear Discriminant Analysis (LDA), K-Means Nearest Neighbour (KNN), Maximum Likelihood Classification (MLC), Support Vector Machine (SVM Linear and Radial Basis Function – RBF), Random Forest (RF)]. A number of methods were implemented just to be ensured the accuracy of model estimates and subsequent interpretation. Coastal wetland ecosystems were subdivided into two subparts, that is, coastal estuary (CE) and inland wetland (IW), to be in line with Costanza et al. (1997, 2014) defined biomes for having the designated equivalent values of these ecosystems. The LULC classification was done in various geospatial tools. SVM and maximum likelihood classification were performed

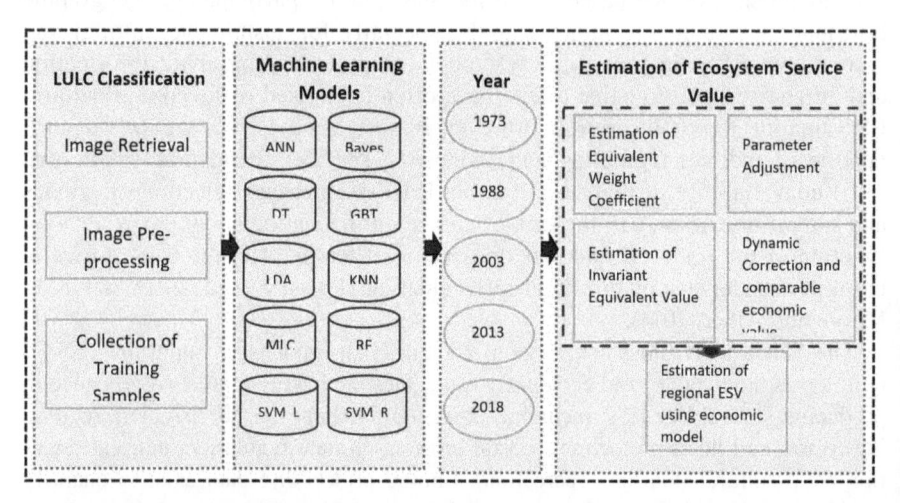

FIGURE 8.3 Methodological flow chart shows the methods and approaches adopted in this study. (ANN – Artificial Neural Network, Bayes – Bayesian, DT – Decision Tree, GBT – Gradient Boosted Tree, LDA – Linear Discriminant Analysis, KNN – K-Nearest Neighbor, MLC – Maximum Likelihood Classifier, RF – Random Forest, SVM L – Support Vector Machine Linear, SVM R – Support Vector Machine Radial Basis Function.)

in ENVI 5.3 software, KNN and LDA were performed in IDRISI TerrSet software, RF was computed in EnMAP-Box suite for ENVI 5.3, ANN, BAYES, DT, GBT, and SVM models were performed in QGIS Orfeo ToolBox (OTB). An average of 50–60 samples were ingested to train the models. Post-classification processing was performed using 4×4 majority filter which has been used for removing isolated pixels from the classification outputs. The accuracy of the model's estimates was examined by various tests, including User's Accuracy (UA), Producer's Accuracy (PA), Overall Accuracy (OA) and Kappa coefficient.

The changes in different ecosystem units were calculated as

$$\Delta LULC_i = \frac{LULC_{End} - LULC_{Start}}{LULC_{Start}} \times 100 \tag{8.1}$$

$$LUCI = \frac{LULC_{End} - LULC_{Start}}{LULC_{Start}} \times \frac{1}{t} \times 100 \tag{8.2}$$

where $\Delta LULC_i$ denotes the change in LULC during the study period i, $LULC_{End}$ and $LULC_{Start}$ refers to the area of LULC at ending and starting time, LUCI denotes the LULC change intensity of category i, t refers to the study period (Sannigrahi et al., 2018; Sannigrahi et al., 2019).

8.2.2 Estimation of Ecosystem Service Values

The updated and adjusted equivalent coefficient values proposed by Costanza et al. (1997, 2014) and Xie et al. (2008, 2017) were utilized for per unit ES valuation of ecosystem units. 17 key relevant ESs, that is, biological control (BC), cultural (CUL), climate regulation (CR), disturbance regulation (DR), erosional control (EC), food production (FP), genetic resources (GEN), gas regulation (GR), habitat (HA), nutrient cycling (NC), pollination (POL), raw materials (RM), recreation (REC), soil formation (SF), water regulation (WR), water supply (WS), and waste treatment (WT), services have been considered for economic valuation and subsequent interpretation.

Using the adjusted equivalent coefficient values, the spatial and temporal dynamics of ESVs were calculated for Sundarbans. The ESVs was computed as follows:

$$ESV_j = \sum_{i=1}^{17} E \times EF_{ij} \times A_j \tag{8.3}$$

$$ESV_i = \sum_{j}^{7} E \times EF_{ij} \times A_j \tag{8.4}$$

$$\text{ESV} = \sum_{i=1}^{17} \sum_{j=1}^{7} E \times EF_{ij} \times A_j \qquad (8.5)$$

where ESV_j, ESV_i, and ESV denote the ESVs (US\$ ha^{-1} year^{-1}) of LULC type j, and ESs i E refers to the equivalent value of food production service of CL (US\$ ha^{-1}), EF_{ij} refers to the adjusted equivalent value coefficient of ES i and LULC j, A_j is the area (ha) of LULC type j, respectively (Sannigrahi et al., 2018; 2020). The spatiotemporal changes in ESVs were calculated as follows:

$$\Delta\text{ESV}_{ij}(\%) = \frac{\text{ESV}_{\text{end}} - \text{ESV}_{\text{start}}}{\text{ESV}_{\text{start}}} \times 100\% \qquad (8.6)$$

$$\Delta\text{ESV}_{ij} = \frac{\text{ESV}_{\text{end}} - \text{ESV}_{\text{start}}}{\text{ESV}_{\text{start}}} \times \frac{1}{t} \times 100\% \qquad (8.7)$$

where ΔESV_{ij} denotes the changes in ESVs of LULC type j ; ES i ESV_{end} and $\text{ESV}_{\text{start}}$ exhibit the ESV of ending and starting year; t represents the time.

Valuation of ESs using global equivalent proxies is always subject to bias and uncertainty. To minimize the valuation uncertainty in the estimation, the Coefficient of Sensitivity (CS) test was performed:

$$\text{CS} = \frac{\left(\text{ESV}_j - \text{ESV}_i\right) / \text{ESV}_i}{\left(\text{VC}_{jk} - \text{VC}_{ik}\right) / \text{VC}_{ik}} \qquad (8.8)$$

where VC denotes value coefficient, i and j refer to the initial and adjusted values, and k is LULC type. It has been observed that ESs turns to be elastic when CS goes above the upper limit (>1); in contrast, the value would turn to be inelastic when the CS would be <1. Moreover, CS = 1 and CS = 0 indicate the complete inelasticity and elasticity (Sannigrahi et al., 2018).

8.3 RESULTS AND DISCUSSION

The LULC classification was performed using ten ML models, and the results of the same are presented in Figure 8.4. All the ML models collectively indicating the prominent changes in few land classes, especially mangrove and inland water surface during 1973–2018. High spatial discrepancies were evident for few algorithms such as MLC, LDA and KNN, whereas a comparably higher spatial accuracy was found for RF, SVM and Bayes algorithm. Besides, the performance of the ten ML models incorporated in this study was examined by different validation approaches including UA, PA, OA and Kappa statistics (Figures 8.5 and 8.6). Among the models, MLC algorithm has the lowest accuracy observed throughout the study periods, except for the year 1973. While the RF and SVM models have performed most accurately.

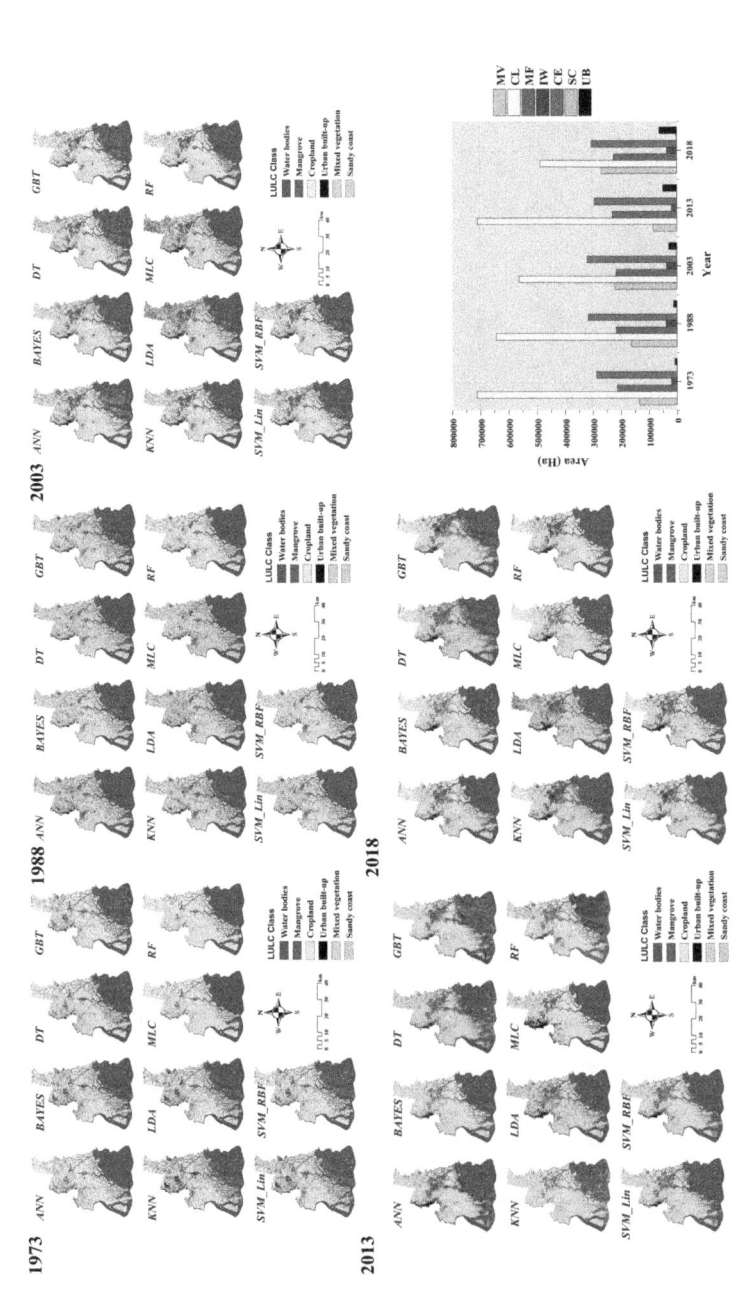

FIGURE 8.4 Spatial distribution and temporal dynamics of different ecosystem types in Indian Sundarbans derived from ten ML models. (ANN – Artificial Neural Network, Bayes – Bayesian, DT – Decision Tree, GBT – Gradient Boosted Tree, LDA – Linear Discriminant Analysis, KNN – K-Nearest Neighbor, MLC – Maximum Likelihood Classifier, RF – Random Forest, SVM L – Support Vector Machine Linear, SVM R – Support Vector Machine Radial Basis Function. MV – Mixed Vegetation, CL – Cropland, MF – Mixed Forest, IW – Inland Wetland, CE – Coastal Estuary, SC – Sandy Coast, UB – Urban Built Up.)

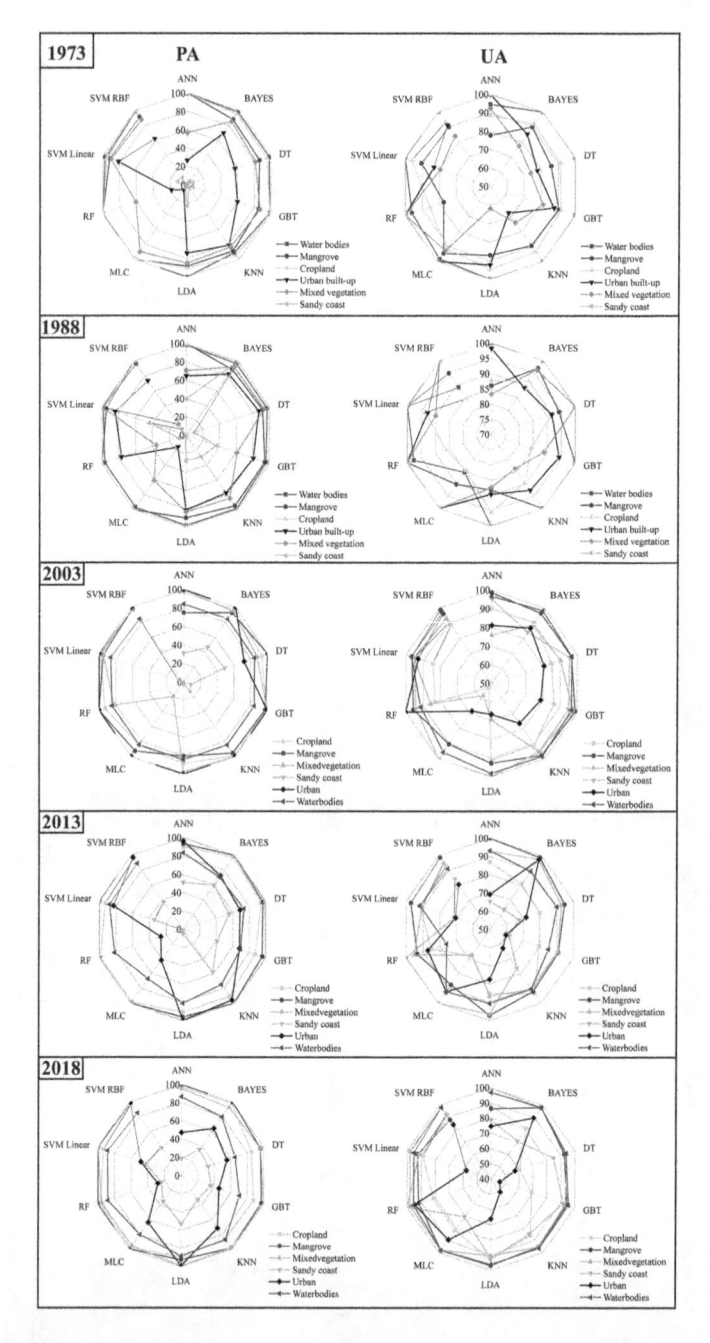

FIGURE 8.5 PA and UA of ten ML models estimated for 1973, 1988, 2003, 2013 and 2018. (ANN – Artificial Neural Network, Bayes – Bayesian, DT – Decision Tree, GBT – Gradient Boosted Tree, LDA – Linear Discriminant Analysis, KNN – K Nearest Neighbor, MLC – Maximum Likelihood Classifier, RF – Random Forest, SVM L – Support Vector Machine Linear, SVM R – Support Vector Machine Radial Basis Function.)

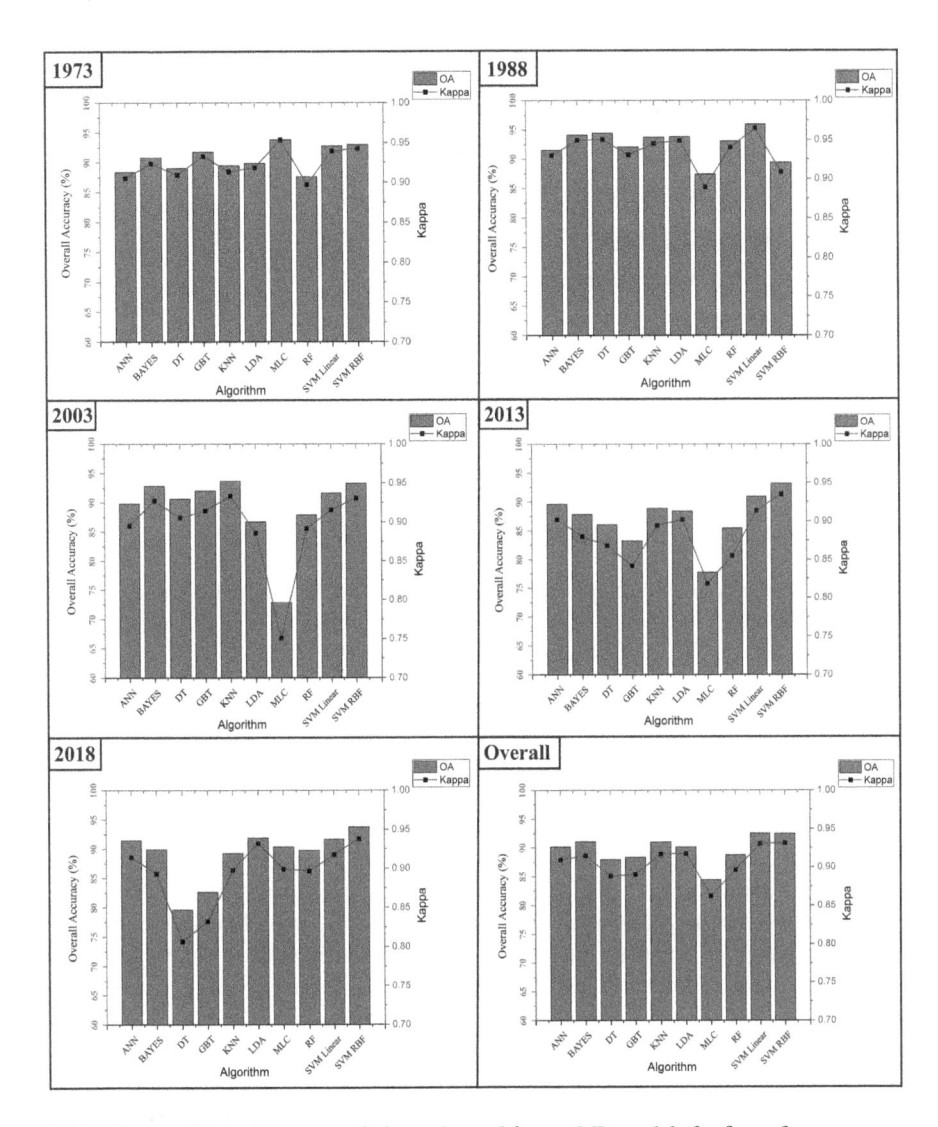

FIGURE 8.6   OA and Kappa statistics estimated for ten ML models for five reference years, 1973, 1988, 2003, 2013 and 2018. (ANN – Artificial Neural Network, Bayes – Bayesian, DT – Decision Tree, GBT – Gradient Boosted Tree, LDA – Linear Discriminant Analysis, KNN – K-Nearest Neighbor, MLC – Maximum Likelihood Classifier, RF – Random Forest, SVM L – Support Vector Machine Linear, SVM R – Support Vector Machine Radial Basis Function.)

To verify the inter-model similarity and consistency among the estimates, the Jaccard coefficient test was performed for all reference years, and results of the same are presented in Figures 8.7 and 8.8. Among the models, high similarity is found for SVM, RF, Bayes, ANN models, while a comparably lower similarity estimates had found for MLC model (Figure 8.7). This suggests the superiority and functional capability of SVM and RF models in capturing the land dynamics more accurately than the remaining ML models considered for the evaluation. The class-wise Jaccard was also

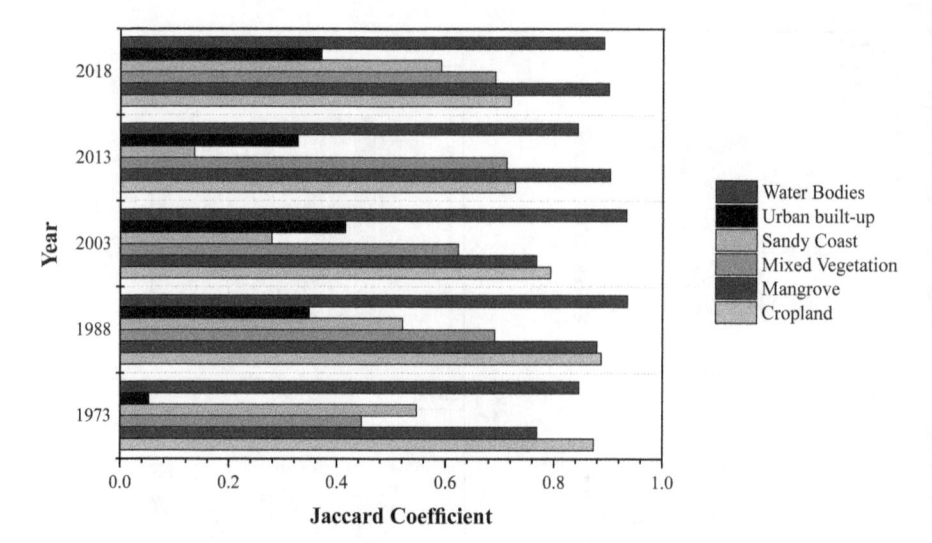

FIGURE 8.7 Jaccard similarity coefficient measured for different LULC classes for five reference years.

computed for each reference years, and it can be seen in the results that similarity value is very high for waterbodies, CL, and mangrove, and the same is found low for more complex classes, such as urban land, mixed vegetation, sand cover, etc. (Figure 8.8). The application of Jaccard similarity is not limited to spatial science research; the same has been utilized in many relevant domains, such as business model (Bag, Kumar and Tiwari, 2019), analyzing characteristics of social network (Bank & Cole, 2008), genome comparisons study (Besta et al., 2020), information retrieval application (Rousseau, 1998). Tomaselli et al. (2013) is the study which explicitly used Jaccard similarity in LULC mapping and comparing land classes among the pairs. This study (Tomaselli et al., 2013) conducted the analysis in five sites, and ten possible pairwise comparisons were made for pairwise evaluation after assuming that high similarity values would be found if substantial overlapping happens among the landscape compositions between two sites. Additionally, it was assumed that the common LULC classes between two sites would only be considered, and subsequently, the similarity scores among the pairs would be recorded. Jaccard value would be 0 when there is no commonality in LULC class distributions among the pairs, and conversely, the value would be 1 when the land-scape compositions of both sites are similar in nature.

The economic values of the key ESs were measured and presented in Figure 8.9. For all reference years, the highest ESVs (million US$) was calculated for waste treatment service, followed by erosion control, habitat, food production, disturbance regulation, genetic, soil formation, water supply, recreation, climate regulation, raw material production, water regulation, cultural, nutrient cycling, biological control, and pollination services, respectively (Figure 8.9). Figure 8.9b shows the ESVs esti-mated for different ecosystem types. Among the major LULC classes, the maximum ESVs (Million US$) was measured for mangrove, followed by coastal wetland, CL, inland wetland, mixed vegetation, urban land, and sandy coast, respectively.

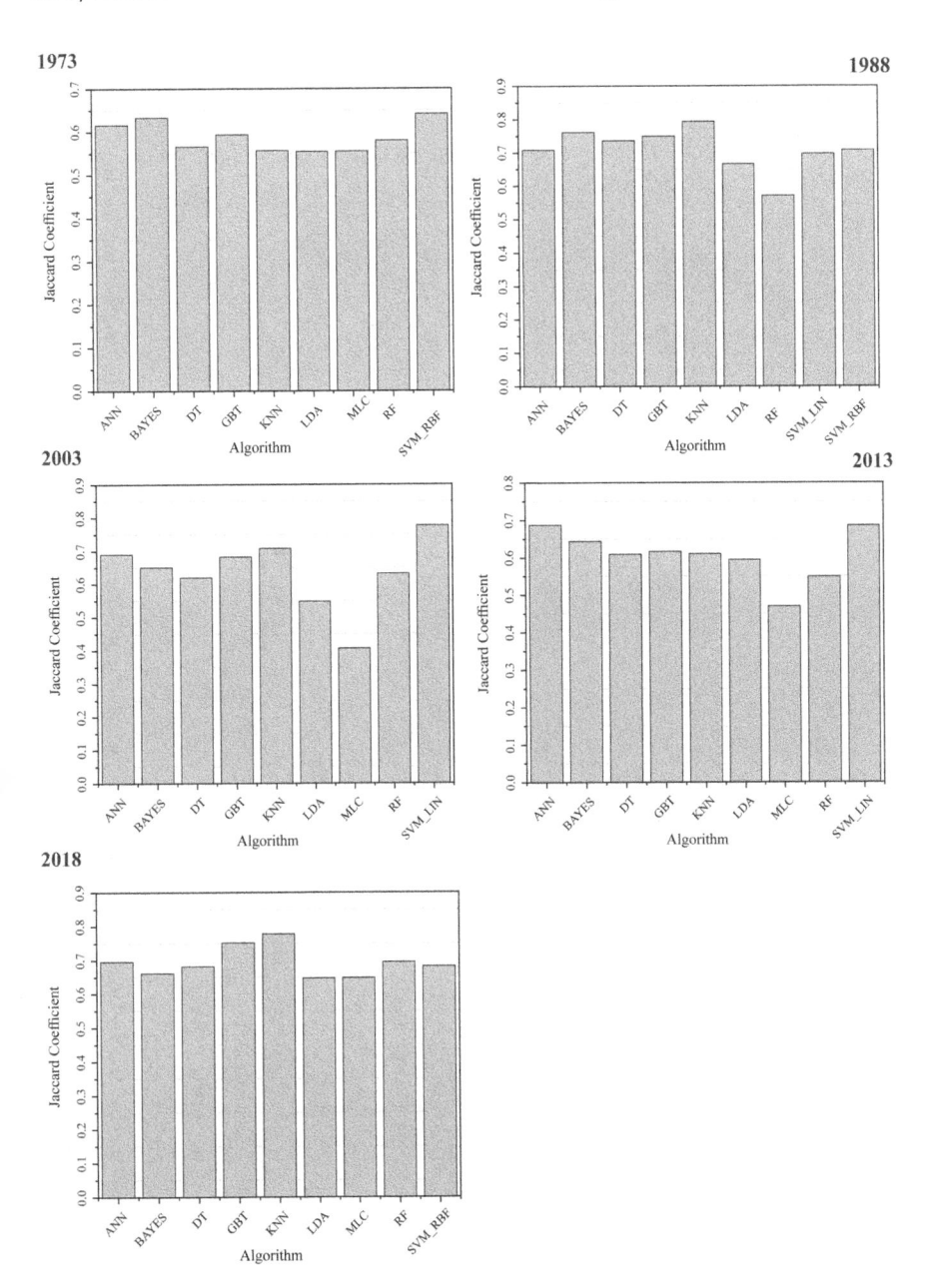

FIGURE 8.8   Jaccard similarity coefficient measured for ten ML models for five reference years. (ANN – Artificial Neural Network, Bayes – Bayesian, DT – Decision Tree, GBT – Gradient Boosted Tree, LDA – Linear Discriminant Analysis, KNN – K-Nearest Neighbor, MLC – Maximum Likelihood Classifier, RF – Random Forest, SVM L – Support Vector Machine Linear, SVM R – Support Vector Machine Radial Basis Function.)

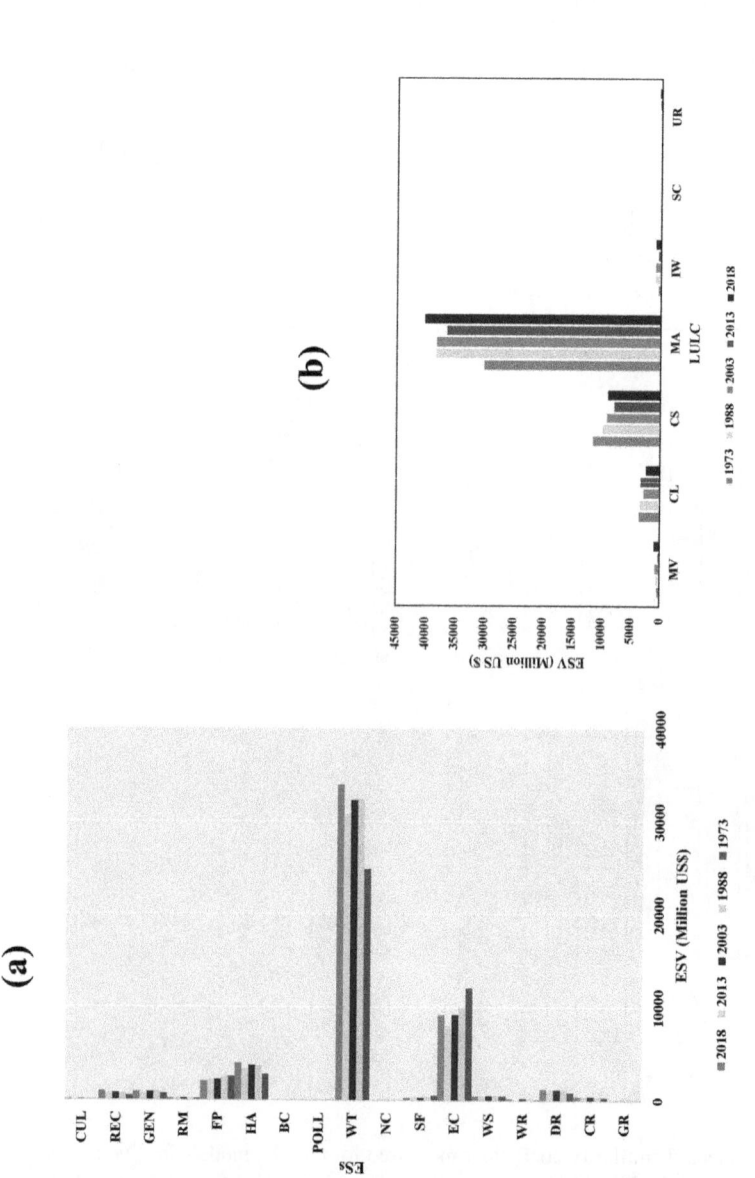

FIGURE 8.9 ES values (Million US $ year) of (a) different ecosystem functions and (b) ecosystem types estimated for the study region. (CUL – Culrural service, REC – Recreation service, GEN – Genetic service, RM – Raw material provision service, FP – Food porduction service, HA – Habitat service, BC – Biological control service, POLL – Pollination service, WT – Waste treatment service, NC – Nutrient control service, SF – Soil formation service, EC – Erosion control, WS – Water supply, WR – Water regulation, DR – Disturbance regulation, CR – Climate regulation, GR – Gas regulation. MV = Mixed forest, CL – Cropland, CS – Coastal estuary, MA – Mangrove, IW – Inland wetland, SC – Sandy coast, UR – Urban builtup land.)

Mangrove ecosystem is providing a substantial amount of ESs services in terms of absorbing gaseous carbon, releasing oxygen, increasing nutrients into the soil, protecting coastal land from sea waves, etc., are few of them (Vo et al., 2012). The degradation of the coastal ecosystem, including mangroves and corals, has profound impacts on community well-being in coastal areas across the world. Uncertainties in livelihood and increasing level of food insecurity are few of the major environmental threats pertinent in the coastal region. In addition to this, habitat degradation and overexploitation of resources destabilized the balance of ecosystems. Burke and Maidens (2004) documented that in the Caribbean, the fisheries production in 2015 would be around 100,000 tonnes with having monetary values 310 million US$ if the current loss of coral reef would have controlled effectively. Uddin et al. (2013) study measured the economic importance of Sundarbans and reported that provisioning and cultural services of Sundarbans could be nearly US$ 744,000 and US$ 42,000 per year in the financial year 2001–2002 to 2009–2010. This study also noted that though this vibrant ecosystem is providing ample amount of ESs, the overexploitation of resources mainly due to the unknowingness of the sustainable limits of resource extraction often build pressure on the system and eventually led the ecosystem less resilient to the concurrent threats the community are dealing across the ecosystems. Therefore, Uddin et al. recommend few must-do options by taking the example of Bangladesh's Sundarbans that defining the sustainable limits up to which the resources can be harnessed, and full inventories of stock and potential of available ecological resources under varied socio-ecological scenarios and climate change could be integrated with conventional sustainable management plans, especially for managing coastal resources.

To verify the sensitivity and elasticity of economic valuation approaches adopted in this study, the CS test was performed and reported in Tables 8.1 and 8.2. Among the 17 ESs, the highest CS values were measured for waste treatment services, followed by erosion control, habitat, food production, disaster reduction, genetic service, genetic service, climate regulation, water supply, soil formation, recreation, water regulation, nutrient cycling, pollination, biological control, raw material, and cultural (Table 8.1). A similar pattern was noted for the remaining four sensitivity levels, that is, CS – 40%, CS – 30%, CS – 20%, and CS – 10% (Table 8.1). Considering the LULC-specific CS estimation, the highest CS score was recorded for mangrove, followed by the coastal estuary, CL and mixed vegetation (Table 8.2). The CS test was implemented in many similar studies, especially in the valuation of ecosystem capitals, to reassure the valuation in given sensitivity levels (Zhao et al., 2004; Kreuter et al., 2001; Li et al., 2007; Tianhong, Wenkai and Zhenghan, 2010). Kreuter et al. (2001) stated that the estimated ESV would be elastic when the ratio between the percentage change in ESVs and percentage change between the coefficient value is greater than unity; conversely, the ESV would become inelastic when the ratio between the above pairs is calculated less than 1. Based on this assumptions, Kreuter et al. (2001) noted that the uses of value coefficient would be more critical when the proportional changes in ESV in respect to changes in proportional changes in value coefficient is higher and vice versa.

TABLE 8.1
CS at Different Scenarios for Five Reference Years

ESF	1973	1988	2003	2013	2018	1973	1988	2003	2013	2018
	VC $\pm$ 50 (%)					VC $\pm$ 50				
CR	0.36	0.40	0.39	0.42	0.38	0.01	0.01	0.01	0.01	0.01
DR	0.93	1.09	1.11	1.08	1.12	0.02	0.02	0.02	0.02	0.02
WR	0.12	0.20	0.20	0.10	0.19	0.00	0.00	0.00	0.00	0.00
WS	0.54	0.49	0.49	0.54	0.45	0.01	0.01	0.01	0.01	0.01
EC	13.04	9.35	8.80	8.24	8.49	0.26	0.19	0.18	0.16	0.17
SF	0.60	0.35	0.29	0.36	0.25	0.01	0.01	0.01	0.01	0.01
NC	0.04	0.07	0.07	0.04	0.07	0.00	0.00	0.00	0.00	0.00
WT	26.92	30.31	30.92	31.49	31.38	0.54	0.61	0.62	0.63	0.63
POLL	0.02	0.02	0.02	0.02	0.02	0.00	0.00	0.00	0.00	0.00
BC	0.04	0.05	0.05	0.04	0.05	0.00	0.00	0.00	0.00	0.00
HA	3.03	3.47	3.60	3.47	3.69	0.06	0.07	0.07	0.07	0.07
FP	2.78	2.38	2.13	2.30	1.89	0.06	0.05	0.04	0.05	0.04
RM	0.20	0.21	0.21	0.21	0.19	0.00	0.00	0.00	0.00	0.00
GEN	0.79	0.85	0.85	0.81	0.82	0.02	0.02	0.02	0.02	0.02
REC	0.54	0.66	0.75	0.81	0.91	0.01	0.01	0.02	0.02	0.02
CUL	0.06	0.10	0.11	0.06	0.11	0.00	0.00	0.00	0.00	0.00
	VC $\pm$ 40 (%)					VC $\pm$ 40				
CR	0.29	0.32	0.31	0.34	0.30	0.01	0.01	0.01	0.01	0.01
DR	0.74	0.87	0.89	0.86	0.90	0.01	0.02	0.02	0.02	0.02
WR	0.09	0.16	0.16	0.08	0.15	0.00	0.00	0.00	0.00	0.00
WS	0.43	0.39	0.39	0.43	0.36	0.01	0.01	0.01	0.01	0.01
EC	10.44	7.48	7.04	6.59	6.79	0.21	0.15	0.14	0.13	0.14
SF	0.48	0.28	0.23	0.29	0.20	0.01	0.01	0.00	0.01	0.00
NC	0.04	0.06	0.06	0.03	0.05	0.00	0.00	0.00	0.00	0.00
WT	21.54	24.25	24.73	25.19	25.11	0.43	0.48	0.49	0.50	0.50
POLL	0.01	0.01	0.01	0.01	0.01	0.00	0.00	0.00	0.00	0.00
BC	0.03	0.04	0.04	0.03	0.04	0.00	0.00	0.00	0.00	0.00
HA	2.42	2.78	2.88	2.78	2.95	0.05	0.06	0.06	0.06	0.06
FP	2.23	1.90	1.70	1.84	1.51	0.04	0.04	0.03	0.04	0.03
RM	0.16	0.17	0.16	0.17	0.15	0.00	0.00	0.00	0.00	0.00
GEN	0.63	0.68	0.68	0.65	0.65	0.01	0.01	0.01	0.01	0.01
REC	0.43	0.53	0.60	0.65	0.73	0.01	0.01	0.01	0.01	0.01
CUL	0.05	0.08	0.09	0.04	0.09	0.00	0.00	0.00	0.00	0.00
	VC $\pm$ 30 (%)					VC $\pm$ 30				
CR	0.22	0.24	0.24	0.25	0.23	0.00	0.00	0.00	0.01	0.00
DR	0.56	0.65	0.67	0.65	0.67	0.01	0.01	0.01	0.01	0.01
WR	0.07	0.12	0.12	0.06	0.11	0.00	0.00	0.00	0.00	0.00
WS	0.33	0.29	0.30	0.33	0.27	0.01	0.01	0.01	0.01	0.01
EC	7.83	5.61	5.28	4.94	5.09	0.16	0.11	0.11	0.10	0.10
SF	0.36	0.21	0.17	0.22	0.15	0.01	0.00	0.00	0.00	0.00

(Continued)

TABLE 8.1 *(Continued)*
CS at Different Scenarios for Five Reference Years

ESF	1973	1988	2003	2013	2018	1973	1988	2003	2013	2018
			VC $\pm$ 30 (%)					VC $\pm$ 30		
NC	0.03	0.04	0.04	0.02	0.04	0.00	0.00	0.00	0.00	0.00
WT	16.15	18.18	18.55	18.89	18.83	0.32	0.36	0.37	0.38	0.38
POLL	0.01	0.01	0.01	0.01	0.01	0.00	0.00	0.00	0.00	0.00
BC	0.02	0.03	0.03	0.02	0.03	0.00	0.00	0.00	0.00	0.00
HA	1.82	2.08	2.16	2.08	2.22	0.04	0.04	0.04	0.04	0.04
FP	1.67	1.43	1.28	1.38	1.13	0.03	0.03	0.03	0.03	0.02
RM	0.12	0.13	0.12	0.13	0.12	0.00	0.00	0.00	0.00	0.00
GEN	0.47	0.51	0.51	0.49	0.49	0.01	0.01	0.01	0.01	0.01
REC	0.32	0.40	0.45	0.49	0.54	0.01	0.01	0.01	0.01	0.01
CUL	0.04	0.06	0.07	0.03	0.07	0.00	0.00	0.00	0.00	0.00
			VC $\pm$ 20 (%)					VC $\pm$ 20		
CR	0.14	0.16	0.16	0.17	0.15	0.00	0.00	0.00	0.00	0.00
DR	0.37	0.44	0.44	0.43	0.45	0.01	0.01	0.01	0.01	0.01
WR	0.05	0.08	0.08	0.04	0.07	0.00	0.00	0.00	0.00	0.00
WS	0.22	0.20	0.20	0.22	0.18	0.00	0.00	0.00	0.00	0.00
EC	5.22	3.74	3.52	3.29	3.40	0.10	0.07	0.07	0.07	0.07
SF	0.24	0.14	0.11	0.15	0.10	0.00	0.00	0.00	0.00	0.00
NC	0.02	0.03	0.03	0.02	0.03	0.00	0.00	0.00	0.00	0.00
WT	10.77	12.12	12.37	12.60	12.55	0.22	0.24	0.25	0.25	0.25
POLL	0.01	0.01	0.01	0.01	0.01	0.00	0.00	0.00	0.00	0.00
BC	0.01	0.02	0.02	0.01	0.02	0.00	0.00	0.00	0.00	0.00
HA	1.21	1.39	1.44	1.39	1.48	0.02	0.03	0.03	0.03	0.03
FP	1.11	0.95	0.85	0.92	0.75	0.02	0.02	0.02	0.02	0.02
RM	0.08	0.09	0.08	0.09	0.08	0.00	0.00	0.00	0.00	0.00
GEN	0.32	0.34	0.34	0.32	0.33	0.01	0.01	0.01	0.01	0.01
REC	0.21	0.26	0.30	0.33	0.36	0.00	0.01	0.01	0.01	0.01
CUL	0.02	0.04	0.04	0.02	0.04	0.00	0.00	0.00	0.00	0.00
			VC$\pm$10 (%)					VC$\pm$10		
CR	0.07	0.08	0.08	0.08	0.08	0.00	0.00	0.00	0.00	0.00
DR	0.19	0.22	0.22	0.22	0.22	0.00	0.00	0.00	0.00	0.00
WR	0.02	0.04	0.04	0.02	0.04	0.00	0.00	0.00	0.00	0.00
WS	0.11	0.10	0.10	0.11	0.09	0.00	0.00	0.00	0.00	0.00
EC	2.61	1.87	1.76	1.65	1.70	0.05	0.04	0.04	0.03	0.03
SF	0.12	0.07	0.06	0.07	0.05	0.00	0.00	0.00	0.00	0.00
NC	0.01	0.01	0.01	0.01	0.01	0.00	0.00	0.00	0.00	0.00
WT	5.38	6.06	6.18	6.30	6.28	0.11	0.12	0.12	0.13	0.13
POLL	0.00	0.00	0.00	0.00	0.00	0.00	0.00	0.00	0.00	0.00
BC	0.01	0.01	0.01	0.01	0.01	0.00	0.00	0.00	0.00	0.00
HA	0.61	0.69	0.72	0.69	0.74	0.01	0.01	0.01	0.01	0.01

(Continued)

TABLE 8.1 (*Continued*)
CS at Different Scenarios for Five Reference Years

ESF	1973	1988	2003	2013	2018	1973	1988	2003	2013	2018
	VC±30 (%)					VC±30				
FP	0.56	0.48	0.43	0.46	0.38	0.01	0.01	0.01	0.01	0.01
RM	0.04	0.04	0.04	0.04	0.04	0.00	0.00	0.00	0.00	0.00
GEN	0.16	0.17	0.17	0.16	0.16	0.00	0.00	0.00	0.00	0.00
REC	0.11	0.13	0.15	0.16	0.18	0.00	0.00	0.00	0.00	0.00
CUL	0.01	0.02	0.02	0.01	0.02	0.00	0.00	0.00	0.00	0.00

CUL, Culrural service; REC, Recreation service; GEN, Genetic service; RM, Raw material provision service; FP, Food porduction service; HA, Habitat service; BC, Biological control service; POLL, Pollination service; WT, Waste treatment service; NC, Nutrient control service; SF, Soil formation service; EC, Erosion control; WS, Water supply; WR, Water regulation; DR, Disturbance regulation; CR, Climate regulation; GR, Gas regulation.

TABLE 8.2
CS at Different Scenarios for the Five Reference Years

		CS					CS (%)				
		1973	1988	2003	2013	2018	1973	1988	2003	2013	2018
MV	VC+50%	0.010	0.012	0.016	0.006	0.019	0.479	0.581	0.787	0.298	0.926
	VC+40%	0.008	0.009	0.013	0.005	0.015	0.384	0.465	0.630	0.238	0.741
	VC+30%	0.006	0.007	0.009	0.004	0.011	0.288	0.348	0.472	0.179	0.556
	VC+20%	0.004	0.005	0.006	0.002	0.007	0.192	0.232	0.315	0.119	0.371
	VC+10%	0.002	0.002	0.003	0.001	0.004	0.096	0.116	0.157	0.060	0.185
CL	VC+50%	0.077	0.063	0.054	0.068	0.045	3.871	3.172	2.704	3.403	2.258
	VC+40%	0.062	0.051	0.043	0.054	0.036	3.097	2.537	2.163	2.722	1.806
	VC+30%	0.046	0.038	0.032	0.041	0.027	2.322	1.903	1.622	2.042	1.355
	VC+20%	0.031	0.025	0.022	0.027	0.018	1.548	1.269	1.082	1.361	0.903
	VC+10%	0.015	0.013	0.011	0.014	0.009	0.774	0.634	0.541	0.681	0.452
CE	VC+50%	0.249	0.185	0.175	0.162	0.167	12.456	9.258	8.771	8.116	8.336
	VC+40%	0.199	0.148	0.140	0.130	0.133	9.965	7.406	7.017	6.493	6.669
	VC+30%	0.149	0.111	0.105	0.097	0.100	7.474	5.555	5.262	4.870	5.002
	VC+20%	0.100	0.074	0.070	0.065	0.067	4.982	3.703	3.508	3.247	3.334
	VC+10%	0.050	0.037	0.035	0.032	0.033	2.491	1.852	1.754	1.623	1.667
MA	VC+50%	0.653	0.720	0.734	0.749	0.746	32.652	36.016	36.688	37.456	37.288
	VC+40%	0.522	0.576	0.587	0.599	0.597	26.122	28.813	29.350	29.965	29.831
	VC+30%	0.392	0.432	0.440	0.449	0.447	19.591	21.610	22.013	22.474	22.373
	VC+20%	0.261	0.288	0.294	0.300	0.298	13.061	14.406	14.675	14.982	14.915
	VC+10%	0.131	0.144	0.147	0.150	0.149	6.530	7.203	7.338	7.491	7.458

(Continued)

TABLE 8.2 (*Continued*)
CS at Different Scenarios for the Five Reference Years

		CS					CS (%)			
		1973	1988	2003	2013	2018	1973	1988	2003	2013	2018
IW	VC+50%	0.010	0.018	0.018	0.009	0.017	0.500	0.899	0.877	0.445	0.830
	VC+40%	0.008	0.014	0.014	0.007	0.013	0.400	0.719	0.702	0.356	0.664
	VC+30%	0.006	0.011	0.011	0.005	0.010	0.300	0.540	0.526	0.267	0.498
	VC+20%	0.004	0.007	0.007	0.004	0.007	0.200	0.360	0.351	0.178	0.332
	VC+10%	0.002	0.004	0.004	0.002	0.003	0.100	0.180	0.175	0.089	0.166
UB	VC+50%	0.001	0.001	0.003	0.006	0.007	0.042	0.075	0.173	0.282	0.361
	VC+40%	0.001	0.001	0.003	0.005	0.006	0.034	0.060	0.139	0.226	0.289
	VC+30%	0.000	0.000	0.000	0.000	0.000	0.025	0.045	0.104	0.169	0.217
	VC+20%	0.000	0.000	0.000	0.000	0.000	0.017	0.030	0.069	0.113	0.144
	VC+10%	0.000	0.000	0.000	0.000	0.000	0.008	0.015	0.035	0.056	0.072

MV, Mixed forest; CL, Cropland; CS, Coastal estuary; MA, Mangrove; IW, Inland wetland; SC, Sandy coast; UR, Urban builtup land; VC, Value Coefficient.

Though the uses of CS is universally approved in many scholarly works, the Aschonitis et al. (2016) study reported few methodological fallacies of sensitivity/elasticity measures in ES valuation estimates. Aschonitis et al. (2016) stated that among the few concerns that raised the validity of the CS approach are (a) the CS approach always assumed to be positive that leads the approach to be unrealistic in real market as it does not follow the "law of demand" and (b) the users can easily manipulate the concept by altering the extent of the study area.

8.4 CONCLUSION

This study thoroughly examined the effects of land cover changes on ESs in Indian Sundarbans. A total of ten ML supervised algorithms were utilized for developing LULC database of the region. Landsat satellite data from 1973 to 2018 were used for LULC classification and subsequent interpretation. Among the ten ML models, the SVM and RF models were found to be the most accurate models, while the MLC model has produced less accurate estimates for all five reference years. In order to verify the similarity of the model estimates, Jaccard similarity test was performed. The results of the Jaccard test indicate that SVM and RF have the highest similarity with other model pairs, and a low Jaccard similarity score was found for MLC. Among the major LULC classes, mangrove and coastal estuaries were found to be the most productive ecosystems of the region. Among the 17 key ESs, the highest ESVs (Million US$) were calculated for waste treatment service, followed by erosion control, habitat, food production, disturbance regulation, genetic, soil formation, water supply, recreation, climate regulation, raw material production, water regulation, cultural, nutrient cycling, biological control and pollination services, respectively. The valuation approaches and methods adopted in this study could be replicated easily for similar research interest across the ecosystems. More domain research in the direction for establishing a robust valuation framework for measuring mangrove ESs would be needed to strengthen the management and preservation of mangroves across the world.

REFERENCES

Aschonitis, V. G., et al. (2016) "Criticism on elasticity-sensitivity coefficient for assessing the robustness and sensitivity of ecosystem services values," *Ecosystem Services*, 20, pp. 66–68. doi: 10.1016/j.ecoser.2016.07.004.

Bag, S., Kumar, S. K. and Tiwari, M. K. (2019) "An efficient recommendation generation using relevant Jaccard similarity," *Information Sciences*, 483, pp. 53–64. doi: 10.1016/j. ins.2019.01.023.

Bagstad, K. J., et al. (2013) "A comparative assessment of decision-support tools for ecosystem services quantification and valuation," *Ecosystem Services*, 5, pp. 27–39. Elsevier. doi: 10.1016/j.ecoser.2013.07.004.

Bank, J., & Cole, B. (2008) "Calculating the jaccard similarity coefficient with map reduce for entity pairs in wikipedia," *Wikipedia Similarity Team*, pp. 1–18. Available at: http://citeseerx.ist.psu.edu/viewdoc/download?doi=10.1.1.168.5695&rep=rep1&typ e=pdf.

Besta, M., et al. (2020) "Communication-efficient jaccard similarity for high-performance distributed genome comparisons," In *2020 IEEE International Parallel and Distributed Processing Symposium (IPDPS)*, pp. 1122–1132. doi: 10.1109/ IPDPS47924.2020.00118.

Braat, L. C. and de Groot, R. (2012) "The ecosystem services agenda:bridging the worlds of natural science and economics, conservation and development, and public and private policy," *Ecosystem Services*, 1(1), pp. 4–15. doi: 10.1016/j.ecoser.2012.07.011.

Brouwer, R., et al. (2013) "A synthesis of approaches to assess and value ecosystem services in the EU in the context of TEEB," *TEEB follow-up study for Europe*, (May), p. 144.

Burke, L. and Maidens, J. (2004) *Burke and Maidens (2004) Reefs at risk in the Caribbean.*

Burkhard, B., et al. (2009) "Landscapes' capacities to provide ecosystem services - A concept for land-cover based assessments," *Landscape Online*, 15, pp. 1–22. doi: 10.3097/ LO.200915.

Comberti, C., et al. (2015) "Ecosystem services or services to ecosystems? Valuing cultivation and reciprocal relationships between humans and ecosystems," *Global Environmental Change*, 34, pp. 247–262. doi: 10.1016/j.gloenvcha.2015.07.007.

Costanza, R., et al. (1997) "The value of the world's ecosystem services and natural capital," *nature. Nature Publishing Group*, 387(6630), pp. 253–260.

Costanza, R., et al. (2014) "Changes in the global value of ecosystem services," *Global Environmental Change*, 26, pp. 152–158. Elsevier.

Daily, G. C., et al. (2009) "Ecosystem services in decision making: time to deliver," *Frontiers in Ecology and the Environment*, 7(1), pp. 21–28. John Wiley & Sons, Ltd. doi: 10.1890/080025.

Duarte, G. T., Ribeiro, M. C. and Paglia, A. P. (2016) "Ecosystem services modeling as a tool for defining priority areas for conservation," *PloS One*, 11(5), p. e0154573. Public Library of Science

Eigenbrod, F., et al. (2010) "Error propagation associated with benefits transfer-based mapping of ecosystem services," *Biological Conservation*, 143(11), pp. 2487–2493. Elsevier Ltd. doi: 10.1016/j.biocon.2010.06.015.

Farrell, T. A. and Marion, J. L. (2002) "The protected area visitor impact management (PAVIM) Framework: A simplified process for making management decisions," *Journal of Sustainable Tourism*, 10(1), pp. 31–51. doi: 10.1080/09669580208667151.

Fisher, B., Turner, R. K. and Morling, P. (2009) "Defining and classifying ecosystem services for decision making," *Ecological Economics*, 68(3), pp. 643–653. doi: 10.1016/j. ecolecon.2008.09.014.

Grêt-Regamey, A., et al. (2016) "Review of decision support tools to operationalize the ecosystem services concept," *Ecosystem Services*, 26, pp. 306–315. doi: 10.1016/j.ecoser.2016.10.012.

Haines-Young, R. and Potschin, M. (2018) "CICES V5. 1. Guidance on the Application of the Revised Structure," *Fabis Consulting*, p. 53.

Kreuter, U. P., et al. (2001) "Change in ecosystem service values in the San Antonio area, Texas," *Ecological Economics*, 39(3), pp. 333–346. doi: 10.1016/S0921-8009(01)00250-6.

Li, R.-Q., et al. (2007) "Quantification of the impact of land-use changes on ecosystem services: A case study in Pingbian County, China," *Environmental Monitoring and Assessment*, 128(1), pp. 503–510. doi: 10.1007/s10661-006-9344-0.

Li, X., et al. (2014) "How important are the wetlands in the middle-lower Yangtze River region: An ecosystem service valuation approach," *Ecosystem Services*, 10, pp. 54–60. Elsevier. doi: 10.1016/j.ecoser.2014.09.004.

Liu, S., et al. (2010) "Valuing ecosystem services," *Annals of the New York Academy of Sciences*, 1185(1), pp. 54–78. John Wiley & Sons, Ltd. doi: 10.1111/j.1749-6632.2009.05167.x.

MEA, 2005. *Ecosystems and Human Well-Being: Synthesis*. Millennium Ecosystem Assessment. Island Press, Washington, DC.

Nelson, E., et al. (2009) "Modeling multiple ecosystem services, biodiversity conservation, commodity production, and tradeoffs at landscape scales," *Frontiers in Ecology and the Environment*, 7(1), pp. 4–11. doi: 10.1890/080023.

Pandeya, B., et al. (2016) "A comparative analysis of ecosystem services valuation approaches for application at the local scale and in data scarce regions," *Ecosystem Services*, 22, pp. 250–259. doi:10.1016/j.ecoser.2016.10.015.

Polasky, S., et al. (2011) "The impact of land-use change on ecosystem services, biodiversity and returns to landowners: A case study in the state of minnesota," *Environmental and Resource Economics*, 48(2), pp. 219–242. doi: 10.1007/s10640-010-9407-0.

Potschin, M. and Haines-Young, R. (2013) "Landscapes, sustainability and the place-based analysis of ecosystem services," *Landscape Ecology*, 28(6), pp. 1053–1065. doi: 10.1007/s10980-012-9756-x.

Richmond, A., Kaufmann, R. K. and Myneni, R. B. (2007) "Valuing ecosystem services: A shadow price for net primary production," *Ecological Economics*, 64(2), pp. 454–462. doi: 10.1016/j.ecolecon.2007.03.009.

Ricke, K., et al. (2018) "Country-level social cost of carbon," *Nature Climate Change*, 8(10), pp. 895–900. Nature Publishing Group

Rousseau, R. (1998) "Jaccard similarity leads to the Marczewski-Steinhaus topology for information retrieval," *Information Processing & Management*, 34(1), pp. 87–94. doi: 10.1016/S0306-4573(97)00067-8.

Sannigrahi, S., et al. (2018) "Estimating global ecosystem service values and its response to land surface dynamics during 1995–2015," *Journal of Environmental Management*, 223. doi: 10.1016/j.jenvman.2018.05.091.

Sannigrahi, S., et al. (2019) "Ecosystem service value assessment of a natural reserve region for strengthening protection and conservation," *Journal of Environmental Management*, 244. doi: 10.1016/j.jenvman.2019.04.095.

Sannigrahi, S., Zhang, Q., Joshi, P. K., et al. (2020) "Examining effects of climate change and land use dynamic on biophysical and economic values of ecosystem services of a natural reserve region," *Journal of Cleaner Production*, 257, p. 120424. doi: 10.1016/j. jclepro.2020.120424.

Sannigrahi, S., Zhang, Q., Pilla, F., et al. (2020) "Responses of ecosystem services to natural and anthropogenic forcings: A spatial regression based assessment in the world's largest mangrove ecosystem," *Science of the Total Environment*, 715, p. 137004. doi: 10.1016/j. scitotenv.2020.137004. Elsevier B.V.

Schmidt, J. P., Moore, R. and Alber, M. (2014) "Integrating ecosystem services and local government finances into land use planning: A case study from coastal Georgia," *Landscape and Urban Planning*, 122, pp. 56–67. doi:10.1016/j.landurbplan.2013.11.008.

Scolozzi, R., et al. (2014) "Ecosystem services-based SWOT analysis of protected areas for conservation strategies," *Journal of Environmental Management*, 146(2014), pp. 543–551. doi: 10.1016/j.jenvman.2014.05.040. Elsevier Ltd.

Snell, M. (2016) "Review of Ecosystem Services Valuation Tools by," (September).

Tianhong, L., Wenkai, L. and Zhenghan, Q. (2010) "Variations in ecosystem service value in response to land use changes in Shenzhen," *Ecological Economics*, 69(7), pp. 1427–1435. doi: 10.1016/j.ecolecon.2008.05.018.

TEEB, 2010. *The Economics of Ecosystems and Biodiversity: Ecological and Economic Foundations*. Kumar, P. (coordinating lead author: de Groot, R.) Earthscan, London & Washington.

Tomaselli, V., et al. (2013) "Translating land cover/land use classifications to habitat taxonomies for landscape monitoring: A Mediterranean assessment," *Landscape Ecology*, 28(5), pp. 905–930. doi: 10.1007/s10980-013-9863-3.

Uddin, M. S., et al. (2013) "Economic valuation of provisioning and cultural services of a protected mangrove ecosystem: A case study on Sundarbans Reserve Forest, Bangladesh," *Ecosystem Services*, 5, pp. 88–93. doi: 10.1016/j.ecoser.2013.07.002.

Vigerstol, K. L. and Aukema, J. E. (2011) "A comparison of tools for modeling freshwater ecosystem services," *Journal of Environmental Management*, 92(10), pp. 2403–2409. doi: 10.1016/j.jenvman.2011.06.040.

Vitousek, P. M., et al. (1997) "Human domination of earth's ecosystems," *Science*, 277, pp. 494–499. doi: 10.1126/science.277.5325.494.

Vo, Q. T., et al. (2012) "Review of valuation methods for mangrove ecosystem services," *Ecological Indicators*, 23, pp. 431–446. doi: 10.1016/j.ecolind.2012.04.022.

Xie, G., et al. (2017) "Dynamic changes in the value of China's ecosystem services," *Ecosystem Services*, 26, pp. 146–154. doi: 10.1016/j.ecoser.2017.06.010.

Xie, G. D., et al. (2008) "Expert knowledge based valuation method of ecosystem services in China," *Journal of Natural Resources*, 23(5), pp. 911–919.

Zhao, B., et al. (2004) "An ecosystem service value assessment of land-use change on Chongming Island, China," *Land Use Policy*, 21(2), pp. 139–148. doi: 10.1016/j.landusepol.2003.10.003.

9 An Appraisal of Tourism Industry as Alternative Livelihood in Indian Sundarbans

Rituparna Hajra
Polba Mahavidyalaya

Tuhin Ghosh
Jadavpur University

CONTENTS

9.1   INTRODUCTION

Tourism industry is well known for its contribution to global Gross Domestic Product (GDP) which is around USD 8.9 trillion (10.3% to total GDP) in the year 2018 (World Travel & Tourism Council 2019). It brings economic value through revenue generation and foreign money exchange to many countries. This sector has significant direct and indirect impact on country's economy. Environment friendly practices in tourism industry also benefit the natural environment along with employment provision and cultural exchange (Ramgulam et al. 2012). The development and prosperity of this economic sector strongly depend on the environmental and socio-cultural resources (Gebhard et al. 2007). Tourism is now considering growing civilian industry which has almost 330 million jobs contribution to employment sector in year 2018 (World Travel & Tourism Council 2019). The 5% increasing trend per annum in tourism industry sector in recent decades is showing that many countries have considered tourism as an important source of income (Trumbic 2005; Williams and Shaw 1988). As a result, tourism is becoming an important element to upgrade and continuously stimulate the economy of local area. Globally tourism industry has attracted capital investment of USD 948 billion (4.3% of total investment) in 2019 (World Travel & Tourism Council 2019). Foreign visitors' spending is the main source of profit in tourism industry. Due to this increasing pressure of tourists many tourist sites developed intense infrastructure, superstructure and facilities which sooner or later, seriously created impact on the environment, thus creating a conflict situation (Casagrandi and Rinaldi 2002). Thus, sustainable tourism is necessary for long-term viability of tourism industry in certain place (Garrod and Fyall 1998). The goal of sustainable tourism was set as the protection, enhancement and improvement of the various components of man's environment for the harmonious development of tourism (Danda 2007). The travel and tourism sector holds strategic importance in the Indian economy providing several socio-economic benefits including provision of employment, income and foreign exchange, development or expansion of other industries such as agriculture, construction, handicrafts, etc. In India tourism industry generated 9.2% of India's GDP in 2018 and supports 8.7% of total employment (World Travel & Tourism Council 2019). West Bengal shares 6.3% and 4.8% of foreign and domestic tourists among India's total (Ministry of Tourism Report 2018).

The world's largest mangrove ecosystem Sundarbans, a UNESCO World Heritage Site, located at the South eastern tip of the 24 Paraganas district is one of the most attractive and alluring places for both foreign and domestic tourists. Sundarbans eco-tourism offers three tourism options including wildlife tourism, beach tourism and religious tourism (Hazra et al. 2014; World Bank Report 2019). People attracted to Sundarbans Tiger Reserve, Kalas Beach, Lothian Island, Bhagabatpur Crocodile Farm, Pakhiralay, Henry Island, Bakkhali beach, etc. There is also non-forest tourism in the Sundarbans which are mainly of religious significance. Among these Sagar Island attracts maximum number of tourists as pilgrimage tourism. There are other places of attraction, like Jatar Deul Temple near Raidighi, Khana-Mihirer Dhibi at Berachampa and Chandraketugarh, etc. (Dasgupta 2008). Traditional Indian cultural heritage gave importance to visit various TIRTHA or pilgrim centre to earn virtue (Kundu 2012). India is famous for its temples and that is the reason that among the different kinds of tourism in India, pilgrimage tourism is increasing most rapidly (Arunmozhi and Panneerselvam 2013). Even though it accounts for more than 90% of domestic tourism in India, it is not readily acknowledged in current public policy (Shinde 2006). This study assesses the tourism profile of 13 blocks of South 24 Parganas which are part of the Indian Sundarbans Delta (ISD) with a special emphasis on case study from Gangasagar. The focus of this study Gangasagar is a famous pilgrim centre where people go to earn virtue and for scenic attraction (Hajra and Ghosh 2014). Gangasagar is situated at the confluence of Ganga river in Sagar Island. Sagar Island is a "CD-Block", an administrative area with a cluster of 42 mouzas or villages which is the largest island in ISD with 2,06,844 population (Census Report 2011) (Figure 9.1). Mythological importance of this region is due to its temple of the great saint Kapil Muni who according to myth was Vishnu had taken birth as per the wish of Kardam muni as his son. In the winter, pilgrims from all over India celebrate the holy dip at the confluence of the River Ganges at Bay of Bengal near Kapil Muni Ashrama on the dawn of the last day of Poush (Makar Sankranti) as

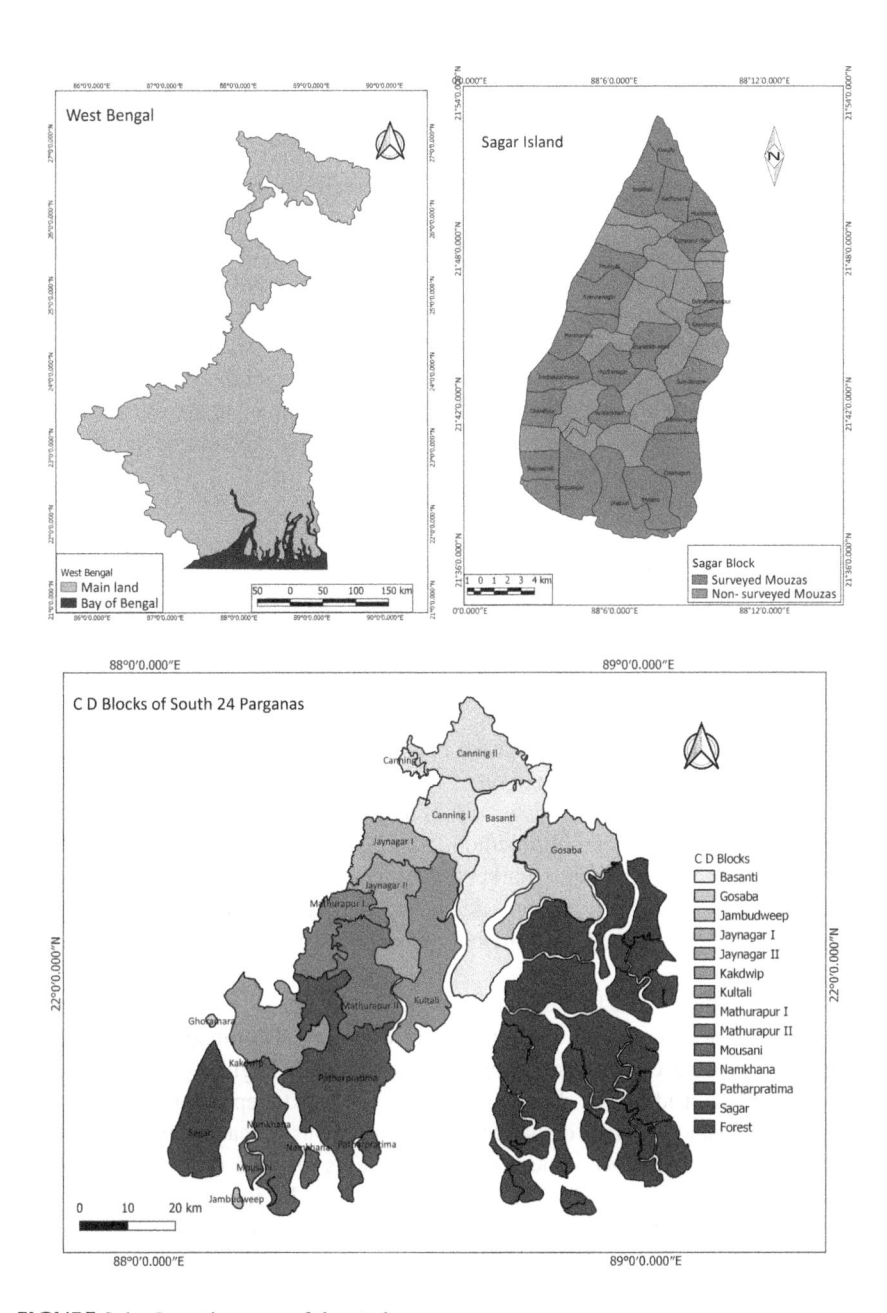

FIGURE 9.1   Location map of the study area.

per Bengali date system (second week of January). Over a few lakhs of pilgrims from all over India visits the place every year. A huge infrastructure has to set each year to cater to this large influx. The local population take advantage of the infrastructural development due to the presence of Kapil Muni temple which includes a large rural hospital and a wide all-weather road running through the length of the island (WWF 2011). Annual monetary involvement during the Gangasagar Mela (GSM) is about 20 million rupees of which only 20% is spent on permanent assets. An estimated 67,500 man-days of jobs are created in the informal sector during the occasion of GSM (Basak 2004). The local people get themselves involved directly or indirectly with the GSM through various types of business such as transport, marketing, food and lodging, etc., which helps to boost up their economy (Hajra, Mitra, and Ghosh 2012). The tourist flow throughout the year is very negligible, while a huge number of pilgrims (tourists) use to visit at a particular time, undoubtedly overshoot the carrying capacity of this island. Tremendous impact of huge number of pilgrims is negatively affecting this coast, which leads to necessity of sustainable tourism to protect the environment from degradation (Hajra, Mitra, and Ghosh 2012). This paper is aimed to assess the scope of tourism as alternative source of income, which can contribute to develop a sustainable economy of this region.

9.2 DATA AND METHODS

This paper uses data from both primary and secondary sources. Secondary data have been procured from published literature, Block level report and District Statistical Handbook. Random sampling was used for this survey. Agricultural production data, workers, tourism-related details for the 13 blocks of South 24 Parganas have been analyzed from secondary source data set. Primary data have been collected through household surveys, hotel survey, market and tourist survey conducted for the purpose of this study through direct interviews within Sagar Island. 617 households have been surveyed using random sampling method from 22 mouzas of Sagar Block. Market survey has been done from major market areas including GSM Prangan, Kachuberia, Chemaguri and Kalibazar. Temporary and permanent shops have been surveyed during and after GSM. Around 50 tourists have been asked questions to assess their satisfaction level.

All the data have been filtered, analyzed and interpreted, along with the cartographic techniques to represent the findings of this study. Dominant and distinctive function analyses have been done using occupational data. Highest percentage share of workers has considered as dominant activity. The next step involves calculation of the mean (X) and standard deviation (SD) for all towns for each functional group. Any town which then shows a percentage of workers of more than mean plus one standard deviation is said to be distinctive in that particular function. The degree of distinctiveness of a function may be expressed in the following way:

$$X + 1SD \text{ to } X + 2SD \qquad (9.1)$$

$$X + 2SD \text{ to } X + 3SD \qquad (9.2)$$

$$\text{over } X + 3S \qquad (9.3)$$

The five-point Likert scale has been used to collect opinion of the respondents regarding different attributes related to the tourism. There are five options for the degree of satisfaction for each question: very poor, poor, moderate, good and very good. The score associated with the degrees of satisfaction is from 1 to 5. Very poor, poor, moderate, good and very good responses are allotted 1, 2, 3, 4 and 5 numbers, respectively. Average satisfaction score (mean score) and some other descriptive statistics (median, mode, standard deviation) for each selected attribute under study are computed. If the mean score of any attribute is 1–2.5 then we can say that the visitors, on an average, are not satisfied with that attributed. On the other hand, if the score lies between 2.51 and 5 then we can assume that visitors are satisfied with that attribute. Ultimately, an overall unweighted or equal weighted composite index, say satisfaction index, based on the mean satisfaction score of each attribute is evaluated by taking an average of the mean scores.

9.3 LIVELIHOOD ISSUES IN THE ISLANDS OF INDIAN SUNDARBANS DELTA

The Ganga-Bramhaputra-Meghna river system carries sediment load composed of sand, silt and clay, and drains through an alluvial plain on its course pouring into the Bay of Bengal, while at the mouth, tidal forces obstruct sediment delivery to the sea, resulting in sediment back-load and deposition along the course, thus helping island formation (Das 2006). The drainage network and dynamic flow pattern of the tidal water, along with erosion and land accretion, make the geomorphological set-up of the area very complex. Agriculture is the main source of income of the inhabitants of ISD. 69% of the non-forest land of this district has been used for agricultural practices in which 39% of total workers is dependent. Aman paddy is the prime produce. Analysis of District Statistical Handbook data shows that there is a decline in paddy production during 2000–2001 to 2012–2013 (Figure 9.2).

Agricultural land loss as a consequence of erosion imposes severe challenges to the farmer community. Cyclone 'Aila' in 2009, 'Sidar' in 2007, 'Bulbul' in 2019 and

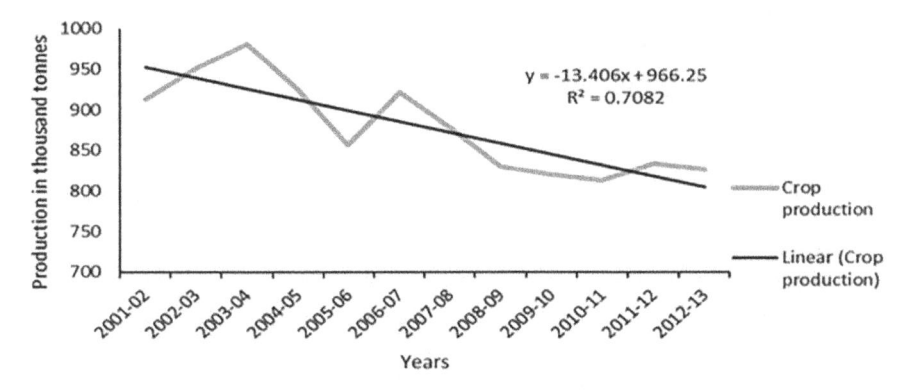

FIGURE 9.2 Trends in paddy production in South 24 Parganas District. (From District Statistical Handbook, South 24 Parganas.)

'Amphan' in 2020 were the most severe climatic disasters in recent time hit the Indian Sundarbans. Respondents often expressed that after the 'Aila' the productivity of the agricultural land decreased to a great extent and, moreover, due to inundation of saline water most of the land became unsuitable for cultivation, even up to 3–4 years. Cultivation was used to be dominant function but now showing a declining trend in most of the blocks of Indian Sundarbans part under South 24 Parganas (Figure 9.3).

9.4 TOURISM PROFILE OF INDIAN SUNDARBANS

The Sundarbans is an archipelago of several hundred islands, spread across 9,630 km^2 in India and 16,370 km^2 in Bangladesh. In India it extends over two districts – 13 blocks in South 24 Parganas and 6 blocks in North 24 Parganas among which the area under South 24 Parganas are the focus of this study. The Sundarbans is of universal importance for globally endangered species including the Royal Bengal Tiger, Ganges and Irawadi dolphins, estuarine crocodiles and the critically endangered endemic river terrapin (Batagur baska). It is the only mangrove habitat in the world for Panthera tigris tigris species (UNESCO). Sundarbans Tiger Reserve encompasses a total area of 2584.89 km^2. Areawise this Reserve is divided into three parts: Sajnekhali Wildlife Sanctuary (362.40 km^2); Sundarbans National Park (1330.10 km^2); Reserve Forest (892.43 km^2); Critical Tiger Habitat (1699.62 km^2); Buffer Area (885.27 km^2) (Sundarbans Tiger Reserve 2017). Tourists from Kolkata can reach Sonakhali via express buses. From Sonakhali, tourist trips are arranged to Sundarbans Tiger Reserve, Kalas Beach, Lothian Island Sanctuary, Bhagabatpur Crocodile Farm, Pakhirala, etc. Watch towers with freshwater pond exist at following places for the facility of wildlife viewing: Sudhanyakhali, Sajnekhali, Netidhopani, Jhingekhali, Burirdabri, Gabbani, Bakkhali, Dhanchi and Lothian Island (Hazra et al. 2014) (Figure 9.4). The Lothian Island Sanctuary (Bhagatpur Crocodile farm) attracts around 30,000 tourists while Sundarban Tiger Reserve gets about 50,000 tourists per year (MOEF 2009). Apart from wildlife tourism religious tourism is also there in Sundarbans including Gangasagar, Jater deul.

9.4.1 NATURE AND WILDLIFE ATTRACTIONS

1. **Sajnekhali Tiger Reserve:** It is one of the most popular spots in Indian Sundarbans. This location has a watchtower, a mangrove interpretation centre, a Bono Bibi temple as well as a crocodile park. The kingdom of birds (Pakhiralaya) at Sajnekhali attracts bird watcher for the sight of seven colourful species of Kingfisher, white bellied Sea Eagle, Plovers, Lap-Wings. Adjacent to the interpretation centre, there exists a tourist lodge managed by West Bengal Tourism Development Corporation (World Bank Report 2019). There are other hotels and private lodges at Pakhiralaya village just opposite to the Sanctuary gateway.
2. **Sudhanyakhali Watchtower:** In the deeper part of forest within sanctuary area on the bank of Sudhanyakhali river this watchtower is situated. This is popular for tiger sighting.

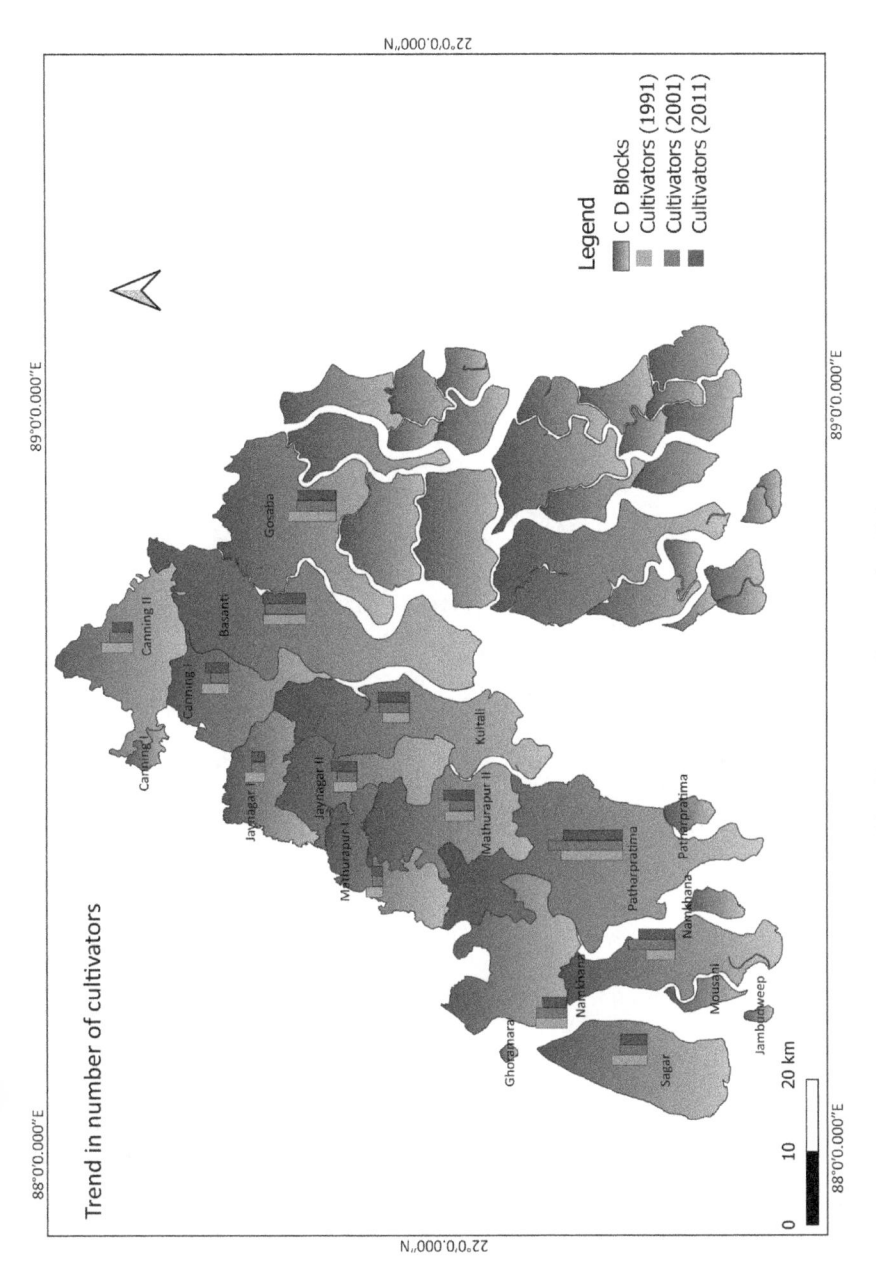

FIGURE 9.3 Trend in numbers of cultivators during 1991–2011. (From Census of India.)

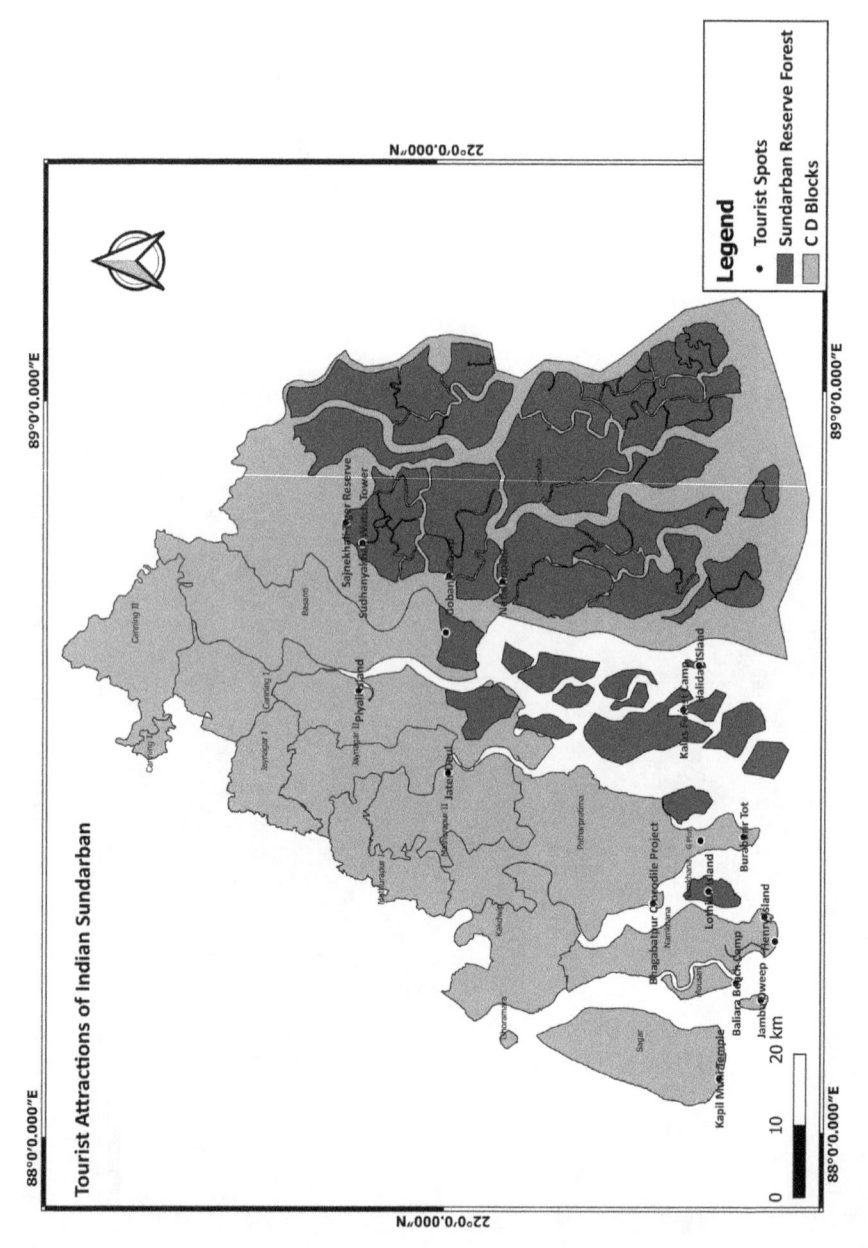

FIGURE 9.4   Tourists attraction of Indian Sundarbans.

3. **Dobanki:** The spot has a watchtower which can be approached by a canopy walk, well protected by fencing, located at a height of 20 ft from the ground. There is a sweet water pond nearby which makes it easy to spot animals. There is a deer park and breeding centre along with a watchtower.

4. **Jharkhali:** This is the first tiger rescue centre of India. There is a Mangrove Interpretation Centre established by Sundarbans Wildlife Society (World Bank Report 2019). Though the Sundarbans Development Board has a two-room rest house here but tourists generally come here for a day visit.

5. **Bhagabatpur Crocodile Project:** This spot hosts India's only crocodile project. There are estuarine species of crocodile and Tortoise. It is situated near Lothian Island and can be reached from Patharpratima village by road. A mangrove interpretation centre, crocodile hatchery, deer park and forest range office are situated here.

6. **G Plot:** This place offers village life within the confluence of river and sea. It is becoming an attractive weekend destination.

7. **Buraburir Tat:** This place is gradually getting popular among tourist due to its pristine beach.

8. **Lothian Island:** It is a wildlife sanctuary which includes estuarine crocodiles, olive ridley sea turtles, spotted deer, jungle cats and rhesus macaques. It is situated at the section of Saptamukhi and Bay of Bengal.

9. **Haliday Island:** A beautiful island with wide beaches and dense mangrove forests is also the place of Sundarbans where barking deer is found.

10. **Kalas Forest camp:** Lying in the estuary of river Matla; this island is a breeding ground for olive ridley turtles during the winter season. It is also popular among bird watchers. Kalas camp is situated near the Kalas islands where tourist hut and forest guard establishments are made. Tourists can spend night here to enjoy nocturnal beauty of nature.

11. **Bakkhali-Frazerganj:** One of the rare beaches that offer both sunrise and sunset view is a favourite attraction for weekend trips; a well-developed and well-connected holiday spot under Namkhana Block.

12. **Henry Island:** Located near to Bakkhali, this island gives tourists both experience of the island and the sea. The place is also home to millions of red crabs that are visible on the beach. Government accommodation is available here.

13. **Baliara Beach:** It is situated at the southern part of Mousani, a small island, and accessible from Sagar and Namkhana. Apart from its peaceful village life and natural beauty this spot offers accommodations in beachside tents which add to the attraction.

14. **Jambu Dweep:** This is an uninhabited island located at a very small distance from Bakkhali. This is being slowly developed as a tourist attraction. However, as of now there are no accommodation facilities.

15. **Piyali Island:** A silted island, situated by Piyali River, is ideal for day-outs. It is located very near to Kolkata as is best reached by road.

9.4.2 Religious Attractions

1. **Netidhopani:** This place was mentioned in Bengali Folklore (Manasamangal) and worshiped for goddess Manasa (the goddess of snakes). The ruins of a 400-year-old Shiva temple and local tales give this place different flavour to the tourists.
2. **Jater Deul:** Located in the village PurbaJata, Archeological Monument was constructed in 975 BC, presently treated as Shiva Temple by the local community. Chaitra Sankranti Mela is famous which attract many locals. This place is failing its tourism popularity due to lack of promotion and management.
3. **Bono Bibi Puja and Mela:** Bono bibi is the guarding deity of the forest and is believed to protect human beings from tiger attacks. She is worshiped by both Hindus and Muslims inhabitants of Sundarbans who are largely dependent on forests for their daily bread. The government is taking initiatives to popularize this festival as an important aspect of Sundarbans tours.
4. **Gangasagar:** The famous Bengali saying 'Sab tirtha bar bar Gangasagar ekbar' amply signifies the importance of the place. The Kapil Muni Ashram is located on Sagar Island. Additionally, every year during Makar Sankranti lakh of pilgrims gather to take holy deep in sea and then visit the temple. This gathering is popularly known as Ganga Sagar Mela which is the second largest congregation after Kumbh mela.

Being rich in natural and cultural significance Tourism in Sundarbans can be an important tool to develop and improve the socio-economic status of the people of Sundarbans as an alternative option of traditional livelihoods. Tourists' arrival is limited mostly during winter for which people here consider tourism-related income as 'additional' revenue (Saville 2001). Sundarbans is an attraction for both domestic and foreign tourists; both are showing an increasing trend in tourist inflow (Hazra et al. 2014). Accommodations are available from West Bengal Tourism Department, private and local organizations. There are also luxury launches run by WBTDC, where tourists can stay at night. Boat trip through the water channels within the forest is the main attraction of Sundarbans. Boat tour and boat carrying population are increasing in number from year 2000 to 2001 (Table 9.1). The estimated trend of total tourist inflow and earnings from tourism industry is increasing during 2000–2013 (Figure 9.5). This improving tourism scenario in Sundarbans can be a potential livelihood for the local people. This paper tries to assess the religious/ pilgrimage tourism potential of Indian Sundarbans focusing on the Gangasagar of Sagar Island and its potential to grow as an industry to support the traditional economy. It is evident from the above discussion that no single livelihood is sufficient to augment a year-long livelihood at Sundarbans, rather people are forced to adopt a mixed pattern of occupation for sustenance.

TABLE 9.1
Progress in Tourism in the District of South 24 Parganas

Year	No. of Launch	No. of Trips Conducted	No. of Tourists Carried	No. of Tourist Lodge
2000–2001	3	94	3621	4
2001–2002	3	146	4690	4
2002–2003	4	108	2649	3
2003–2004	4	121	2992	3
2004–2005	5	316	5673	3
2005–2006	4	110	3667	3
2006–2007	4	N/A	3500	3
2007–2008	4	180	4000	3

Source: Deputy Director of Tourism, Govt. of WB; West Bengal Tourism Development Corp. Ltd.

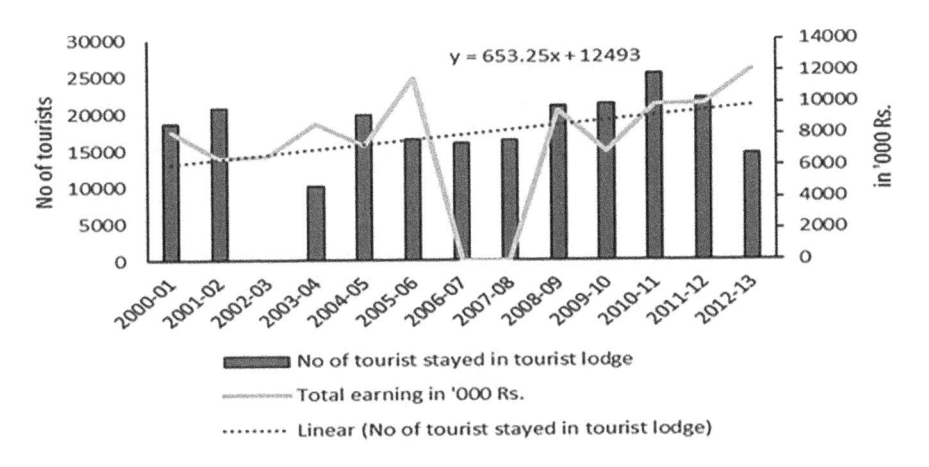

FIGURE 9.5 Tourist inflow and earnings from tourism industry, Indian Sundarbans (District Statistical Handbook) (Data are not available – Total Earning for the year 2006–2007 & 2007–2008; No of tourist year 2002–2003).

9.5 PILGRIMAGE TOURISM: AN ALTERNATIVE LIVELIHOOD

GSM is the largest fair in West Bengal and this activity involves almost all the Government departments, local bodies, scores of non-government voluntary organizations for a period of over a fortnight. It is worth mentioning that one of the most major sources of the total community from the annual "GSM". In the Sagar Island larger percentage of local residents are in opinion that the tourism is a lucrative source of income. Dominant and distinctive function analysis has been done to assess the occupational structure which shows that daily/agricultural labourer outnumbers the other services (Figure 9.6). Though agriculture is dominant, but order of distinctive function shows that transport and tourism and tourism-related

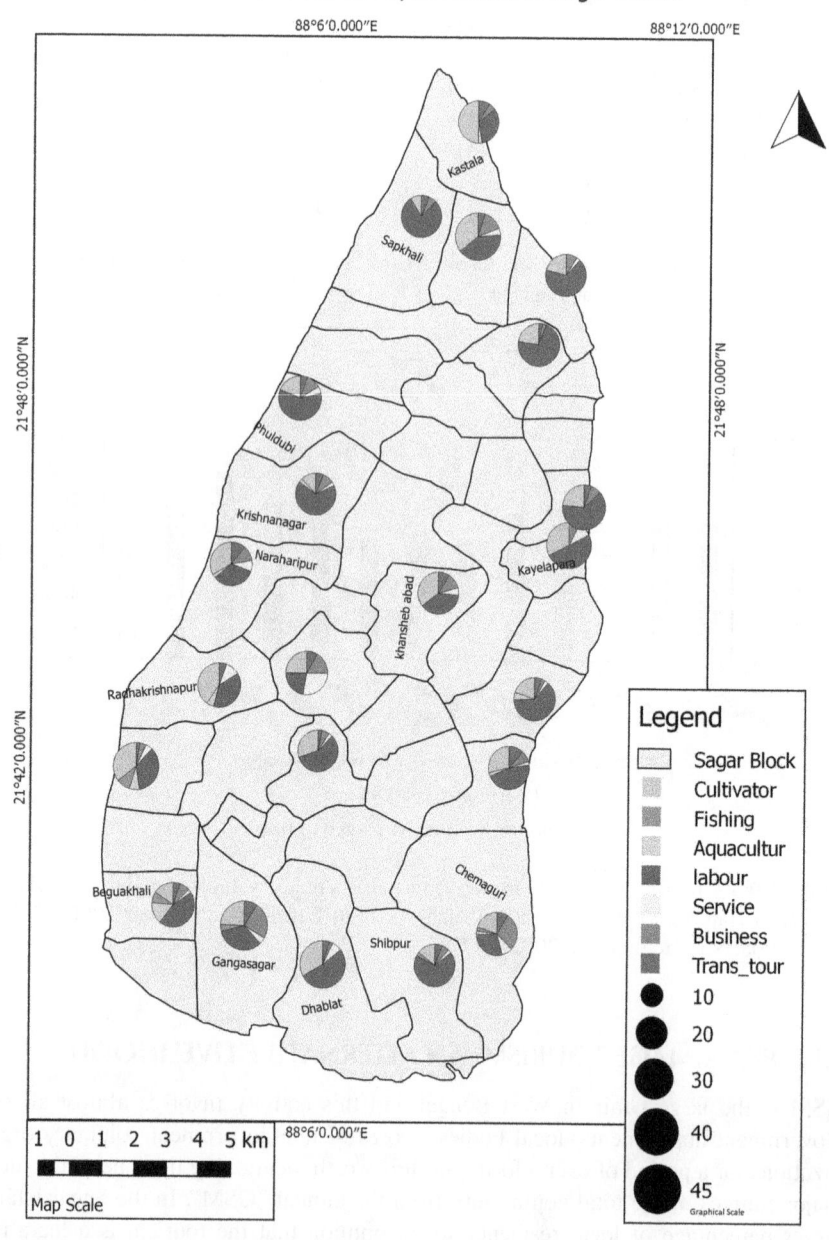

FIGURE 9.6   Dominant Functions in surveyed mouzas of Sagar Island. (Primary data.)

business occupation have significant impact on some parts of Sagar Island including Gangasagar, Kachuberia. Table 9.2 is showing the functions under three distinctive orders from 22 surveyed mouzas of Sagar Island; derivation of scale is mentioned in method section.

The local people are appointed as skilled, unskilled and semi-skilled workers in construction of road, Yatrinibas, and temporary residence like Hoghla cottage, latrine, bathroom, etc., with enhanced opportunity to earn cash. Villagers get involved in construction work at preparatory phase. They earn USD 62 per month in average during that phase. It appears from the survey that most of of business circuits are concentrated near Kochubria, Chemaguri, GSM Prangan, Sridham and Kali Bazar during the festival. Kochuberia and Chemaguri are two entry points of Sagar. GSM Prangan is the main place where the festival occurs whereas Sridham, Kali Bazar, is the marketplace situated nearer to the main function. There are several temporary tea stalls and rice hotels set-up by locals during the Ganga Sagar Mela (Figure 9.7). In the main centres of attraction, the number of temporary stall/'dala-mala dokan' is the highest. Gangasagar-Bakkhali Development Authority (GBDA), established in 2013, has now prepared a separate complex for dala-mala shops and tried to make these temporary constructions into a permanent one. The owners have informed that the tourist and pilgrims are in search of hotels and tea stalls especially just after offering 'puja' to the temple. To meet the demand gradually some temporary hotels and tea stalls are usually set-up by the local villagers and the people of the nearby villages get benefitted by cash earning during that period. Survey statistics show that during mela week average business of each temporary shop is near about Rs. 5000 (Figure 9.8). Due to the huge number of pilgrims permanent business units can also earn extra money during GSM festival. A comparison between the monthly income and the income of this particular period has shown an increase of five times in this particular time than the rest of the year (Figure 9.9).

Accessibility and transport linkages are most influential factor behind the spread of any tourist centre. Gangasagar can be approached from Kolkata on road and railway but there is no direct link; three steps are required to reach Gangasagar. It is around 90 km from Kolkata through NH 6 and SH 79 up to Harwood Point near Kakdwip where ferry service is available for crossing over the Muriganga River to reach Kachuberia (the northern tip of the island). Kachuberia to Gangasagar is linked by road where bus and cars are the available transportation. Gangasagar bus terminus is situated at Sridham near mela road no 1. It is found from the field survey that around 20 buses are there in service throughout the year and the number increases up to around 40/50 during the fair. An additional ferry ghat has been used during mela for VIP services. One of the major hindrances behind the growth of tourism in Sagar Island is its location and poor accessibility. From the ground survey it has found that the ferry services now become limited due to the siltation in Muriganga. Only high tide time is favourable for ferry which has shrunk to 4–5 hours per day. Construction of bituminous road from Benuban link to Chemaguri crossover by GBDA has now makes Benuban to Namkhana ferry service more accessible. The road map has been prepared to show the accessibility of Gangasagar (Figure 9.10). Recently a helicopter service has started from Kolkata to Gangasagar; a helipad has been made near mela ground. A leisure cruise trip has also started very recently from Babughat, Kolkata to Kachuberia along the Hooghly river.

TABLE 9.2
Distinctive Order of Functions in Sagar Island

Scale of Distinctiveness of Different Functions

Mouzas	I Order Distinctiveness	II Order Distinctiveness	III Order Distinctiveness
Kastala	Aquaculture, Transport and Tourism	Cultivator	-
Sapkhali	Transport and Tourism	Daily Labour	-
Kachuberia	Transport and Tourism, Service, Business	Cultivator	-
Muriganga	Daily Labour and Business	-	-
Companir char	Transport and Tourism	Daily Labour	-
Phuldubi	Daily Labour, Business, Transport and Tourism, Service	-	-
Debimathurapur	Daily Labour, Business, Transport and Tourism	-	-
Kayelapara	Cultivator, Daily Labour, Business	Service	-
Sumatinagar	Aquaculture, Transport and Tourism, Daily Labour	-	-
Bankimnagar	Aquaculture, Business	-	Transport & Tourism
Rudranagar	Transport and Tourism	Business	Service
Khansheber Abad	Cultivator, Sservice, Business	Transport and Tourism	-
Krishnanagar	Aquaculture, Daily labour, Business, Transport and Tourism	-	-
Naraharipur	Cultivator, Service, Business, Aquaculture	Transport and Tourism	-
Radhakrishnapur	Aquaculture	Cultivator, Service, Transport and Tourism	-
Chandipur	Service	Cultivator, Aquaculture	Fishing
Beguakhali	Business, Transport and Tourism	-	Fishing, aquaculture
Gangasagar	Cultivator, Service, Transport and Tourism	Business	Fishing
Chemaguri	Fishing, Aquaculture, Service	Transport and Tourism	Business
Shibpur	Fishing, Daily labour, Transport and Tourism	-	-
Dhablat	Cultivator, Daily labour, Service	Transport and Tourism	-
Krittankhali	Daily labour	-	-

Source: Primary survey.

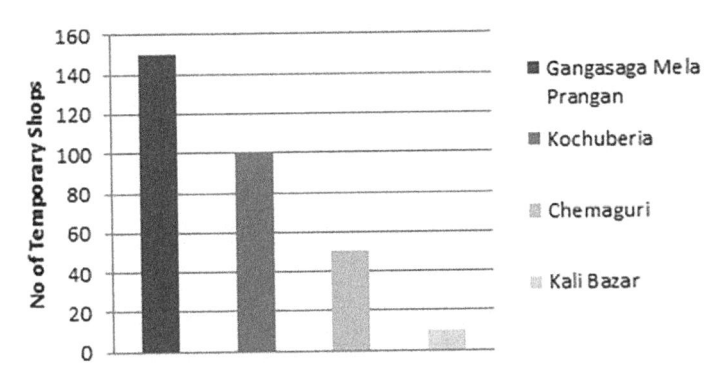

FIGURE 9.7 Number of temporary shops at four main points. (Primary survey.)

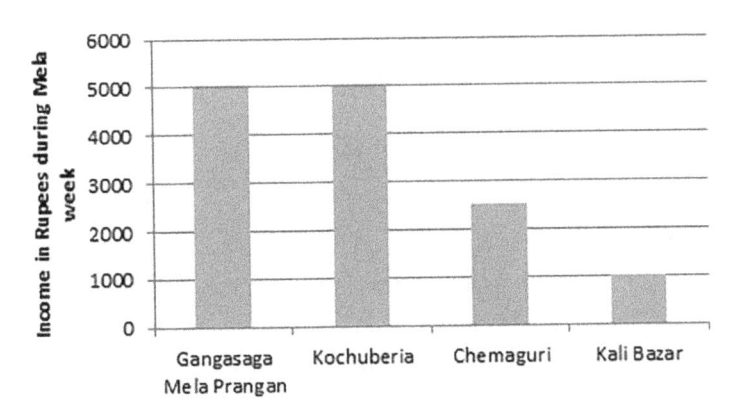

FIGURE 9.8 Income from temporary business units during Mela week. (Primary survey.)

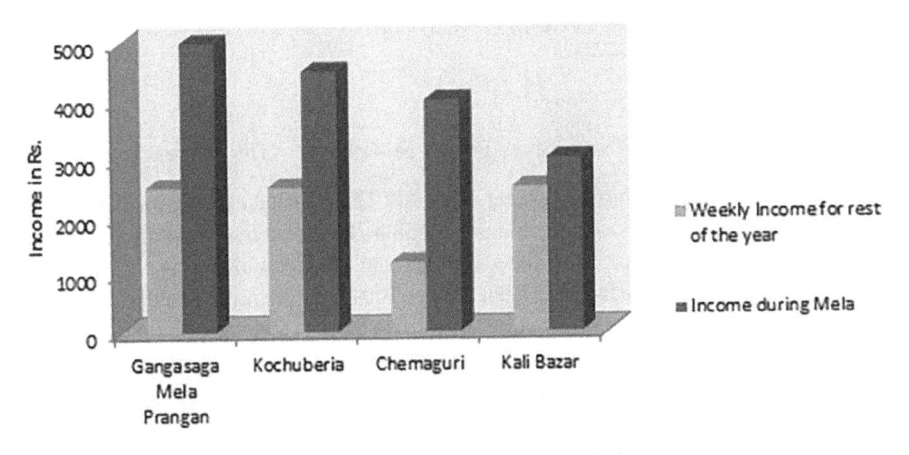

FIGURE 9.9 Comparison of weekly income and income of permanent business units during Mela week. (Primary survey.)

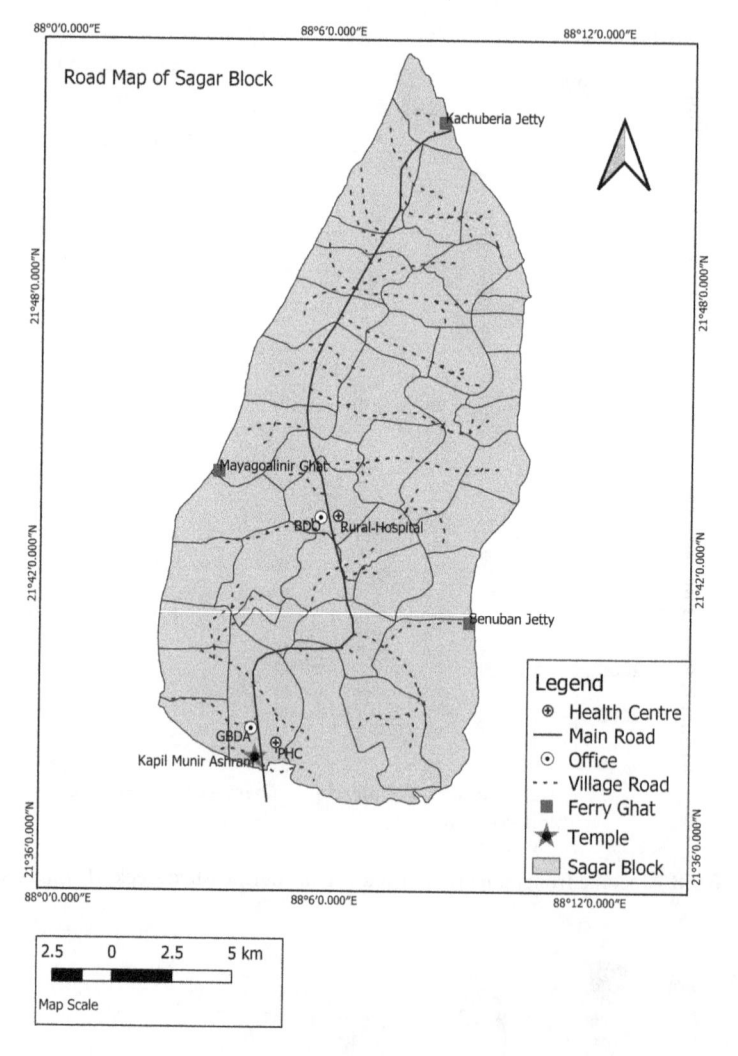

FIGURE 9.10 Road map of Sagar block. (From Sagar Community Development Block Office.)

The accommodation in Gangasagar is mostly Dharmashala maintained by different religious and business organizations including Bharat Sevashram Sangha, Ganga Bhaban, Haryana Bhavan. They meet the demand of accommodating thousands of pilgrims coming here during GSM but except Bharat Sevashram Sangha others are mainly restricted to specific community. As tourism here is mela oriented most of these Dharmashalas remain vacant or shut down at other time. The seasonal nature of tourism hampers the establishment of new hotels/ guest houses. There are few guest houses from Government sectors like Zilla Parisad Bungalow, Iggrigation Department Bungalow, and Public Health Engineering Department Bungalow, etc. But these guest houses are not available for commoners during the festival. One has to book these lodges from Kolkata through the concern department which is very complex process itself. Temporary lodges were created by Government, private owners, NGOs and other voluntary organizations.

Irrigation Department established Hotel Sinchan and the Public Health Engineering department (PHE) has three rest houses. Hotel Larika had private ownership before but in 2010 it has been undertaken by the State Tourism Department of Govt. of West Bengal in 2010 renaming it to Gangasagar Tourist Lodge. In recent years the number of lodges and guest houses is increasing since Sagar Island is getting popularity in leisure and weekend trips also. Table 9.3 is showing the major lodges and their booking details are based on the hotel survey.

Other infrastructures have developed owing to the increase in tourist arrival. To help the crowd during this festival, the West Bengal Government takes initiative to arrange for the accommodation, sanitation, food and drinking water, and all the infrastructure facilities for the tourists and pilgrims. GBDA has established a number of facilities including toilet, market complex, drinking water, rest room, beautification, lighting of the temple ground-sea beach to attract tourists. Till October, 2011, there was no grid electricity in Sagar but later electric connection helps to grow the tourism activities. A survey on the tourist infrastructural facilities has been done at three points of Sagar-Kachuberia (the main entry point), Rudranagar (capital centre) and Gangasagar (main temple and mela prangan) (Table 9.4). The analysis shows that

TABLE 9.3
Accommodation in Gangasagar

Dharamshala/Lodges/ Guest Houses/Hotels	Booking Centre	Availability	Target Community	Tariff per Room per Day (Approximately)
Bharat Sevashram Sangha	Onsite and through trustee board	Whole year	Open to all	Rs. 250
Ganga Bhavan	Through trustee board	Limited	Specific community	N/A
Haryana Bhavan	Through trustee board	Limited	Specific community	N/A
Irrigation Bungalow	Kolkata through concern department	Whole year but restricted for VIPs during mela	Officials mainly	Rs. 700– Rs. 750
PHE Bungalows	Kolkata through concern department	Whole year but restricted for VIPs during mela	Officials mainly	N/A
Zilla Parisad Bungalow	Kolkata through concern department	Whole year but restricted for VIPs during mela	Open to all	Rs. 700– Rs. 750
Hotel Larica/Gangasagar Tourist Lodge	Onsite	Whole year	Open to all	Rs. 750
Youth Hostel	Online	Whole year	Open to all	N/A
Wooden Bungalow	GBDA	Whole year	Officials	Rs. 2000–Rs. 2500
Bamboo House	District Magistrate	Whole year	Officials	Rs. 2500

Source: Field/ Hotel survey.

TABLE 9.4

Tourist Infrastructural Facilities

Places	Restaurants	Toilets	ATM	Petrol Pump	Electricity
Kachuberia	Very few	Yes (one)	No	Yes	Yes
Rudranagar	Very few	Yes (one)	No	Yes	Yes
Gangasagar	Good in number	Yes	Yes	Yes	Yes

Source: Field survey.

the development of tourist facilities could influence the tourist arrival throughout the year and spread of tourism.

Tourist satisfaction level has been estimated by descriptive statistic using ten criteria from primary survey data. The satisfaction level based on arithmetic mean has been classified into two classes – not satisfied (0–2.5) and satisfied (2.5–5). Table 9.5 indicates the satisfaction level of these criteria. As can be said from descriptive statistic tourists are not satisfied with accommodation, transportation and other points of attraction rather than Kapil Muni temple having values less than mean 2.5. As discussed above limited transportation link and accommodation hinder the growth of tourism in this study area. There are very few other attraction points that have been developed in this island which can be extended using its scenic beauty. The potentiality of expansion of tourism could be affirmed from other criteria whish shows satisfied values. Kapil Muni temple, sea beach, market and souvenir are having high satisfaction values. Respondents gave positive response for leisure/weekend trip here which indicate the possibility in increase of tourist arrival with proper promotion and management. The assessment has found overall satisfaction value is 3.066 which

TABLE 9.5

Tourist Satisfaction Level in Gangasagar

Criteria	Mean	Median	Mod	SD	Satisfaction Level
Transportation	2.30	2	2	0.78	Not satisfied
Accommodation	2.06	2	2	0.76	Not satisfied
Kapil Muni Temple	4.20	4	4	0.60	Satisfied
Market complex	3.92	4	4	0.74	Satisfied
Food	2.96	3	3	0.82	Satisfied
Tourist facilities (toilet, rest room, health centre, ATM)	2.84	3	3	0.95	Satisfied
Sea-beach	3.78	4	4	0.83	Satisfied
Other points of attraction	2.04	2	2	0.72	Not Satisfied
Crime	3.20	3	3	0.60	Satisfied
Leisure/weekend tourism	3.36	3.5	4	0.99	Satisfied
Satisfaction index (equal weighted)		3.066			Satisfied

Source: Primary data.

indicates that tourists are satisfied with this place. It can be said from the analysis that Gangasagar has the potential to establish as a popular tourist centre with useful utilization of its cultural heritage and aesthetic value.

9.6 DISCUSSION

The unique coastal ecosystem of the Sagar Island is characterized by resource combination at the interface of land and sea offering beaches, scenic beauty, rich terrestrial and marine biodiversity, diversified cultural and historic heritage, etc. (Hajra and Ghosh 2014). Gangasagar has its heritage value to the religious minded people all over India and already established as famous pilgrimage spot (Kundu 2012; Hajra, Mitra, and Ghosh 2012; Dasgupta, Mondal, and Basu 2006). Sagar Island has great aesthetic and traditional heritage value which has potential to attract the tourists for leisure and holiday-trips. This study intends to assess the potentiality of this island to grow as tourist spot and expansion of tourism activity as alternative livelihood to local inhabitants. It is evident from the declining crop production and number of cultivators that agriculture practice does not promise much in terms of attaining better well-being for the local community (Hajra and Ghosh 2016, 2018). In line with the study Kabra (2003) this study shows that the loss in agricultural production leads to increase labour migration to informal sector affecting the sustenance of the economic structure. This study assessed the economic benefit of local community from yearly GSM which shows almost five times income hike during mela period than rest of the year. This finding is similar with Basak (2004), Dasgupta, Mondal, and Basu (2006) and Shinde (2006). A huge infrastructural set-up is made to cater to this large influx which the local population can take advantage of throughout the year. This includes a large rural hospital and a wide all-weather road running through the length of the island (WWF 2011). Study has assessed the tourist facilities available in this island which seems not to be adequate to spread tourism activities whole year. This study found that with proper management strategy tourism industry might be expand as alternative livelihood for local community in Sagar Island which is in line with previous literature Kundu (2012). Distinctive function analysis shows that transport and tourism related activities have significant impact in some mouzas of Sagar Island. Tourist satisfaction level has been assessed which indicates that tourists have positive response towards this pilgrimage. This study tries to suggest an implementable and comprehensive strategy to formulate a sustainable coastal tourism according to the guidelines of Global Sustainable Tourism Criteria (GSTC).

9.7 STRATEGY FOR TOURISM INDUSTRY AS GSTC

The GSTC baseline criteria for destinations were drafted on 7th of March 2012 and updated on 4th of June 2012 (GSTC 2012) to set minimum global standards for sustainable practices. United Nations' Millennium Development Goals such as poverty alleviation, gender equity and environmental sustainability, including climate change, are the main concerns of this criterion. The GSTC was developed based on the recognized approaches including the UNWTO Destination Level Indicators, GSTC Criteria for Hotels and Tour Operators, and all other widely accepted principles and guidelines, certification criteria and indicators (GSTC 2012). The GSTC for Destinations is considered as a process to make sustainability the standard practice in all kinds f of tourism. GSTC can be considered as a basic guideline for sustainable tourism for specific location but there may be circumstances due to which a criterion is not applicable to a specific tourism destination and few criteria may have limitations depending on the nature of the location (Hajra and Ghosh 2014).

This study tries to find the applicable criteria from GSTC for Sagar Island to expand tourism here. It is found that major emphasis should be given on tourism strategy, infrastructural development and climate change adaptation, crisis management, tourist satisfaction and promotion from the GSTC (2019). First, proper infrastructure is needed to cater the tourist flow whole year. This may include increase of ferry service, buses, accommodation option, food shops, ATM, etc. The available accommodations mostly do not have onsite booking facility and complex booking process often discourage the weekend tourists. Second, delivering local economic benefits criteria should be assessed and accordingly work divisions and engagement of labour could be implemented by authorities. Appropriate measures may include levels of visitor volume, visitor expenditure, employment and investment and the distribution of economic benefits. Third, the destination encourages the retention of tourism spending in local economy through supporting local enterprises, supply chains and sustainable investment. It promotes the development and purchase of local sustainable products. Fourth, the policy should be formulated to evaluate, rehabilitate, and conserve cultural heritages and landscape. Fifth, the waste by millions of pilgrims during GSM causes an undesirable change in the physical, chemical and biological characteristics of the air, water and land which harmfully affects the health, survival or activities of humans or other living organisms. There is a need to check the growing pollution to save the environment on the part of the State Govt. as well as the local people. The waste management, water treatment, cleaning of beach and temple ground must be done regularly. Sixth, the study area is prone to natural hazards including erosion, cyclone and tidal surge. Risk reduction, crisis management and emergency response plan are needed to mitigate the adverse impact of environment on tourism industry. Seventh, the main attraction of Sagar is to visit 'GSM' during the month of January and Kapil Muni Temple. Along this religious spot there are many other temples like Mansa Mandir, old Shiva Temple, light house, casuarina forest, Hooghly river confluence, sand dune which can be promoted for leisure and weekend trips also. Recently a number of initiatives have been taken by GBDA which might be a firm step towards growth of pilgrimage tourism in the Sagar Island.

REFERENCES

Arunmozhi, T. and Panneerselvam, A., 2013. Types of tourism in India. *International Journal of Current Research and Academic Review*, 1(1), pp. 84–88.
Basak, C.M., 2004. Pilgrimage on the ocean-development of Sagar Island, Bay of Bengal. In *Oceans'04 MTS/IEEE Techno-Ocean'04* (IEEE Cat. No. 04CH37600) (Vol. 2, pp. 954–958). IEEE.
Casagrandi, R. and Rinaldi, S., 2002. A theoretical approach to tourism sustainability. *Conservation Ecology*, 6(1), p. 13.
Census Report, 2011. Govt. of India.

Danda, A.A., 2007. Surviving in the Sundarbans: threats and responses. Unpublished Ph.D., University of Twente.

Das, G.K. 2006. Sundarban: Environment and ecosystem. Sarat Book Distributors.

Dasgupta, R., 2008. Assessment of vulnerability & adaptation strategies for the Sundarbans island system, West Bengal, India, in the perspective of climate. Unpublished Ph.D. thesis, Jadavpur University.

Dasgupta, S., Mondal, K. and Basu, K., 2006. Dissemination of cultural heritage and impact of pilgrim tourism at Gangasagar Island. *The Anthropologist*, 8(1), pp. 11–15.

Garrod, B., and A. Fyall. 1998. Beyond the rhetoric of sustainable tourism? *Tourism Management* 19, pp. 199–212.

Gebhard, K., Meyer, M. and Roth, S., 2007. Criteria for sustainable tourism for the three biosphere reserves Aggtelek, Babia Gora and Sumava. Ecological Tourism in Europe (ETE) and UNESCO.

GSTC. 2012. Global Sustainable Tourism Criteria. Global Sustainable Tourism Council.

GSTC. 2019. Global Sustainable Tourism Criteria. Global Sustainable Tourism Council.

Hajra, R., Mitra, R., and Ghosh, T., 2012. Impact of Gangasagar Mela on sustainability of Sagar Island, West Bengal, India. *International Journal of Research in Chemistry and Environment*, 2 (1), 140–144.

Hajra, R. and Ghosh, T., 2014. Formulation of methodological approach for sustainable tourism using 'GSTC'criteria: A case study of Sagar Island, India. *International Journal of Innovative Research and Development*, 3(1), pp. 305–309.

Hajra, R. and Ghosh, T., 2016. Migration pattern of Ghoramara Island of Indian Sundarban-identification of push and pull factors. *Asian Academic Research Journal of Social Sciences & Humanities*, 3(6), pp. 186–195.

Hajra, R. and Ghosh, T., 2018. Agricultural productivity, household poverty and migration in the Indian Sundarban Delta. *Elementa: Science of the Anthropocene*, 6(1), p. 3.

Hazra, S., Das, I., Samanta, K., and Bhardra, T., 2014. Impact of climate change in Sundarban Area West Bengal, India. In Earth Science and Climate Book-9326/ 17.02.00. Caritus India, SCiAF.

Kabra, K.N., 2003. The unorganised sector in India: Some issues bearing on the search for alternatives. *Social Scientist*, 31, pp. 23–46.

Kundu, S.K., 2012. A study of tourism potentialities and problems in Sagar Island of West Bengal. *Golden Research Thoughts*, 1(8), pp. 984–984.

Ministry of Tourism Report. 2018. India Tourism Statistics at a Glance, 2018. Government of India.

MOEF. 2009. Environmental and Social Assessment. In World Bank Assisted Integrated Coastal Zone Management Project. MOEF- ICZM Project Report, Govt of India: Centre for Environment and Development, Thiruvantapuram.

Ramgulam, N., Mohammed, K.R. and Raghunandan, M., 2012. The quest for sustainable business tourism: An examination of its economic viability in Trinidad. In *Global Conference on Business & Finance Proceedings* (Vol. 7, No. 2, p. 331). Institute for Business & Finance Research.

Saville, N. M. 2001. "Practical Strategies for Pro-Poor Tourism: Case study of Pro-poor Tourism and SNV in Humla District, West Nepal". PPT Working Paper No 3. ODI, IIED, ICRT.

Shinde, K., 2006. Religious tourism: Intersection of contemporary pilgrimage and tourism in India. In Journeys of Expression V: Tourism and the Roots/Routes of Religious Festivity Conference Proceedings CD-ROM. Sheffield Hallam University.

Sundarban Tiger Reserve. 2017. "Indian Sundarban." http://www.sundarbantigerreserve.org/str/, Accessed 25/08/2020.

Trumbic, I., 2005. Tourism carrying capacity assessment in the Mediterranean coastal tourist destinations. In *Proceedings of the 14th Biennial Coastal Zone Conference New Orleans*, Louisiana.

Williams, A. M. and Shaw, G., eds. 1988. *Tourism and Economic Development: Western European Experiences.* London, UK: Belhaven.

World Bank Report. 2019. Conceptual Plan for Integrating Community-Based Tourism Along the Bangladesh-India Protocol Route for Inland Navigation: Third Draft Report (English). World Bank Group, Washington, D.C.

World Travel & Tourism Council. 2019. The Economic Impact 2019, World. World Travel & Tourism Council. London, UK.

WWF. 2011. Indian Sundarban Delta: A Vision. edited by A. Srikanthan, Danda, G. WWF.F

10 An Analysis of Small-Scale Fisheries Management Status by Focusing on Degrading Fisheries Resources in the Sundarbans

Indrajit Pal and Afshana Parven
Asian Institute of Technology

Md. Ashik-Ur-Rahman
Khulna University

Mohammad Sofi Ullah
University of Dhaka

Khan Ferdousour Rahman
Asian Institute of Technology

CONTENTS

10.1 INTRODUCTION: BACKGROUND AND MOTIVATION

Sundarbans Mangrove forest (SMF) has a complicated and diverse ecosystem. The study presents the empirical results based on the secondary information available about the mangrove Fisheries and Aquaculture interlinked with the life and livelihoods. Sundarbans Mangrove Fisheries (SMF) plays a vital role for the mangrove dependent people. Degraded Fisheries management and frequent natural disasters made the life of the people extreme vulnerable. People sufferings have no limit due to the lack of sufficient income, work and resilient livelihoods. Due to the over exploitation of natural resources, illegal shrimp larvae harvesting, overfishing, use of destructive fishing gear, poison fishing, lack of policies and management, lack of institutional capacity and social acceptance of new knowledge adaptation for sustainable fisheries management results in severe environmental degradation in the study area that affects small-scale fishers food security (Gain et al., 2019a, Gain et al., 2019b). With other factors, natural disaster triggers the losses of deforestation that extinguish the fish stocks and habitat for other aquatic animals. Impacts of natural disaster and banning of fishing area jeopardize the resilience of fishing communities. To promote the stakeholders' empowerment sustainable management of mangrove fisheries needs to be linked with coastal governance. Involvement of civil societies including the local communities in decision-making process can help to restore, conserve and manage the mangrove ecosystem.

The Sundarbans Mangrove Forest (SMF) area is a deltaic wetland formed by silt transported over time by one of the largest Ganges–Brahmaputra–Meghna (GBM) river system and among the total area 60% lies in Bangladesh and the rest in India (Habib et al. 2020). The SMF is in the south of the Tropic of Cancer, to the northwest of the Bay of Bengal (21°30′–22°30′N, 89°12′–90°18′E) (Mozumder et al. 2018). The amount of land of Sundarbans is 5,950 km² in which 1696.99 km² is a sanctuary area. Numerous aquatic species (Approximately 400 species) use these mangrove marshes as a breeding and nursery grounds (Gain and Das 2014). The temperature, salinity and physiological parameters make this mangrove forest as an ideal environment for them. Fishes, mollusks and crustaceans use Sundarbans and its nearby marine and brackish water as their breeding, nursery and feeding environment (Muhibbullah et al. 2005). Furthermore, the estuary in the SMF is a major food source for other fishes and prawns that spend most of their lives in freshwater and descend annually to the estuary for spawning (Rainboth 1990). The fisheries management authorities need to have sufficient of the Sundarbans Mangrove Forest (SMF) ecosystem, including fish stocks and fishing household living strategies, to incorporate such knowledge in the process of management planning (Mozumder et al. 2018). Otherwise, fish stocks will end up both in quantity and quality. Unfortunately, fishermen in this region have reported that they are spending more time and efforts to capture fewer and smaller fish (Hoq 2014). And impacts on the fisheries resources seem to have an effect to the small-scale fishermen (Rainboth 1990).

The Sundarbans Mangrove Forest (SMF) has a distinct management history. To maintain the ecological balance and sustainable utilization of resources, the Government of Bangladesh (GoB) has formulated and implemented different management policies and action plans (Hoq 2014).

Focusing only on inland freshwater fisheries management over the recent decades resulted in a constant lack of management strategies of SMF fisheries in Bangladesh (Mozumder et al. 2018). Fisheries resource depletion can be visible due to many reasons, such as the poor resource users over-exploit the fish stock out of sheer necessity, and the resultant degradation further aggravates poverty in turn. Furthermore, non-compliance with fishing rules and regulations and the attempts of small-scale fishers to support their livelihoods by any means possible, result in increasing fishing pressure. Their use of destructive fishing methods and gears, and a tendency to fish whatever is available, including larvae and juveniles, again reduce the fish stock (Murshed-e-Jahan, Belton, and Viswanathan 2014).

Management based on a guessing game will deteriorate the sustainability of fisheries resources. High biomass means high productivity and this management strategy has proved as a failure in the past and will be in the future. Any serious attempts to build knowledge must come from a strong foundation of basic research (Rainboth 1990). The growing threats to sustainable fisheries arise from excessive exploitation of the resources, indiscriminate fishing disregarding specificity fishing gears, locality and season and destructive fishing like use of poisons in the closed creeks and khals, reduction of freshwater flow, intrusion of salinity in the elaborate parts of the forest leading to severe environmental degradation; these are but the prime concerns for the sustainability of the resources (Hussain 2014). Management strategies should be based on these constraints and maintained properly with strict law enforcing agency.

Sundarbans protects the estuarine embankments from disasters like cyclone. People in that area are highly dependent on the Sundarbans ecosystem for their life and livelihoods. Disaster affects the mangrove forest and tremendous damage to wildlife. Disaster also impacts the artisanal fisheries in the mangrove area where the fishers are relatively poor and has less capability for the investment in fisheries sector where people in that area have highly interactions in marine and aquatic resources (Allison et al. 2009). More than 3 million people are directly or indirectly involved in various livelihood activities but the forest is reducing terrifyingly by various factors such as natural disasters (e.g. Cyclone Aila, Sidr, etc.) and over-exploitation of natural resources (Shah, Huq, & Rahman 2010). The objective of the study is to understand the Mangrove Fisheries Management in the Sundarbans area where people are dependent on the Mangrove ecosystem. It is also important to know that which factors are affecting the mangrove ecosystem as well as the artisanal fisheries and what are the existing rules and regulations for the small-scale mangrove fisheries management to maintain the biodiversity and the livelihoods of the poor people in that area. The research targets in this paper to interlink the various factors, disaster management and mangrove fisheries management policy adjustment with the adaptation measures for resilient fishers' community in Sundarbans area.

10.2 PRESENT STATUS OF FISHERIES RESOURCES OF THE SUNDARBANS

10.2.1 AVAILABLE FISH SPECIES

The Sundarbans, located in the south-west of Bangladesh, is the largest mangrove forest and consisting of a group of plants, coastal waters, fishes, shellfish and crustaceans. The Sundarbans ecosystem supports rich fisheries diversity and constituents of 177 species of fishes (Mustafa, Ahmed, and Ilyas 2019). The wetlands of Sundarbans are very rich in Biodiversity. It is one of the most biologically productive of all-natural ecosystems with great economic importance. The primary aquatic resources of the Sundarbans are as follows (Pomery et al. 2016):

1. **Hilsa:** is the single most important species of Sundarbans. It is caught form the estuarine area and the rivers inside the SRF. The lower part of the SRF is essential for the hilsa harvest. The Baleswar has been reported by fishers to be highly crucial for hilsa inside SRF.
2. **Whitefish:** there are many species of white and finfish. The Sundarbans Biodiversity Conservation project reported 204 bony fish species. Some commercially important species are datney, taposhi, vetki, pangus, poma, churi, kain, tengra, etc.
3. **Prawn and Shrimp:** The adult golda and bagda are highly essential for commercial harvest. These are export items and have a high price, although the catch amount is not much. There are 26 species of prawn and shrimps in the SRF.
4. **Cartilaginous fish:** These are non-bony fish. They are mainly marine but available in the estuary and SRF. There about 20 species of cartilaginous, and some of them have economic value. *Himantura* sp. is sold in the market, and people eat them. Shark has economic value as well.
5. **Reptiles:** The turtles are of economic importance. Others are crocodile and snake.

6. **Crab:** The mud crab is a highly essential and export item. There are about 44 species of crab.
7. **Mollusks:** 36 species. In some cases, mollusk meat is used in shrimp farms as food. Its shell is also used to produce lime and decorative and ornamental items. Mollusks can be an export item if necessary steps are taken by the government and other authorities concerned.
8. **Dolphin:** The mammals. Dolphin is one of the most attractive marine mammals that can contribute to the expansion of the tourism business. It does not have much economic value in terms of food.
9. **Post-Larvae of Prawn and Shrimp:** The Post-Larvae (PL) of Bagda and golda has high economic value.
10. **Dry Fishery:** At Dublar chard. Mainly white fishes of all species are dried.

10.2.2 SECTORAL CONTRIBUTION TO REVENUE

Despite continued degradation, the Sundarbans contributes 3% to the country's gross domestic product out of 5% contribution of the country's forestry sector (Habib et al. 2020).

The fisheries of Sundarbans are significant for the local economy and livelihoods of thousands of poor people living around and outside the landscape area. There are many other stakeholders. It produces 2%–5% of the total capture fisheries. In 2003–2004 the Forest Department (FD) production estimate was 433,000 Metric ton (MT) (Hoq 2008). IPAC PRA finding shows that an average of 47% of households within the 5 km area in the landscape in Bagerhat & Satkhira district are engaged in fishing.

Approximately 40,000–70,000 boats operate in the Sundarbans Reserve Forest (SRF) for fishing. FD revenue collection data have been considered for representing the value of different groups of fish. It is noticeable from the following figures that shrimp fry was providing higher revenue (4–16 million Bangladesh Taka [BDT]) than any other group of fish during the last 5 years. Whitefish and Golda (Machrobrachium rosenbergii) were second in rank (1.7–2.7 million Taka) for revenue, whereas dried fish (list of dry fishes in Table 10.1), Illish (Tenualosa illisa) and crab (Scylla Serrata) were in the third position. Bagda (Penaeus monodon) and

TABLE 10.1
Extinct and Declining Fish Species

Extinct and Declining Fish Species	Present Status
Kaiful	Extinct
Goda	Extinct
Kakshil	Declining rapidly
Kanmas	Declining rapidly
Tepa	Declining rapidly
Harina	Declining rapidly

Source: USAID (2010).

Golda (M. rosenbergii) are the most valuable shellfish species. Among the white fish, Plotosus canius, Lates calcarifer, Polynemous indicus, Pampus Chinensis and Liza parsia are more valuable than the others. The other shrimps, *P. indicus*, Metapenaeus Monoceros, and mud crab (S. Serrata), are also valuable (USAID 2010).

10.2.3 FISH EXPLOITATION TRENDS

The assumption of fishers in the Sundarbans is around a 50% decrease in the fish caught during the last 10 years, and the reduction is somewhat less in the lower part and more in the upper part of the SRF (Hoq 2014). In recent years, fishing with poison (organophosphate agro-chemicals) in the canals of the Sundarbans has increased. It is recognized to be highly detrimental to the aquatic resources of the SRF. Moreover, there is an increasing trend in the number of fishers as well as fishing efforts observed (Hoq 2014). A recent report on coastal fisheries in Bangladesh shows that catch per unit fishing effort is falling, and several species of marine shrimp and fish stocks are in decline (Murshed-e-Jahan, Belton, and Viswanathan 2014).

4-year forecasting on the yield of Sundarbans fisheries (based on landed data) shows marked decline of white fish, Scylla serrate and fry catch (Mustafa, Ahmed, and Ilyas 2019). Sundarbans fishing communities in Bangladesh are at higher threat due to frequent natural disasters, which damages natural resources systems. The Sundarbans water supports 208 species of fish and crustaceans belonging to 84 families, a higher total than that of other tropical mangroves. The mean fish biomass is 39 kg ha^{-1} (Hoq 2007). A prominent feature of the SRF fishery is that its size has been gradually expanding. The present production is about 12,000 MT, almost double of them in 10 years back. 4-year forecasting on the yield of Sundarbans fisheries (based on landed data) shows a marked decline of white fish, S. Serrata, shrimp and crab residuals, and Penaeus monodon fry catch (Hoq 2007).

Bangladesh has achieved remarkable progress in the fisheries sector since its independence in 1971. This sector is contributing a very significant role in socioeconomic development and deserves the potential for future development in the agrarian economy of Bangladesh. It contributes 3.57% to our national GDP and around one-fourth (25.30%) to the agricultural GDP. This sector provides a major (60%) share of animal protein.

Bangladesh is blessed with vast and rich fisheries resources. The diversified fisheries resources of the country are divided into two groups as Inland and Marine fisheries. Again, Marine fisheries include Industrial (Trawl) and Artisanal fisheries.

According to the definition of the sector of fisheries, Sundarbans Fisheries are confined in Sundarbans only, and it comprises flowing rivers and a mangrove area. The water area for Sundarbans fisheries is estimated at 177,700 hectares (DoF 2018). But the annual fish production in Sundarbans area is declining (Figure 10.1). Figure 10.2 also shows average fish production in the Sundarbans area in 2004–2019.

10.2.3.1 Vegetation and Waterbody Changes Trend in the Sundarbans

According to satellite data analysis, the Sundarbans is about 6,004 km^2. The Sundarbans is a natural forest area, where there are a lot of branches of rivers and streams. The ecology, biodiversity and ecosystem depend on the Sundarbans

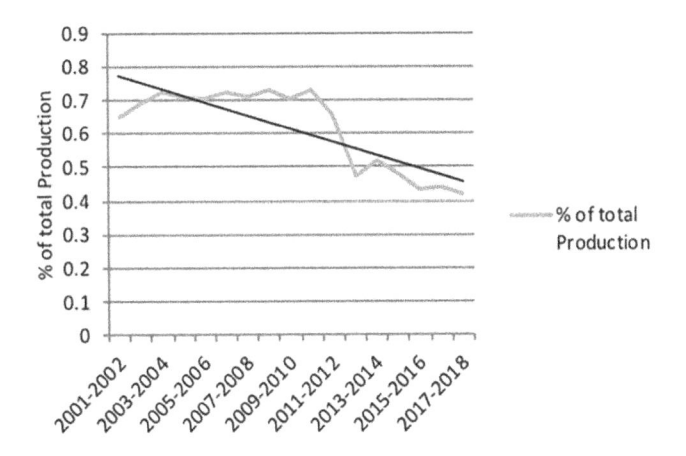

FIGURE 10.1 The declining trend of Sundarbans fisheries production. (From DoF 2018.)

waterbody and vegetation coverage. Because of this, 2-decades satellite data from 2000 to 2019 were analyzed to find out the Sundarbans waterbody and vegetation coverages, which is related to fisheries resource management. The satellite data analysis result shows that the Sundarbans is increasing gradually; it had 30.06% wetlands in 2000 which stood at 30.54% in 2010 and again in 31.25% in 2019 (Figure 10.1 and Table 10.2) and in 2029 it will be 31.82% (Table 10.2). Similarly, the Sundarbans vegetation coverage is decreasing gradually, she had 69.94% vegetation coverage in 2000 which stood at 69.46% in 2010 and again in 68.75% in 2019 and in 2029 (Figure 10.3 and Table 10.3) it will be 68.18% (Table 10.4).

The medium-resolution Landsat data sets are sufficient for land use and land cover classification. In this study, amongst the Landsat images, the Landsat TM, Landsat ETM+ and Landsat OLI/TIRS spectral channels were chosen specifically to map vegetation type and water features of Sundarbans, the present research. The attempt was only classifying land cover to show the changes in the Sundarbans because the dimension of the change gives lots of clues about biodiversity level conditions in Sundarbans. Furthermore, the image analysis accuracy assessment is given in the below table (Table 10.3).

Table 10.3 shows the accuracy of the image analysis. The overall Kappa coefficient for the year 2000, 2010 and 2019 had been found at 88.68%, 86.92% and 83.49%, respectively. Conversely, overall accuracy was at 94%, 93% and 91%.

Both waterbody and vegetation coverage's increasing and decreasing rates are steady; on an average about 35 km² waterbody is increasing per decade in the Sundarbans (Figures 10.4 and 10.5).

The Sundarbans ecosystem is decreasing with many endangered species (Flora and Fauna) due to the excessive use of natural resources, disease and increase salinities changing in vegetation system and coastal erosion (IUCN 2016). The Sundarbans is decreasing due to the highest pressure of human settlements near the border of the Sundarbans (IUCN and UNESCO 2016). Sundarbans are also becoming increasingly deforested because the local people are drastically cutting down the trees in the forest. This real picture was found by analyzing satellite images of the area.

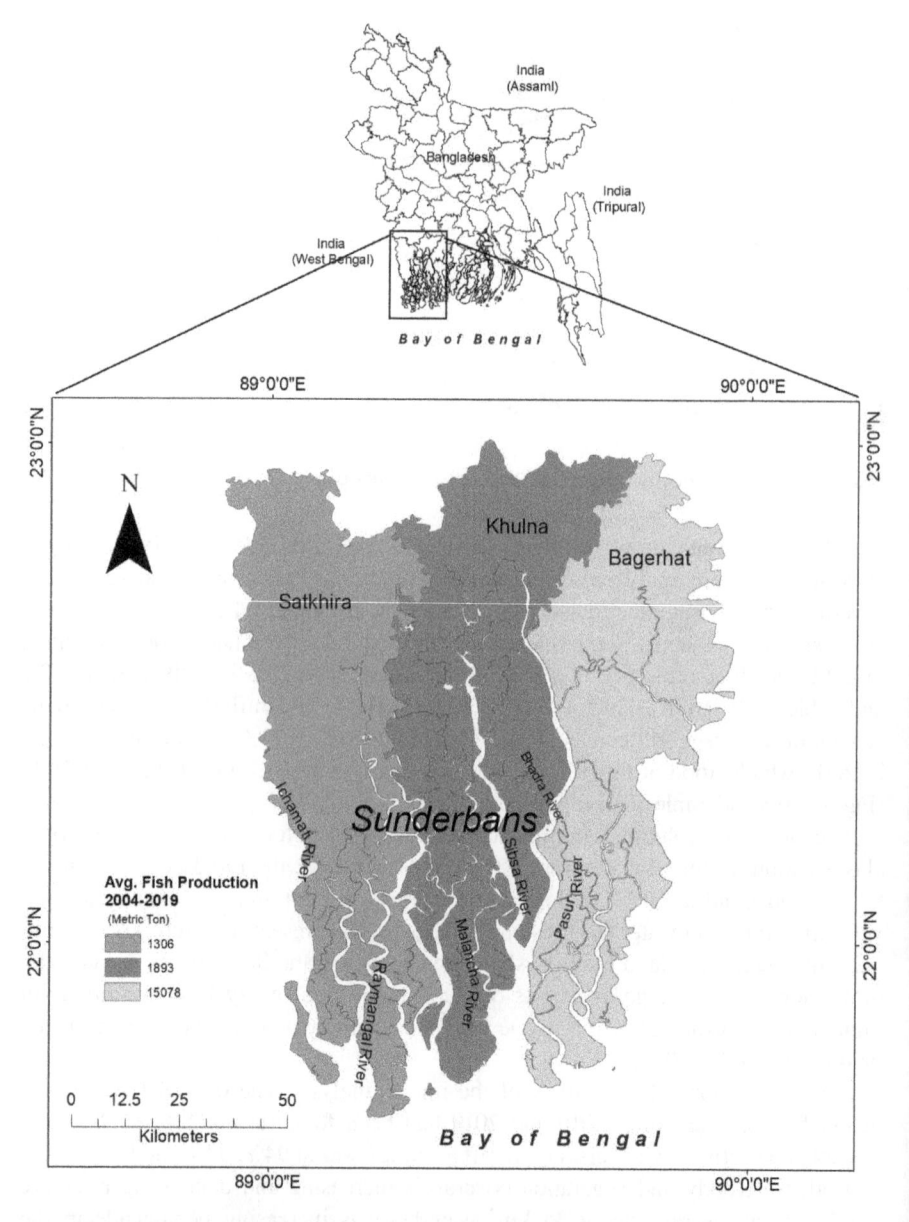

FIGURE 10.2   Average fish production MT in Sundarbans area.

The satellite analysis has shown that about three (3) km^2 of the forest area has been depleted every year and converted to wetland and water body in the area (Figure 10.3). To enrich biodiversity, there needs to be a balance between the presence of forests coverage and wetlands. It is difficult to imagine the balance of land cover now, natural and human pressure. However, since the forests in the Sundarbans are gradually declining, therefore, biodiversity is also gradually declining and this is reality.

TABLE 10.2
Annual Fish Production of the Sundarbans Fisheries

Year	Production MT	% of total Production	Productivity (Kg ha^{-1})	Growth Rate
2017–2018	18225.24	0.42	103	0.7685
2016–2017	18086.04	0.44	102	7.20
2015–2016	16,870	0.43	95	−4.03
2014–2015	17,580	0.48	99	−4.27
2013–2014	18,366	0.52	103	15.18
2012–2013	15,945	0.47	90	−26.21
2011–2012	21,610	0.66	122	−3.74
2010–2011	22,451	0.73	126	9.85
2009–2010	20,437	0.70	115	10.697
2008–2009	18,462	0.73	104	1.7134
2007–2008	18,151	0.71	102	2.25
2006–2007	17,751	0.72	100	8.08
2005–2006	16,423	0.70	92	4.44
2004–3005	15,724	0.71	88	3.16
2003–2004	15,242	0.72	86	9.78
2002–2003	884	0.69	78	12.46
2001–2002	12,345	0.65	69	-

Source: DoF (2018).

In addition to deforestation, the human footprint is hindering the development of biodiversity in the Sundarbans. Satellite data analysis has shown that the waterbody and wetland have been increased in the Sundarbans in the last two (2) decades. What is the reality of the fisheries sector in the current situations? We know that all types of mechanical vessels are moving freely in the rivers of the Sundarbans; therefore, natural breeding of fish is decreasing. On the other, oil leakage from vessels pollutions and sedimentation from upstreams have affected the fisheries sector. The river basin in the coastal area is getting wider; on the other hand, its depth is decreasing due to sedimentation which has to affect the fisheries sector also.

10.2.4 Impacts on Biodiversity

The Sundarbans is highly productive, supports immense Biodiversity and provides a wide range of ecosystem services including flood mitigation, fish breeding and production, livelihoods, climate change mitigation and adaptation, and other socioeconomic and recreational functions (BFD 2010). The existence of Sundarbans, forming an ideal mangrove ecosystem, supports large groups of fish, shrimp, edible crab, and also supplies food and cash to the coastal communities. One-third of the country's population is dependent on Sundarbans (Mozumdar et al. 2018). With over 3.5 million people from the surrounding areas depend directly or indirectly on the Sundarbans for their livelihood, the forest has been reducing alarmingly day by day (Hoq 2007). During the past 10 years, the availability of most desired Penaeus

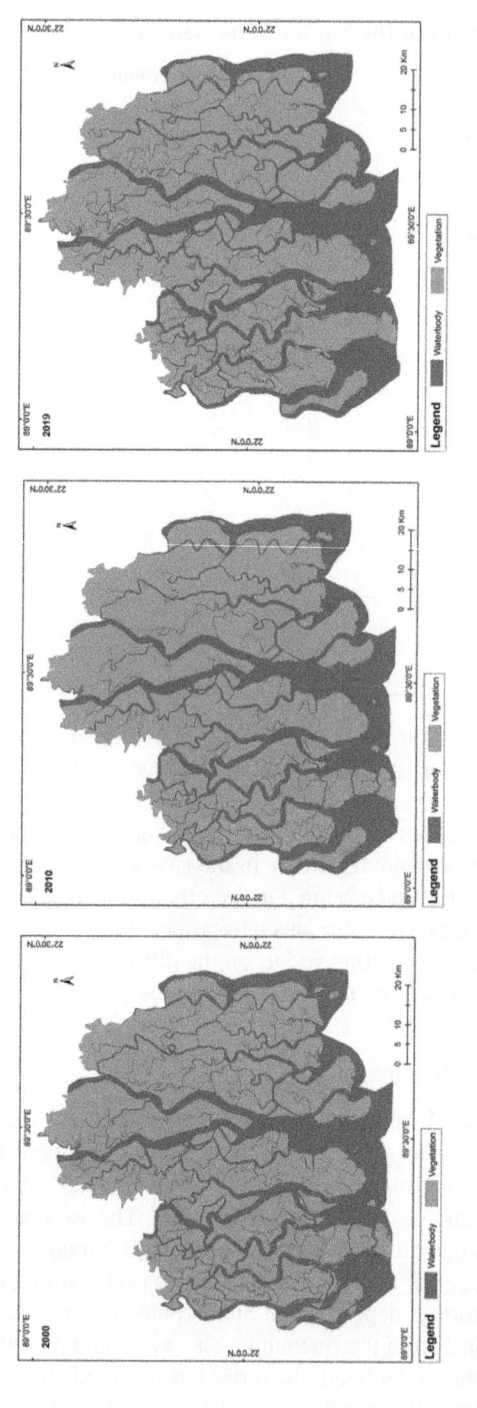

FIGURE 10.3   Land cover changes in the Sundarbans from 2000 to 2019. (Satellite Data from https://glovis.usgs.gov/.)

TABLE 10.3
Overall Accuracy Assessment for Land Cover Analysis

Year	Overall Accuracy (%)	Kappa Coefficient (T)
2000	94	88.68
2010	93	86.92
2019	91	83.49

TABLE 10.4
Changes in Land Cover in the Sundarbans from 2000 to 2029

	2000		2010		2019		Forecast (2029)	
Landcover	Area (km²)	%	Area (km²)	%	Area (km²)	%	Area (km²)	%
Waterbody	1804.633	30.06	1833.64	30.54	1876.17	31.25	1910.631365	31.82
Vegetation	4199.746	69.94	4170.73	69.46	4128.21	68.75	4093.745646	68.18
Total	6004.379	100.00	6004.37	100.00	6004.38	100.00	6004.377011	100.00

Source: https://glovis.usgs.gov/, 2000, 2010 and 2019.

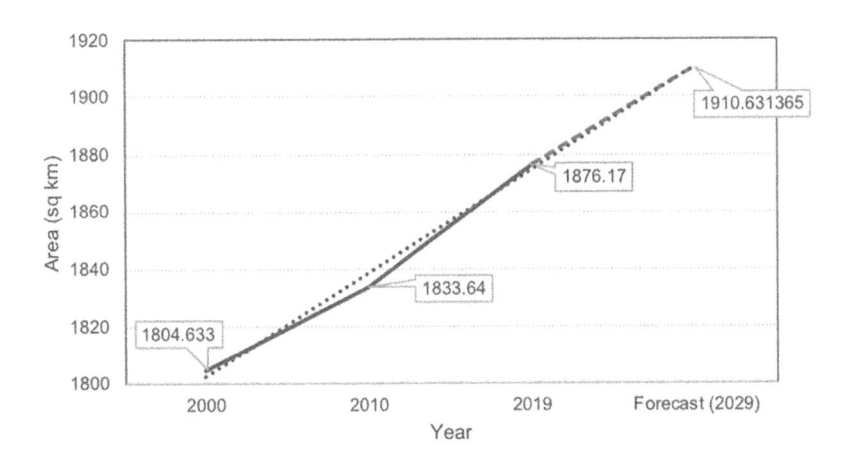

FIGURE 10.4 Trends of the waterbody changes in the Sundarbans from 2000 to 2029.

monodon PL has been gradually declined (Hoq 2007). If the nursery grounds are destroyed, or the juveniles are captured, the chance for the juveniles to return to the sea will be reduced, leading to the scarcity of mature mother stock. This will result in less availability in the sea and the estuary for breeding. The desperate attitude of the fishermen contributes to the degradation of both wood and non-wood resources in SRF. They usually cut the mother trees of gewa (Excoecaria agallocha) and sundri

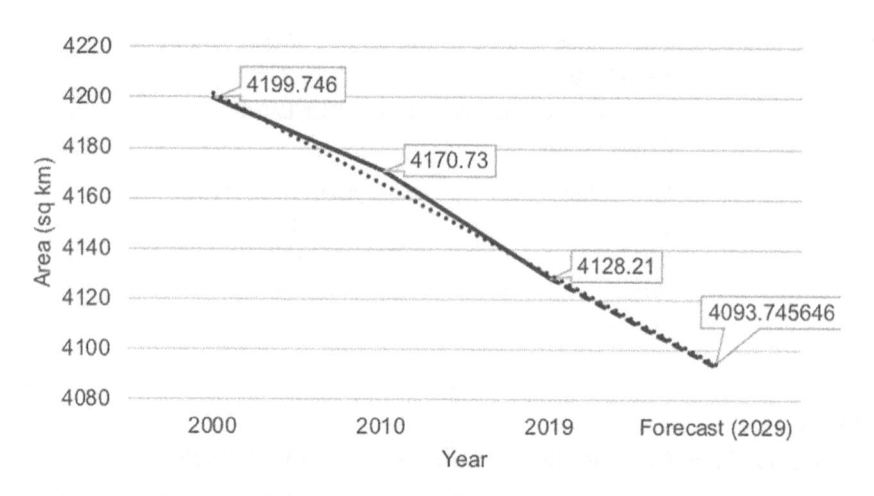

FIGURE 10.5 Trends of the vegetation changes in the Sundarbans from 2000 to 2029.

(Heritiera fomes). The Sundarbans ecosystem is a famous nursery and breeding area for key fisheries of coastal and marine waters of the Bay of Bengal. Despite continued degradation, the Sundarbans contributes 3% to the country's gross domestic product out of 5% contribution of the country's forestry sector (Roy and Alam 2012). In recent years, population pressure, economic development and unsustainable management practices have resulted in the rapid depletion and degradation of the Sundarban's resources, threatening its Biodiversity and the livelihood of the local community (Hoq 2014).

Fishers of the Sundarbans use different kinds of harmful nets and gear for catching fish, which cause damage to aquatic lives; for example, monofilament gill nets (called current Jal) are responsible for the killing of different aquatic animals and small-sized fishes. Fine-meshed set bag nets (locally called Behundi Jal), pull and push nets (Thela Jal), fine-mesh mosquito nets (Chingri Pona Jal), longshore nets (Khuti Jal) have been identified as the most destructive among all the fishing gears in the Sundarbans. Catch mortality is very high for these nets. Set bag nets used for collecting shrimp fry in the estuary and rivers of the Sundarbans also catch eggs, spawn and larvae of all species along with adult fish. It is highly detrimental to declining fish diversity (Habib et al. 2020).

10.2.5 LAND-USE CONFLICTS

The total land area of SMF is 4,143 km², and the remaining water area of 1,874 km² encompasses rivers, small streams and canals. The existence of the SFM, forming an ideal mangrove ecosystem, supports large groups of fish, shrimp, edible crab, and supplies food and income opportunities to the coastal communities. The forests are being reduced alarmingly day by day due to immense pressure from usage by around 3.5 million people who depend directly or indirectly on the SMF for their livelihood (Mozumder et al. 2018).

10.2.6 DISASTER IMPACTS ON SUNDARBANS AND MANGROVE FISHERIES

Due to the climate change, frequency and intensity of natural disaster is increasing (Guha & Roy 2016) in Sundarbans. Biophysical aspects of Sundarbans have intensively studied but very fewer studies predict what could be the alternative development pathway for the mangrove management in the changing climate change situation (Roy el al. 2020). Natural hazards tremendously affect the wildlife in the mangrove forest including trees, fish stocks and animals. Fodder scarcity happens due to the broken trees that limits the movement of the wild animals. On the other hand, freshwater canals are infested with saline water that causes the scarcity of drinking water for the animals.

Operative disaster management approaches are required to reduce the disaster risk and loss in Sundarbans area. Pal and Shaw (2018) reveals that operative disaster risk governance strategies save lives as well as minimize the damage and extra burdens to the economy and increase the resilience in emergencies (Lwin et al. 2020). Due to the natural disaster, mangrove forest-dependent people and artisanal fishers lose their jobs that adversely affect the economy as well as the socio-economic condition of the people. The GoB has comprehensive disaster management plan for early warning system (Hossain 2011; Izumi and Shaw 2015) but lack of coordination between different sectorial management that affects the plan and adaptation policy for a development of a resilient community (Pal et al. 2017). The disaster has cascading effects to the local fishers as they face multilevel natural disasters so disaster risk analysis in the study area is important to reduce the multi-hazard risk reduction that has been followed in different papers (Pal et al. 2018).

10.3 WHY FISHERIES MANAGEMENT IS VITAL TO CURB THE RESOURCE DEPLETION

Managing common pool resources has always been a challenging task. In terms of 'fisheries management', it is an integrated process of information gathering, analysis, planning, consultation, decision-making, and implementation with the enforcement of rules and regulations by different institutions. The principles and goals of fisheries management have been changing over time to fulfil the different needs of the ecosystem and the community. A key aspect of fisheries resource management and sustainable use is an assessment of the present status of resources or the information gathering on this resource (Williams 2014). Unfortunately, there is inadequate monitoring of fish stocks in the SRF, and fisheries resources have not been inventoried (Faridah-Hanum et al. 2013). The previous section focused on giving an idea of this resource status over the past years. However, more accurate inventories are needed, i.e. GPS-mapped fishing grounds for operations emanating from Satkhira and Khulna Range (Thompson et al. 2016). The state government is primarily in charge to control and manage SMF and the water bodies inside this forest through the FD (Hoq 2007). Although to protect fisheries and mangroves, there are several management policies, that are not being implemented correctly always (Mozumder et al. 2018). Socio-economic and cultural heterogeneity should be addressed in structuring management systems and policy implementation.

Furthermore, local politics, religion, education level and influence of the local authority on the management plan should be reflected. Because, as all these issues play direct or indirect roles in policy implementation (Mozumder et al. 2018). For the sustainable utilization of the SMF resources, eventually contributing to the improvement of the small-scale fishers' livelihoods in the Sundarbans region, there is an urgency to update existing policies and management issues (Hoque Mozumder et al. 2018).

10.3.1 SUSTAINABLE FISHERIES MANAGEMENT: A TWO-FOLDED MANAGEMENT PRACTICE

Local fishermen have noted that they are spending more time and effort to capture fewer and undersize smaller fish (Islam and Chuenpagdee 2013). Along with traditional management options, sustainable fisheries management is in practice in the SMF to overcome the issues of unsustainable fish catch. The following two-fold requirements are suggested for the sustainable management of the SRF fisheries resources:

i. **Resource Conservation Measures:** Maintain the fisheries resources to a level that does not degrade from the present level (i.e. ensuring sustainable harvest) by controlling the number of fishers and checking the type of gears used.

ii. **Resource Improvement Measures:** Improve fisheries resources through different management and conservation interventions.

Major initiatives taken under this sustainable fisheries management are as follows:

10.3.1.1 Fishing Area Ban

Khal restrictions and protected sanctuaries were also acceptable, though the prospect of increasing protected area coverage is less. Restricted khals may be easier to build awareness about and to monitor and enforce than protected sanctuaries. However, both of these approaches require robust science and constant monitoring to ensure that the protection of that area will have an ecological impact. Furthermore, establishing new closed areas will likely stretch the already limited enforcement capabilities of the FD, so it would perhaps be more efficient to focus efforts on those protected areas that already exist (Thompson et al. 2016). Year round fishing ban may be practiced, like:

i. All water bodies in the existing three and proposed wildlife sanctuaries will be banned for fishing. Besides, a 1-km wide zone on the northern periphery areas of the sanctuary may also be included under the fishing area ban.

ii. A permanent fishing ban by FD to be enforced in the 18 canals that have been declared closed in the buffer zone.

iii. Canals less than 25 feet wide within 3 km area of FD permanent Camp office/Petrol Office located throughout the Sundarbans.

10.3.1.2 Fishing Ban during Breeding Season

Size limits and small fish release regulations were also relatively acceptable but to a lesser extent. The small fish release might be another way to protect *S. argus*, but setting an overall size limit for all species would mean some sexually mature small fish would be released unnecessarily, and by setting a limit for *S. argus* only at its size of sexual maturity, this creates sorting work for the fisher. Size limits through bans on the least selective gears would be a less complicated way forward. There is, however, amounting literature disputing the assumptions underlying the selective fishing paradigm, particularly in broad, adaptive and low-tech fisheries like those in the Sundarbans (Table 10.2). Small fish plays an essential role in food security in developing countries, and when multi-species interactions are considered, size selectivity can reduce total yields (Thompson et al. 2016).

Bangladesh Government is committed to reduce the poverty in the country. The Government has many social-safety net programs in the country like Vulnerable Group Feeding (VGF) Program in Bangladesh, Fund for climate change, Disaster and climate resilient fund, comprehensive disaster management program, Fishermen ID Card and Fisheries Project, etc., to protect the poor fisherman and disaster affected people (MOF, Bangladesh). The Government of Bangladesh shows massive inventiveness towards SDG 14 that is conservation and sustainable use of marine resources. The government has taken initiatives to improve hilsa fish in the country, to ensure the supply of a treasured and captivating fish species that has political interest as well. The decision has immense positive response that hilsa fish caught larger than previous years with a high price and fishermen receive financial benefit in local market. It can be described as conservation with economic benefit initially. Yet the costs of these regulations are directly affecting the poor, illiterate and in everlasting debt to the small-scale fishermen. The government provides a small amount of payment for ecosystem services (PES) for alternative income generation, good enough to receive something than nothing during the banning period (Porras et al. 2017). The artisanal fishermen have no bargaining power in the policy and decision-making design that actually affects their life and livelihoods. Informal markets are always neglected in policy-making due to the lack of clarity of the market. There are initiatives for hilsa fishermen but other workers who are involved in shrimp and other aquatic sector are neglected and not included in this PES. But, they may be included in VGF, disaster and climate resilient fund for the development project, a kind of local political influence made the task harder (Islam et al. 2016).

Building resilience is not the task of government alone. The multi-stakeholder characteristics of resilience are highlighted in all the global frameworks. Sendai framework suggests society engagement and institutionalized partnership through multi-stakeholder disaster risk reduction at various levels and declares the responsibilities across public and private stakeholders, including business and academia, to ensure mutual outreach, partnerships and complementarities in roles and accountabilities (Pal & Bhatia 2017).

10.3.1.3 Ban of Fishing Gear and Mesh Size Control

In the Sundarbans waters at least 14 different gears operate (USAID 2010). These gears are grouped into three broad categories, such as single-species single-gear fishery; single-species multi-gear fishery; and multi-species multi-gear fishery. Based on the gears they are again classed into groups based on whether they are nets of different types, hook and lines for fishing fin fishes and crabs, etc.; nets that are operated setting at fixed position in the estuaries or rivers called set bag nets or hand nets used for shrimp PL collection from the estuarine rivers (USAID 2010).

Compared with size limits and species bans like *S. argus*, gear bans were some of the most acceptable fishing regulations. This 'gear ban' is easy to impose among fisheries communities of SMF (Thompson et al. 2016).

10.3.1.4 Boat License Certificate Limit

Boat license certificates (BLCs) are issued for every year in July (at the beginning of the Govt. financial year) and are valid for only 1 year. The pre-requirement for a BLC is a certificate and attested photograph from a local public representative, i.e. Union Parisad Chairman. After having a BLC, fishermen have to pay annual fees for that BLC which is based on the 'Maundage' of their boat. Maundage (M) is calculated as follows (Hoq 2014):

$$M = L = W \times H \times 0.356$$

where L = boat length (in feet), W = maximum boat width (in feet), H = boat height (in feet) and for dry fish, M (dry fish) = $(L \times W \times H \times 2)/10$.

To prevent over fishing, the number of BLCs provided by BFD to allow fishermen for catching fish inside Sundarbans was limited. The maximum number of annual BLC issuance has been 12,000. The first priority in issuing BLC is given to those boat owners who live within 5 km area around the Sundarbans. The maximum limit of permits for a month is given for three times and 5–7 days fishing is allowed under one permit (Habib et al. 2020) (Table 10.5).

10.3.2 Major Resources Management Plans to Protect Fisheries Resources of the SMF

The SMF comprises three sanctuaries such as Sundarbans East, Sundarbans South and Sundarbans West (IUCN 1997). In August 1997, a 5-year Conservation Management Plan was prepared for the three wildlife sanctuaries. After that a 12-years Integrated Forests Management Plan was developed in January, 1998, for the period of 1998–2010 for the same three wildlife sanctuaries (Razzak et al. 2020). Prior to this plan, Forest Resource Master Plan was developed in 1993 to cover the periods of 1993–2012 which did not provide sufficient guidance to address the present issues regarding resources depletion (Figure 10.1). At the same time, an agreement was reached with USAID (United States Agency for International Development), FD, EU and IPAC in preparing a Strategic Management Plan (SMP) as recommended in the SEALS (Sundarbans Environmental and Livelihood Security) project preparation report. The IRMP is formed based on the SMP. This IRMP for the SRF and its surrounding landscape is thus the result of numerous discussions and meetings with the FD, DoF and Department of Environment; local stakeholders including Co-Management Organizations (CMOs); civil society members; USAID; USFS (United States Forest Service); and IPAC project (Razzak et al. 2020).

TABLE 10.5
Current Monitor and Conservation Measures Taken by Bangladesh Forest Department

Measures Taken	Implementation Periods
Fishing ban in water bodies and wildlife sanctuaries	All over the year
Fishing ban in specific 18 declared canals in the buffer zone	All over the year
Fishing ban canals less than 25 feet wide throughout the Sundarbans	All over the year
Fishing ban in all canals	July-August
Fishing ban in beels and canals	February–March
Complete ban of using monofilament gills net (current jal), set bag net (behundi jal), push net (thela jal), channel stake net (khalpeta net)	All over the year
No fishing by poison, insecticide and de-watering	All over the year
No fishing by the net with mesh size more than 01 inch to 15 mm	All over the year
Fishing ban three finfish species viz. Pangas (*Pangasius pangasius*), Sea bus (*Lates calcarifer*) and Kain magur (*Plotosus canius*)	May–June
Ban on Hilsha (*Tenualosha ilsha*) fishing for 22 days	October
Catching of Hilsha (*Tenualosha ilsha*) and Pangas (*Pangasius pangasius*) below 23 cm	November–June
Boal (*Wallago attu*) lower than 12 inch	April–August
Ban of fishing of the species Shilon (*Silonia silondia*), Vola (*Johnius argentatus*) and Air (*Bagarius bagarius*) lower than 12 inch	February–June
Ban of fingerling and fish fry collection	All over the year

Adopted from Habib et al. (2020).

10.3.2.1 Current Co-Management Practices

Co-management and its implementation are in its early stages in the country, but may be the key to alternative fisheries resource management in future (Hossain et al. 2006). At present there is only one CMO in the study area, named, Sarankhola Co-management Organization (SCMO). SCMO contributes to sustainable natural resource management and enhanced biodiversity conservation in targeted forest and wetland PAs with the goal of preserving the ecological bio-diversity while promoting equitable economic growth and strengthening environmental governance. The total numbers of council members are 56 and committee members are 24. Total numbers of village conservation forums are 21 and forum members are 3,354. Since its formation, SCMO has taken up several co-management activities with the aim of sustainable natural resource management in the area. SCMO's working area is about 31,227 ha, spreading over one Range Office, three Station Offices, one Upazila, two Unions and 20 villages/paras (Razzak et al. 2020). Policies supporting co-management approaches to natural resource management include the Bangladesh Forest Department's Nishorgo Vision 2010, which focuses on co-management and community partnerships as strategies for strengthening the management of protected areas (Razzaque 2017). Another example is the USAID-funded

Integrated Protected Area Co-management Project (IPAC), 65 managed by the FD, which seeks to engage local stakeholders through a participatory co-management process, empowering them with decision-making rights and positive incentives, thus promoting their interest in, and commitment to, the protection of biodiversity resources (Razzaque 2017).

10.3.2.2   Role of Forest Department

Scientific management of forest resources was first initiated in the Sundarbans in the 1870s when a Forest Management Division was established exclusively for the management of Sundarbans in the Gangetic Delta. The FD exercises its control on fisheries resources in SRF through collection of tolls, taxes and revenue from fishing boats, fishermen and fishery products (Hoq 2007).

10.3.2.3   Regulations Related to the Sundarbans Fisheries

The on-going depletion of fisheries jeopardizes human food security, resilience of fishing communities and livelihood options for current and future generations (Lam and Pitcher 2012), as well as impacting marine species, habitats and ecosystems beyond the targeted species (Hobday et al. 2011). To reduce these threats, fishing regulations are drafted and enforced to control among other things: the area and time of usage; size, sex and species captured; gear use; and fishing effort (Thompson et al. 2016). The Department of Forest (FD) presently faces in managing the Sundarbans (Table 10.6).

10.3.2.4   Gradual Changes in the Principles of Fisheries Management

A number of regulations have been enacted for the conservation of the resources and ecosystem, but yet to rigorously enforced (Hoq 2007). Management process initiated long ago for the SMF since the pre-British time and it shaped to present through several changes. Different objectives at different times acted as driving forces behind this change (Table 10.7).

10.3.3   Stakeholders Analysis

Existence of ten livelihood groups in the SMF is found, such as Bawali (wood cutter), Nypa collectors (golpata used as roof materials), Mawali (honey and bee wax collector), Jele (Fisher), Majhi (Boatman), Crab collector, Medicinal plant collector, Shrimp fry collector, Chunery (oyster and snail collector) (USAID 2010). The average monthly income among the fishermen is about 3,750 BDT of which 70% comes from fisheries and the rest from non-fishery activities. While about 48% had some secondary occupation, only 24% of their income was derived from it. Interestingly, only 4% of their secondary occupations involved harvesting resources from the Sundarbans. A few womenfolk (16%) earned additional income through animal husbandry, bawali, manual labor, fishing, handicrafts, poultry, and tea vending, and tailoring. The fishers use a variety of fishing gears that include: Cast net, Crab line, Crab trap, Creek net, Drag net, Drifting gill net, Fixed floating gill net, Long line, Hook and rod, Long-shore net, Otter fishing, Set-bag net, Post-larvae box net, Post-larvae hand drag net, Post-larvae hand push net, Post-larvae pole net and Post-larvae

TABLE 10.6

Existing Fisheries Management and Conservation Rules in SMF

Legislation	Major Outcomes	Implementing Agencies
Indian Forest Act, 1878	(i) Empowers the FD to manage the inshore and offshore fisheries in the Sundarbans and near shore 20 km marine waters	FD
Hunting and Fishing Rules, 1959	(i) A fishing permit is required to fish in reserved or protected forests; (ii) Royalty may be charged on fish caught in tidal waters of reserved and protected forests; (iii) It is illegal to use poison, explosives or fixed engine fishing gears, or to dam or bale water in reserve and protected forests	FD
Major Fisheries Regulations for SRF: Khal Closure Regulation (1989); Collection and Export of Live Crab Regulation (1995); Closed Season Regulation (2000)	(i) Closes 18 khals permanently for fishing to ensure natural fish breeding; (ii) Closes the entire SRF for crab fishing from December to February to ensure crab breeding; (iii) Closes fishing in the entire SRF for five species (*P. pangasius, P. canius, L. calcarifer, M. rosenbergii, S. serrata*) during 1st May–30th June to ensure natural breeding.	FD
Wildlife Sanctuary Regulations, 1999	(i) Fishing is permanently prohibited in the three wildlife sanctuaries of SRF	FD
Other Regulations for Fisheries in SRF	(i) It is illegal to place nets across a Khal and thereby completely block it; (ii) it is illegal to sting a rope transversely across a Khal.	FD

Adopted from Hoq (2007).

set-bag net. Of these, gillnets and set-bag nets are the most commonly used gears in the wildlife sanctuary (GoB 2018).

10.3.3.1 Conflicts over Resource Exploitation due to Poor Management

Different users of fishery resources are involved in fisheries management to balance the competing demands. Due to differences in power, interests, values, priorities, and manner of resource exploitation conflicts among fisheries stakeholders arise (Murshed-e-Jahan, Belton, and Viswanathan 2014). Intra-group users can also be involved in conflicts, such as conflicts arise as small-scale fishers, who are present in millions, interact with stakeholders including other fishers. Failure of authorities is also liable for this type of conflicts. The sector suffers further due to a lack of inter-agency coordination among the various governments in situations with jurisdiction over fisheries (Murshed-e-Jahan, Belton, and Viswanathan 2014). Conflicts of this type occur when a group of fishers asserts that their fishing operations and rights are negatively affected by the action of another group of fishers or stakeholders. The study found that disputes gravitate around competing claims on fishing grounds mostly between active gears such as Small Mesh Drift Nets (SMD), but also occur between active and passive gears such as SMD and

TABLE 10.7
Historical Changes in the Principles of Fisheries Management

Periods	Legal Initiatives	Major Objectives	Major Outcomes
Pre-British rule (before 1757)	No management	Resource extraction	Resource extraction
British rule (1757–1947)	Charts of India Forest, 1855	Conservation idea generation. Controlling the resources	Awareness with important realization and first regulations regarding felling trees for revenues
Indian forest act 1894	Declaration of 'Reserve forest' 1875–1876 under Indian Forest Act, 1894	Introduction of formal forest policy to be administrated. Targeting benefits with commercial management of wood and non-wood forest products for public at large and for the local people under regulations and rights.	Resource extraction
Pakistan Rule (1947–1971)	Forest Policy of Pakistan, 1955 Revised Forest Policy of Pakistan, 1962	Classifications of the Sundarbans on the basis of its utility and objectives. Acceleration of timber harvesting. Speed up regeneration for increased harvesting. Ignorance of the principal of sustainable forest use and rights of local people	Over-exploitation of forest resources from the Sundarbans. Protection of wildlife and habitats. Realization of overuse, ecological degradation.
Bangladesh Rule (1971–present)	National forest policy, 1979	Qualitative improvement based on modern trend and technology for extraction and utilization of forest resources. Coastal mangrove plantation	There were inconsistencies as conservation leaves little incentives to expand forest-based industries and becomes detrimental to forest health by increasing degradation through illegal harvesting. Inappropriate land tenure agreement caused illegal feelings of mangrove trees and encroachment of the land.
	Revised national forest policy, 1994	Multidimensional use of its resources including water and fish. Keeping the bio-environment intact and consideration of global warming and climate change for its existence. Use of appropriate extraction technology. Identification of protected areas. Ensuring participation of local people	Sustainable management

Adopted from Hoq (2014).

Marine Set Bag Nets. When two parties fishing in the same area accidentally drift into each other and become entangled the nets may need to be cut, thereby also resulting in conflicts between the two parties. Conflicts of this type can also happen between fishers and boat owners when the latter refuse to pay fishers' according to their earlier commitments, or are reluctant to provide safety equipment before the fishing voyage. Boat owners who were interviewed admitted that this often causes conflicts with fishers. However, owners stated that fishers did not always provide them with the true figures of fish catches. They suspected some fishers under their employ illegally sold fish at sea in order to gain extra benefits. According to owners, this is the main reason for conflict with the fishers they employ. Fishers and boat owners also reported conflicts with fish traders due to the nature of market governance structures. Conflict arises when local fish traders create a syndicate and force the fishers or boat owners to sell their catch directly to them, preventing traders from other areas from competing. Fishers reported that they never received the perceived 'true' market value from these fish traders. Conflict also happened between money lenders and fishers when the latter failed to repay their loans (Habib et al. 2020).

10.4 ANALYSIS AND DISCUSSION

The research plan was to understand the mangrove Fisheries Management (SMF) and their associated ecosystem services dependent people life and livelihoods. The small-scale fishermen from the mangrove area are poor, uneducated and has less bargaining power to involve in decision-making process for the mangrove management activities of the government. In stakeholder analysis they have less involvement but more interest. But they are neglected when the policies are making to protect the Sundarbans. Not only the small-scale fisheries but also the agricultural activities should be addressed with other sectors. People in that area involved in many other activities to cope up the adverse climate change situation. They faced extreme level of post-disaster crisis in their life and livelihoods. Biodiversity composition changes due to the saline water intrusion, sea level rise, cyclone that affects the wild life composition in the mangrove forest which affects the small-scale poor fisherman and mangrove-dependent people. With other factors, natural disaster triggers the losses of deforestation that extinguish the fish stocks and habitat for other aquatic animals. Impacts of natural disaster and banning of fishing area jeopardize the resilience of fishing communities, and they face the lack of resources and enough food consumption. To cope up with the adverse living mechanism and living style of the people, money lending mechanism, lack of resources people are engaged with Shrimp fry collections which has negative social and ecological impacts. Estuarine or mangrove fisheries are part of broader marine environment. Hilsa fishery has more economic value and showed the cultural identity in Bangladesh. So, Government has more attractive policy to restore the stocks of Hilsa fishery by applying the PESs but other fishermen are in negligence. Shrimp aquaculture is a prominent aquaculture in the study region where land-use conflicts are very common. Large shrimp farm owners and rich political people force the marginal land owner to sale the land forcefully. There are also small-scale shrimp farmers. About 80%–90% people are engaged in shrimp farming in that area. They are engaged in from wild shrimp larvae catching to shrimp processing industries. Women and children are also involved in this industry. If the gher (small shrimp pond) is small-scale they repair dykes, clean ponds and stocking shrimp larvae so that they have a sense of ownership as well. However, the study area is highly productive biodiversity and that needs to be protected to protect people in that area. So, the government has taken initiatives to support the costal fisheries management for protection the habitat, maximize the benefit of utilization of resources in a sustainable way, and minimize the conflicts between the stakeholders. Government and non-Government organizations are working together to get the maximum outcome from a project based coastal management. But, lack of coordination between different sectors create jeopardize of the initiatives for the co-management of natural resources. Role of civil society is also unclear and their involvement also negligible. Though Government has taken initiatives to manage the mangrove fisheries and aquaculture in that area, natural disasters make the community vulnerable as they are failure to develop their coping capacities. Lack of public access to science based information, lack of stakeholders involvement including civil society, squalor of home-grown knowledge on hazard, livelihood transformation, gendered based economic activities and inadequate adaptive capacity are one of the major causes for the failure of the management practices in coastal fisheries management. However, effectiveness of a policy must be evaluated from the point of view of the social and institutional mechanisms that bring about changes in the behaviour of actors and organizations thereby bringing about the intended outcomes (Pal et al. 2017).

So, the government should focus not only ecosystem management or co-management of fisheries resources in the area but also include the disaster management policy to reduce the critical gap in response and recovery to recognize the vulnerability of the community to multiple hazards for developing appropriate capacity building for communities to measure and achieve the risk understanding so that they can appropriately adjust with the adverse and become more resilient before the hazardous event occurred (Esraz-Ul-Zannat et al. 2020). For example, Mangrove for future project that promotes investment in Coastal ecosystem conservation for sustainable development with IUCN and UNDP and use of the resources by addressing the challenging issues to coastal ecosystem and livelihood that promotes an ocean-wide approach for coastal management and help building the ecosystem dependent coastal communities.

10.5 CONCLUSION

The study focused on the small-scale fisheries management status in the Sundarbans mangrove forest where people are dependent on the agriculture, shrimp culture and natural resources for their life and livelihoods. Human-natural resource use conflict must be minimized for sustainable natural resources management. Good governance must be initiated including all stakeholders with clear rights of the stakeholders and disaster-affected people. The poor fishers, small-scale farmers, women, widow and minor communities in that area should be highly prioritized. The GoB-integrated disaster risk reduction, adaptation strategies and coastal zone management for most vulnerable region of Sundarbans area and increase the social safety nets to minimize the exogenous shocks in response to local demand and donor initiative projects (Hassan et al., 2013) and more initiatives should be taken to strengthen the community resilience. Human rights organization can take steps to make people more aware about the rights during and after the disaster to bring a sustainable coastal resource use by the locals. Besides, political commitment for biodiversity conservation, formal legal arrangement, strengthening forest protection law, sustainable education and awareness, research, monitoring, national and international collaboration may help to develop good governance for developing a resilient community for climate change, disaster adaptation for a sustainable mangrove fisheries and forest.

REFERENCES

Allison, E.H., Perry, A.L., Badjeck, M.C., Neil Adger, W., Brown, K., Conway, D., Halls, A.S., Pilling, G.M., Reynolds, J.D., Andrew, N.L. and Dulvy, N.K., 2009. Vulnerability of national economies to the impacts of climate change on fisheries. *Fish and Fisheries*, 10(2), pp. 173–196.

BFD (2010) 'Integrated Resources Management Plans for The Sundarbans (2010–2020) Vol-1', Nisorgo Network, Bangladesh Forest Department (BFD), Ministry of Environment and Forest, People Republic of Bangladesh, 1(December), p. 281. Available at: http://www.nishorgo.org/tbltd/upload/pdf/0.12860600 1357814923_SRF_IRMP_Volume1_16.1.2011-Revised Version2_6–9–11.pdf.

DoF. (2018). Yearbook of Fisheries Statistics of Bangladesh, 2017–18. Fisheries Resources Survey System, Department of Fisheries. Bangladesh: Ministry of Fisheries, 2018. Volume 35: p. 129.

Esraz-Ul-Zannat, M., Abedin, M.A., Pal, I. and Zaman, M.M., 2020. Building resilience fighting back vulnerability in the coastal city of Khulna, Bangladesh: A perspective of climate-resilient city approach. *International Energy Journal*, 20(3A), pp. 549–566.

Faridah-Hanum, I., Latiff, A., Hakeem, K.R. and Ozturk, M. eds., 2013. *Mangrove Ecosystems of Asia: Status, Challenges and Management Strategies*. Springer Science & Business Media, Berlin.

Gain, A.K., Ashik-Ur-Rahman, M. and Benson, D., 2019a. Exploring institutional structures for tidal river management in the Ganges-Brahmaputra Delta in Bangladesh. *DIE ERDE–Journal of the Geographical Society of Berlin*, 150(3), pp. 184–195.

Gain, A.K., Ashik-Ur-Rahman, M. and Vafeidis, A., 2019b. Exploring human-nature interaction on the coastal floodplain in the Ganges-Brahmaputra delta through the lens of Ostrom's social-ecological systems framework. *Environmental Research Communications*, 1(5), p. 051003.

Gain, D. and Das, S.K., 2014. Present status and decreasing causes of shellfish diversity of Passur river, Sundarban, Bangladesh. *Aquaculture, Aquarium, Conservation & Legislation*, 7(6), pp. 483–488.

Go, B. 2018. "Community-Based Resources Management Plan (CBRMP) of the Wildlife Sanctuaries for Dolphins in Bangladesh Sundarbans," no. July: 102.

Guha, I. and Roy, C., 2016. Climate change, migration and food security: Evidence from Indian Sundarbans. *International Journal of Theoretical and Applied Sciences*, 8, pp. 45–49.

Habib, K.A., Neogi, A.K., Nahar, N., Oh, J., Lee, Y.H. and Kim, C.G., 2020. An overview of fishes of the Sundarbans, Bangladesh and their present conservation status. *Journal of Threatened Taxa*, 12(1), pp. 15154–15172.

Hassan, R., Islam, M.S., Saifullah, A.S.M. and Islam, M., 2013. Effectiveness of social safety net programs on community resilience to hazard vulnerable population in Bangladesh. *Journal of Environmental Science and Natural Resources*, 6(1), pp. 123–129.

Hoq, M.E., 2007. An analysis of fisheries exploitation and management practices in Sundarbans mangrove ecosystem, Bangladesh. *Ocean & Coastal Management*, 50(5–6), pp. 411–427.

Hoq, M.E., 2014. Management strategies for sustainable exploitation of aquatic resources of the Sundarbans Mangrove, Bangladesh. In *Mangrove Ecosystems of Asia* (pp. 319–341). Springer, New York.

Hossain, M.F., 2011. Disaster management in Bangladesh: Regulatory and social work perspectives. *Journal of Comparative Social Welfare*, 27(1), pp. 91–101.

Hossain, M.M., Islam, M.A., Ridgway, S. and Matsuishi, T., 2006. Management of inland open water fisheries resources of Bangladesh: Issues and options. *Fisheries Research*, 77(3), pp. 275–284.

Hussain, M.Z., 2014. Bangladesh Sundarban Delta Vision 2050: A first step in its formulation-document 2: A compilation of background information. IUCN, International Union for Conservation of Nature, Bangladesh Country Office, Dhaka, Bangladesh, pp.1–192.

Islam, M.M. and Chuenpagdee, R., 2013. Negotiating risk and poverty in mangrove fishing communities of the Bangladesh Sundarbans. *Maritime Studies*, 12(1), p. 7.

Islam, M.M., Mohammed, E.Y. and Ali, L., 2016. Economic incentives for sustainable hilsa fishing in Bangladesh: An analysis of the legal and institutional framework. Marine policy, 68, pp.8–22.

IUCN, 1997. "Sundarban Wildlife Sanctuaries Bangladesh." World Heritage Nomination IUCN Techn.

IUCN, 2016. The Sundarbans Assessment report. https://worldheritageoutlook.iucn.org/explore-sites/wdpaid/145580#:~:text=The%20ecosystem%20is%20showing%20degradation, going%20coastal%20erosion%20and%20retreat.

IUCN and UNESCO, 2016. Report on the Mission to The Sundarbans World Heritage Site (Bangladesh). Gland, Switzerland and Paris, France: IUCN and UNESCO World Heritage Centre.

Izumi, T. and Shaw, R., 2015. *Disaster Management and Private Sectors*. Springer, Tokyo.

Lwin, K.K., Pal, I., Shrestha, S. and Warnitchai, P., 2020. Assessing social resilience of flood-vulnerable communities in Ayeyarwady Delta, *Myanmar. International Journal of Disaster Risk Reduction*, 51, p. 101745.

Muhibbullah, M., Amin, S.N. and Chowdhury, A.T., 2005. Some physico-chemical parameters of soil and water of Sundarban mangrove forest, Bangladesh.

Mozumder, M.M.H., Shamsuzzaman, M., Rashed-Un-Nabi, M. and Harun-Al-Rashid, A., 2018. Socio-economic characteristics and fishing operation activities of the artisanal fishers in the Sundarbans Mangrove Forest, Bangladesh. *Turkish Journal of Fisheries and Aquatic Sciences*, 18, pp. 789–799.

Murshed-e-Jahan, K., Belton, B. and Viswanathan, K.K., 2014. Communication strategies for managing coastal fisheries conflicts in Bangladesh. *Ocean & Coastal Management*, 92, pp. 65–73.

Mustafa, M.G., Ahmed, I. and Ilyas, M., 2019. Population dynamics of five important commercial fish species in the Sundarbans ecosystem of Bangladesh. *Journal of Applied Life Sciences International*, 22, pp. 1–13.

Pal, I., Ghosh, T. and Ghosh, C., 2017. Institutional framework and administrative systems for effective disaster risk governance–Perspectives of 2013 Cyclone Phailin in India. *International Journal of Disaster Risk Reduction*, 21, pp. 350–359.

Pal I., Shaw R. (2018) Disaster Governance and Its Relevance. In: Pal I., Shaw R. (eds) Disaster Risk Governance in India and Cross Cutting Issues. Disaster Risk Reduction (Methods, Approaches and Practices). Springer, Singapore. https://doi.org/10.1007/978-981-10-3310-0_1

Pal, I., Tularug, P., Jana, S.K. and Pal, D.K., 2018. Risk assessment and reduction measures in landslide and flash flood-prone areas: A case of Southern Thailand (Nakhon Si Thammarat Province). In: Samui, P., Kim, D., and Ghosh, C. (eds.), *Integrating Disaster Science and Management* (pp. 295–308). Elsevier, Amsterdam.

Pomery, R., Thompson, P. and Courtney, C.A., 2016. Marine Tenure and Small-Scale Fisheries: Learning from the Bangladesh Experience and Recommendation for the Hilsa Fishery. USAID Tenure and Global Climate Change Program, Washington, DC.

Porras, I., Mohammed, E.Y., Ali, L., Ali, M.S. and Hossain, M.B., 2017. Power, profits and payments for ecosystem services in Hilsa fisheries in Bangladesh: A value chain analysis. *Marine Policy*, 84, pp. 60–68.

Rainboth, W.J., 1990. The fish communities and fisheries of the Sundarbans: Development assistance and dilemmas of the aquatic commons. *Agriculture and Human Values*, 7(2), pp. 61–72.

Razzak, M.A., Ahsan, M.E. and Nahar, N., 2020. The impact of co-management on the Sundarbans fisheries: Evidence from Sharankhola, Bagerhat, Bangladesh. *Resource*, 10(2), pp. 41–50.

Razzaque, J., 2017. Payments for ecosystem services in sustainable mangrove forest management in Bangladesh. *Transnational Environmental Law*, 6, p. 309.

Roy, A., & Alam, K. (2012). Participatory forest management for the sustainable management of the Sundarbans Mangrove Forest. *American Journal of Environmental Sciences*, 8, 549–555.

Roy, J., Islam, S. T., and Pal, I., 2020. White Paper on Energy, Disaster, Climate Change: Sustainability and Just Transitions in Bangladesh, International Energy Journal, Vol 20, Special Issue 3A, October 2020, Bangabandhu Chair Special Issue (Volume 1), Energy, Disaster, Climate Change: Sustainability and Just Transitions in Bangladesh. Pp. 579–580.

Shah, M.S., Huq, K.A. and Rahman, S.M.B., 2010. Study on the conservation and management of fisheries resources of the Sundarbans. Integrated Protected Area Co-Management (IPAC), Bangladesh.

Thompson, B.S., Bladon, A.J., Fahad, Z.H., Mohsanin, S. and Koldewey, H.J., 2016. Evaluation of the ecological effectiveness and social appropriateness of fishing regulations in the Bangladesh Sundarbans using a new multi-disciplinary assessment framework. *Fisheries Research*, 183, pp. 410–423.

Williams, R., 2014. The socio-cultural impact of industry restructuring: Fishing identities in Northeast Scotland. In: Urquhart, J., Acott, T., Symes, D., and Zhao, M. (eds.), *Social Issues in Sustainable Fisheries Management* (pp. 301–317). Springer, Dordrecht.

Section 4

Vulnerability of Sundarbans through Geospatial Analysis

11 Biophysical Vulnerability Assessment of Indian Sundarbans Mangrove

SayaniDatta Majumdar, Niloy Pramanick,
and SugataHazra
Jadavpur University

CONTENTS

11.1 INTRODUCTION

The mangroves are considered as one of the most productive and diverse ecosystems that have definite adaptations to flourish in the tidal environment of tropics and subtropical coastlines. These littoral forests are found in sheltered estuaries and along the river banks and lagoons extending for several kilometers inland [1]. Mangrove forests may be found either as isolated patch of stunted trees – in highly saline conditions or as luxurious forests with a canopy reaching 30–40 m in height under suitable environmental conditions. These are generally grouped into two types – (i) true mangroves and (ii) mangrove associates [2,3]. True mangrove species like *Heritiera fomes, Bruguiera gymnorrhiza, Avicennia alba,* and *Rhizophora mucronata* thrive the intertidal zones, whereas mangrove associates like *Hibicustilisaceus, Suaeda nudiflora, and Thespesia populnea* can endure both littoral and terrestrial environmental conditions [4].

Mangrove ecosystems have significant ecological and economic values. They are the significant source of food, fuel, and fodder for coastal communities and play a decisive role in the marine food chain by acting as breeding grounds for different types of amphibians, fishes, prawn, shellfishes, and crustaceans [5,6]. It protects the shoreline and island areas from several natural hazards like tropical cyclones and tsunamis by acting as a natural barrier. Not only do they maintain the water quality by trapping runoff sediments and nutrients from the polluted coastal water but also prevent coastal erosion [7,8]. Recently blue carbon is globally treated as an important climate change mitigation tool because of their potential to trap and store huge amount of organic carbon in the sediments for millennia, and mangroves along with sea grasses are notable Blue Carbon ecosystems [9,10]. Moreover by serving as ecotourism hotspots and providing aquaculture, fuel, honey, traditional medicines, etc., the mangroves further contribute to the human livelihood directly or indirectly [11,12].

Covering around 75% of the world's coastline the spatial extent of mangroves is mostly restricted within the tropics and subtropics [13,14]. Globally, total mangrove cover is 156,220 km^2 [15]. Almost all the mangroves throughout the world occur in patches, especially along the sea coasts and deltaic regions, the only exception being the Sundarbans, which alone marks the single largest patch of mangrove forest shared between India and Bangladesh covering ~10 million ha [16]. Mangrove forests of India constitute only 0.67% of total forest cover in the country [17,18]. The narrow and steep slope attributed by the presence of Western Ghats and higher salinity levels due to lack of major water flowing rivers have restricted the growth of mangroves mostly in the funnel shaped estuaries and backwaters of the western coastal margins [19]. In contrast to these particular geomorphic settings of western coasts, the prolonged coastlines, plenty of fresh water along with sediments brought in by the large rivers like Ganges, Godavari, Mahanadi, and Krishna have facilitated luxuriant growth of mangroves in the deltaic areas of Eastern coasts [20]. The island mangrove habitats predominantly thrive in the intertidal zones along the coast [20].

During recent decades, mangroves have been injudiciously exploited by coastal population for timber, fodder, fuel, charcoal and honey [21,22]. Around 35% of the mangrove forest area has disappeared globally since 1980 [22,23]. Bangladesh Sundarbans, mangroves, have lost 45% of its coverage due to logging, shrimp farming and various natural catastrophes [24]. Apart from the increased anthropogenic pressure, relative rise in sea-level concomitant with climate change has been noted as a potent threat to the survival of mangroves [25–28]. Reduced mangrove area and health increase the threat to safety of coastal population as well as shoreline development from coastal hazards like storm surges, tsunami, hurricanes, and erosion [7,26,29].

11.2 VULNERABILITY AND ITS COMPONENTS

According to the Intergovernmental Panel on Climate Change (IPCC) second assessment report, vulnerability is defined as the degree to which a system is exposed to the adverse effects of climate change [30,31]. It has three essential dimensions – Exposure, Sensitivity, and Adaptive Capacity [31–33].

Exposure refers to magnitude of stresses that a system may encounter [31,33,35]. While sea level and rainfall variations are some of the potential exposure components that are directly related to climate change [34,36,37], there are also some exposure factors which being independent of climate change can lead to definite mangrove vulnerability. The tidal ranges, fluvial sediment supply, variability in wave energy are such factors which govern a particular geomorphic setting of mangroves [33].

Sensitivity refers to the extent by which a system is affected by various exposure components [33,38]. It is governed by the system's inherent characteristics to withstand the various impacts of climate change in the form of, flooding owing to rise in sea level, elevated sea surface temperature, decreased fresh availability due to rainfall variations, frequent occurrence of storm, etc. (IPCC 2014). Deteriorating mangrove health, mangrove forest loss, decline in forest productivity, and biodiversity are some of the specific sensitivity indicators for determining the vulnerability [39,40].

Adaptive capacity for a mangrove ecosystem can be determined by its adaptation to migrate landwards when the rate of relative sea level rise exceeds net vertical accretion [26,41]. Response of individual species to different impacts as well as human response and their participatory actions can be some of the significant adaptive strategies to reduce a system's susceptibility to climate change exposures [33].

11.3 VULNERABILITY ASSESSMENT METHODOLOGY

Vulnerability assessment of mangroves of Indian Sundarbans was carried out in response to climatic change and associated sea level rise in the Northern part of Bay of Bengal. A spatial vulnerability assessment model was adopted by taking into consideration the bio-physical parameters in spatial and non-spatial formats as used by Ref. [42] in his study of mangroves of Western Niger Delta. The data collection, pre-processing and quantification of various natural indicators that contribute to the mangroves increasing susceptibility to climate changes were aided by integrated remote sensing applications and suitable Geographical Information System (GIS) platforms (here Arc GIS 10.3.1).

11.3.1 STUDY AREA

The Indian Sundarbans delta encompassing an area of 9,630 km^2 is bordered by "Dampier-Hodges line" in the north, Bay of Bengal in the south, Harinbhanga-Raimangal river in the east and Hooghly River in the west [43–45]. The name "Sundarbans" was named after "Sundari" (*Heritiera fomes*), a dominant mangrove tree species [43,46].

Hugli, being a tributary of river Ganga, acts as the main artery of the Sundarbans mangrove ecosystem. The region is well drained by the freshwater flown of river

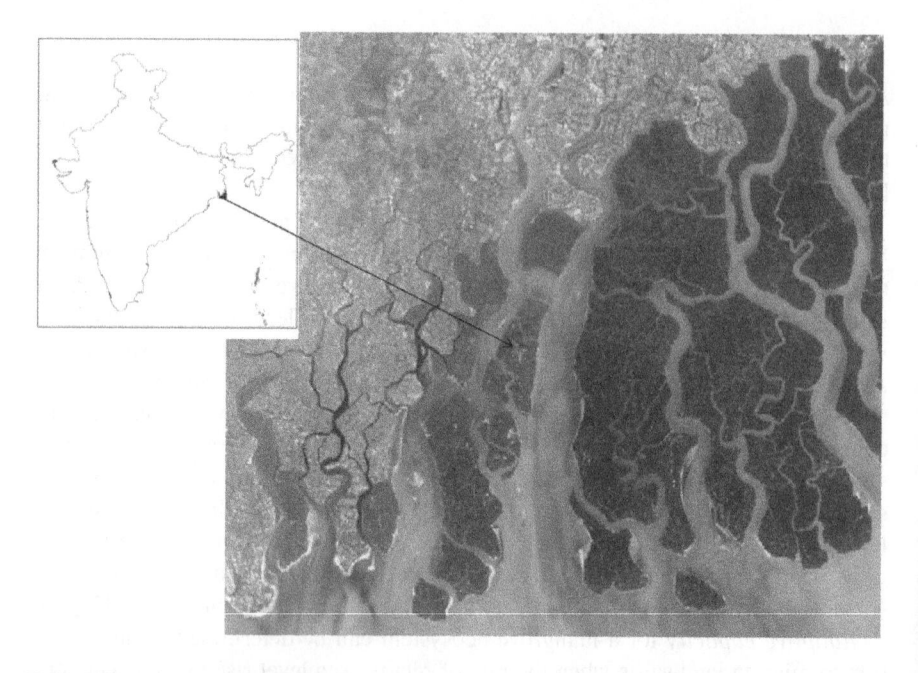

FIGURE 11.1 The study area.

Ganga, Brahmaputra and Meghna, which reveal high seasonal variation in their discharge [16].

The region of Indian Sundarbans (Figure 11.1) has a humid, tropical, maritime climate with 80% of annual rainfall occurring in the summer monsoon (southwest monsoon) months. The central and northern part of the region gets an annual rainfall of about 1,650–1,800 mm whereas 2,790 mm of downpour is received by the outer coast. During June and July the mean maximum temperature is 29°C, whereas mean minimum temperature is 20°C. Heavy rainfall in the monsoon seasons leads to high river discharge that ranges between 2,952 and 11,897 m^3 s^{-1} which steadily declines during non-monsoonal months ranging between 900 and 1,500 m^3 s^{-1}) [47].

The mangrove forests of Indian Sundarbans covering an area of 2,400 km^2 are located in the districts of 24 Parganas (North and South) within 21° 32′–22° 40′N latitudes and 88° 05′–89°51′E longitudes. Accounting for 62% of the country's total mangrove forests [46,48], Indian Sundarbans region as a rich species diversity characterized by a wide variety of flora and various species of fish, birds, reptiles, and benthic invertebrates (like arthropods, molluscs, etc.), phytoplankton, zooplankton, bacteria, fungi, etc. [18].

11.3.2 CLIMATE CHANGE AND HEALTH OF INDIAN SUNDARBANS MANGROVES

A. **Rise in temperature:** According to [49,50], earth is anticipated to warm 2°C–6°C by 2,100 owing to uncontrolled human forcings. After a thorough analysis of surface air temperature anomaly data of Sundarbans and adjoining Bay of Bengal by [51] it was found that there had been an increasing trend in rise in temperature due to the existing global warming phenomena.

The recent analysis of various climatic parameters of Indian Sundarbans by [52] revealed an increasing trend in surface water temperature by 6.14% and 6.12% in western and eastern sectors of the region, respectively. An increase in temperature may impact the mangroves by altering the seasonal forms of reproduction or changing the length of time between flowering and fall of mature propagules of the plants [53,54].

B. **Rise in Sea-level:** The combined effect of global warming and melting of glacial ice has resulted in eustatic sea level rise by 10–20 cm due to thermal expansion of ocean [55]. The scientists have estimated that the twentieth century witnessed a global sea level rise at the rate of 1.0–2.0 mm and by the end of twenty-first century, projected rise in mean sea level projections would range from 0.09 to 0.88 [49]. Sea level rise is considered as a potent factor behind recent land loss in Indian Sundarbans coupled with shoreline retreat (maximum of 2.8 km between 1968 and 2014) as witnessed by the southern islands. While Sagar Island has shrunk by 15%, the other three other islands, Lohachahara, Suparibhanga and Bedford, have completely disappeared.

C. **Effect of salinization:** Ericson et al. (2006) reported about the world's highest rate of sea level rise in the Bay of Bengal of >10 mm year^{-1}. It is a well-established fact that the survival of mangroves would be threatened if their landward migration is restricted by dearth of space to spread out supported by higher rate of sea level rise at the expense of their migration [56–58]. The healthy mangroves require a daily regime of freshwater-seawater flux [58–61] which further provide perfect salinity conditions and any sorts of variations in salinity profile of the adjacent water column or soils result in the differential growth of above ground biomass of salinity sensitive mangrove species [62]. Consequent upon silting up of several distributaries of River Ganges the supply of freshwater to the Sundarbans is severely limited [63]. Thus the western sector of the Sundarbans which otherwise received water from Farakka Barrage is experiencing a decline in salinity, whereas all the distributaries of Ganges acting as arms of the sea either submerges the mangroves during high tide or the retreat of water during low tide facilitates very scanty or no input of freshwater at all from the upstream in the eastern sector [64]. Such ongoing salinity imbalances is responsible for the most tenacious flora of the Sundarbans, i.e. *Heritiera fomes* to become threatened [63,64].

D. **Changing pattern of tropical cyclones:** In the recent times during the post-monsoon season (October–November) some of the deadliest tropical cyclones have originated over the Bay of Bengal [65]. During 1981–2010 the intensity of major tropical cyclones increased with greater wind speed of 49 m s^{-1} in the post-monsoon season. While increased sea surface temperature induced by global warming and upper ocean thermal layers facilitated intensification of the cyclones, heightened convective instability in the atmosphere further promoted their growth [65,66]. According to Islam [67] storm-inflicted mangrove forests of the Sundarbans take a period of 25 years for full recovery. Due to recent episodes of frequent storm visit (1988, 1991, Sidr in 2007, Nardis in 2008 and Aila in 2009) and the Asian tsunami of 2004 the regeneration ability of mangroves has weakened significantly.

E. **Change in precipitation and temperature regime:** El Niño Southern Oscillations (ENSO)-induced climatic variations have profound effect on rise in sea surface temperature as well as irregularities of monsoon rainfall and intensification, coupled with flooding events, in Southeast Asia [68]. With gradual increase in severity of the decadal ENSO events, the frequency of occurrence of heat spells and rainfall irregularities has increased throughout many parts of the world. Such anomalies are in turn adversely affecting the biodiversity of wetland ecosystem and primary productivity of species of Sundarbans in the form of variations in the tidal regime, fluctuations in riverine discharge, estuarine sediment dynamics [69].

F. **Erosion and Accretion with respect to sea level change:** Sea level rise is solely responsible for the recent land losses in the Sundarbans region which further triggered significant loss of mangrove cover in the coastal stretches. Sundarbans have lost about 284 sq.km of its land in the last 50 years at an increasing rate of 2.85–5.5 km^2 while the rate of accretion has been only 84 km^2 in the past 50 years [70,71]. Within the time period of 1989–2014, Indian Sundarbans mangrove region witnessed coastal erosion at the rate of 5.70 km^2 per year [15]. While Lohachahara, Suparibhanga, and Bedford have completely disappeared, Sagar Island has shrunk by 15% and Jambudwip by 50% [72,73].

11.3.3   SELECTION OF INDICATORS FOR VULNERABILITY ASSESSMENT

By implementing the definition of vulnerability laid down in IPCC AR IV framework, a vulnerability index (VI) was developed based on the indicators selected from the aspects of exposure, sensitivity and adaptive capacity. The criteria for selection of components were based on indicator's inherent characteristics of reflecting the processes and consequences of impacts of climate change on mangrove ecosystems. The various components of exposure sensitivity and adaptive capacity selected are the following listed in Table 11.1 for thorough understanding.

11.3.4   DESIGN OF VULNERABILITY INDEX

Following the methodology of [90], our GIS-based assessment of mangrove vulnerability consisted of four basic modules: (i) data gathering, (ii) data input and pre-processing, (iii) data storage and processing, and (iv) data output. The first step included collection of spatial and non-spatial data from extensive literature survey, remote sensing products, and visual interpretation of cartographic data sets as well as field work. After selecting the appropriate parameters of exposure sensitivity and adaptive capacity, all the relevant data sets pertaining to these three fields were accumulated and pre-processed. In the next step all the individual data sets were integrated using the Geo-Relational model in GIS environment [91] where all the attribute information that describe the real-world spatial data sets were depicted in the form of points, lines, and polygons. Using the Composite Map model the attribute values of various geographical features were stacked one above the other to form a new layer and a new attribute to express [91].

Thus using Arc GIS 10.3.1 version all the identified components of vulnerability were depicted in appropriate layers and evaluated.

TABLE 11.1

Factors, Components and Their Measurement Techniques

Factors	Components	Measurement	Units	References
Exposure	Tidal range	Local tidal gauge records	Meters	
	Precipitation	TRMM rainfall data	mm	Kumar et al. [75], Shrestha et al. [74]
	Air and Sea surface temperature	IMD time series temperature data, Future temperature data from available climate projections	Degree Celsius	Clough and Sim [76], Cheeseman et al. [77,78]
		MODIS SST	Degree Celsius	Field [34], Alongi [36]
	Earth skin temperature	NASA AGRO CLIMATE DATA	Degree Celsius	Buyadi et al. [79], Islam and Islam [80], Thakur et al. [81]
	Specific and relative humidity	NASA AGRO CLIMATE DATA	Na	
		NASA AGRO CLIMATE DATA		Irabor et al. [42]
	Current speed, wind speed	Frequency and wind speed of storms – Ocean Motion	Square kilometers	Cahoon [82], Alongi [36], Faraco et al. [83]
		NASA AGROCLIMATE DATA	Square kilometers	
Sensitivity	Mangrove forest health	Forest health assessment using Remote Sensing and GIS Techniques (EVI, NDVI, SAVI, TNDVI etc.)	Na	Fiu et al. [84], Ghosh and Mukhopadhyay [85]
	Soil Salinity Slope Elevation	Field data collected SRTM DEM	Ppt	Ellison and Zouh [86], Ellison and Strickland [87]
	Seaward edge retreat	Erosion and accretion mapping using RS GIS techniques.	Square meters	Fiu et al. [84], Ellison and Zouh [86], Punwong et al. [88], Ellison and Strickland [87]
	Reduction in mangrove area	Forest cover change mapping from temporal satellite images.	Square kilometers	Ellison and Zouh [86]
	Saline blank	Land use land cover	Square meters	Ellison and Zouh [86]
Adaptive Capacity	Regeneration rate after an extreme event	Existing Literature	Square meters	Marimuthu et al. [89]

1. **Erosion and accretion in Sundarbans with respect to sea level fluctuations-induced shoreline change:** During the recent years it has been observed that the islands of Hoogly Malta estuary are suffering from significant land loss owing to relative sea level changes and the rate of erosion and accretion

phenomena is strongly correlated with the sea level fluctuations in different island systems [51]. In order to estimate this sea-level variation, simulated tide data of Diamond Harbour and Sagar stations were obtained for the period 1990–2018 from X Tide web portal. This data were further verified with the data obtained and examined by Hazra et al. [51] from the tide gauge data of Sagar island observatory (21°31′N 88°03′E) at 6 hours interval. As used by Hazra et al. [51] the high and low tides of the months provided the monthly mean sea level of that month. The mean monthly tide values were further averaged to provide annual mean sea level. The least square regression line through the tidal values plotted provided the requisite value of annual mean sea level. Tidal range was also calculated as the difference between the highest high tide data for the month of September and lowest low tide of the month for March. For further analysis of sea surface height TOPEX-POSEIDON, Jason series Satellite-Altimeter data were used and calibrated. The global average sea level trend generated from Integrated Multi Mission Ocean Altimeters TOPEX/Poseidon Jason-1 and OSTM/Jason-2 Version 4.2 data of resolution 360° × 132° was obtained from PO.DAAC (Physical Oceanography Distributed Active Archive Centre) of NASA's Jet Propulsion Laboratory for this purpose.

Since shoreline is one of the dynamic landforms in the coastal area accurate detection and monitoring of it is necessary for understanding other coastal processes like erosion and accretion. For shoreline change detection all the satellite scenes from 1990 to 2018 (at 5 years interval) were pre-processed and geo-referenced as suggested by Nayak [92]. Following the methodology of Mujabar and Chandrashekhar [93], the satellite scene 1990 was taken as the base year for delineating mean high tide line as the land and sea boundary. All the shorelines were manually digitized to near perfection and the shape files were exported to suitable GIS platforms (here Arc GIS 10.5) for further analysis. The extracted shorelines were overlaid over the base year to produce the shoreline change map. The erosion and accretion due to shoreline change were further analyzed from the map (Figure 11.2).

2. **Assessing meteorological data**
    i. **Sea surface temperature:** Moderate Resolution Spectroradiameter (MODIS) level 3 binned monthly night time SST products were obtained for the North Indian Ocean from NASA Ocean Color Web portal at 4 km spatial resolution for 15 years from 2003 to 2018. The downloaded data were further processed for analysis using the Sea DAS software package of Ocean Color web. Averaged monthly means of 12 months (Jan–Dec) were computed and the seasonal temperature variability of the last two decades was observed.
    ii. **Precipitation:** Tropical Rainfall Measurement Mission rainfall data were analyzed for determining the bio-physical characteristics of mangroves which tends to change with respect to seasonal variability [75]. Monthly precipitation data products (TRMM_3B43_v7) of the study region were obtained from NASA GIOVANNI web portal (https://giovanni.gsfc.nasa.gov/giovanni/) for a period of 28 years (1990–2018). The data were further processed and suitable statistical inferences were drawn using the R programming language.

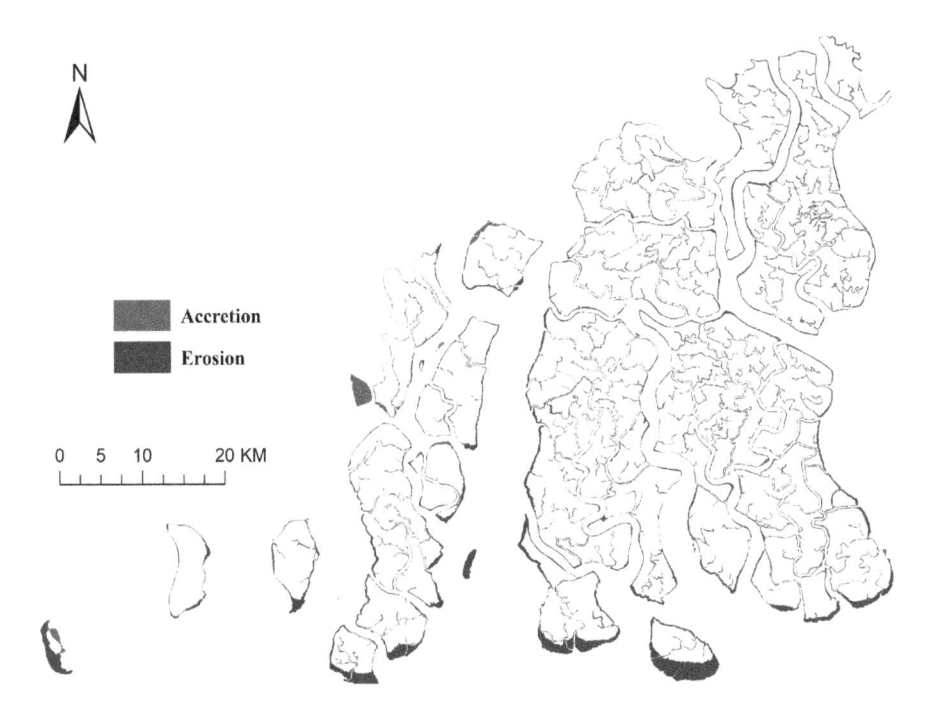

FIGURE 11.2 Erosion and accretion in Indian Sundarbans Region.

iii. **Air temperature**: Monthly air temperature of the study region was incorporated from Indian Meteorological Department (IMD) and analyzed for the same time frame of 28 years, i.e. 1990–2018.

iv. **Other parameters**: Earth's skin temperature, wind speed, specific humidity and relative humidity data of the Sundarbans region were acquired from POWER Project Data Sets portal [95] of NASA which provide solar and meteorological data sets in various formats. The data regarding frequency and wind speed of various storm occurrences within the time frame of 1990–2018 were obtained from NASA's Ocean Motion and Surface Current web portal and further analyzed in Arc GIS 10.5 platform.

3. **Determining the areas of mangrove cover:** The areal extent of mangroves has significantly changed along with extensive land loss and land gain by erosion and accretion of sediments owing to sea level fluctuations. Remote sensing-based multitemporal approach was used for mapping the mangrove areas as well as to monitor changes underwent by the forest cover over a span of 28 years (Figure 11.3). After applying necessary image pre-processing techniques like radiometric calibrations and atmospheric corrections over the same multi-spectral data sets (as used in mangrove health assessment). Normalized Difference Vegetation Index (NDVI) algorithm was performed for estimating vegetation cover using near red and infra-red bands. Such ratio indices help to convert multispectral information into a single index. Moreover NDVI is a necessary pre-classification step to separate mangrove vegetation from other vegetation cover. An unsupervised classification on

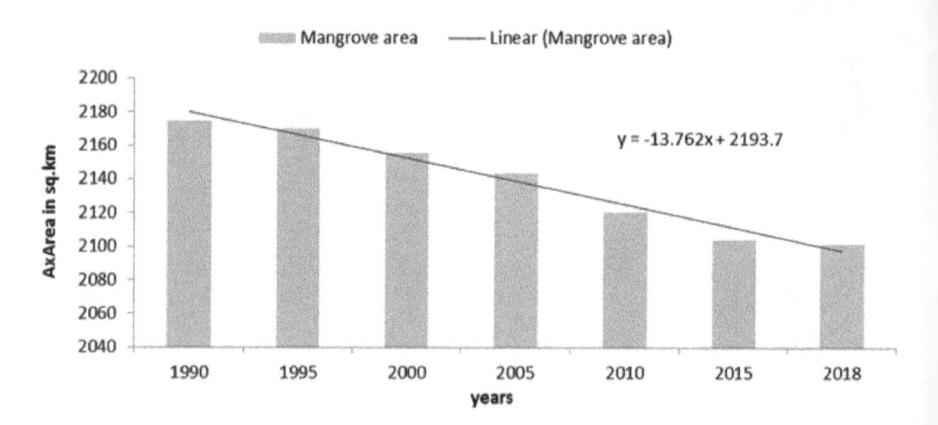

FIGURE 11.3   Temporal change in area of mangroves of Sundarbans.

the NDVI images using K-means clustering algorithm generated 150 classes at the convergence level of 95%. All the mangrove classes were identified with reference to the field data and merged into a single mangrove category. For identifying changes in mangrove area, multi-date data classification technique based on single analysis of multi-temporal data sets [94] was adopted where the classes where changes have occurred showed statistical differences from the classes where no change has taken place.

4. **Defining health of Indian Sundarbans mangroves:** To analyze the health of the Indian Sundarbans mangroves stress maps were prepared on a desired time scale. Following the methodology used by Rahman et al. [97] and Ghosh and Mukhopadhyay [85], National Oceanic and Atmospheric Administration (NOAA) Advanced Very High Resolution Radiometer (AVHRR)-based Global Vegetation Index of 16 km resolution was used for assessment of mangrove health. Global Vegetation Index (GVI) refers to a combination of Vegetation Condition Index (VCI) and Temperature Condition index (TCI). While VCI indicates plants greenness in terms of long term ecosystem changes induced short-term weather-related NDVI variations [96], Temperature Condition Index reflects the thermal conditions in terms of brightness temperature. GVI is based on NDVI and brightness temperature (BT) AVHRR products, which are obtained from the visible (VIS, 0.58–0.68 µm, Ch1), near infrared (NIR, 0.72–1.00 µm, Ch2) and (thermal) and infrared (IR, 10.3–11.3 µm, Ch4) AVHRR channels.

$$VCI = 100 \left(NDVI - NDVI_{min} \right) / \left(NDVI_{max} - NDVI_{min} \right)$$

$$TCI = 100 \left(BT_{max} - BT \right) / \left(BT_{max} - BT_{min} \right)$$

While calculating GVI, high frequency noises were removed from NDVI and BT products, rainfall and associated weather variations, and fluctuations in mean seasonal cycles were also estimated [98]. Finally Vegetation Health Index was calculated as a linear combination of VCI and TCI.

$$VHI = 0.5\,VCI + (1 - 0.5)\,TCI$$

All the three indices were scaled from 0 (highly stressed vegetation health) to 100 (very healthy vegetation) [85,97]. The spatial average values of Vegetation Health Indices were calculated by averaging the GVI product, over land pixels of Indian Sundarbans region. The area specific-weighted average vegetation health indices were finally estimated for the entire study site. The VHI values were further classified and composite stress map was prepared to compare the vegetation health at temporal scale (Figure 11.10).

5. **Identifying saline blanks from land cover map:** The saline blanks are a common feature in the mangrove swamps of Sundarbans region. During high tide when water intrudes the land it gets accumulated in the low land areas. After the water evaporates the salts and sediments are left behind whose gradual accumulation increases the salinity and inadaptability of the region [99]. Thus the mangrove vegetation gets depleted around these shallow salt pools creating bald white patches [51]. For delineating such geomorphic features the land cover map of the years 1990 and 2018 was prepared. Multi-spectral satellite Landsat 5 data for the year 1990 and Landsat OLI (Operational Land Imager) for 2018 were used. The images underwent atmospheric correction where surface radiance is converted to surface reflectance through proper recalling coefficients obtained from product metadata file. Additionally sun glint correction was performed on the Landsat OLI image. The same classification techniques as used in mangrove area mapping were performed over the corrected images to generate four classes, namely, (i) water, (ii) saline blanks, (iii) mangroves, (iv) beach and mudflat (Figure 11.8).

6. **Preparation of digital elevation model (DEM):** Slope or elevation is considered to be one of the most significant parameters to assess the vulnerability of coastal areas and transformation of coastal vegetation transformation with respect to rise in sea level [100]. For this study Shuttle Radar Topography Mission (SRTM) DEM of the years 1990 and 2018 of the region was collected from United States Geological Survey (USGS) website for evaluation of regional slope, erosion prone regions and the change of mangrove habitat with rise in sea level (Figure 11.4).

7. **Determining soil salinity of Sundarbans region:** Thorough literature survey has been done for assessing soil salinity of the islands. According to Joshi and Ghose [101] and Gopal and Chauhan [16] the soil salinity of Sundarbans region decreases from the coast towards inland as well as from west to east. Thus Bangladesh Sundarbans is oligohaline (0.5%–5% salinity) and Indian Sundarbans is polyhaline with 18%–30% salinity. Soil samples were collected during GPS-based field survey. On the basis of soil samples collected from the different locations a vectorized point map of those locations was prepared. Further Kriging geospatial interpolation technique was adopted in Arc GIS 10.5 platform to interpolate the soil sample locations in the island boundary shape file layer to prepare a raster soil salinity map of the region (Figure 11.7).

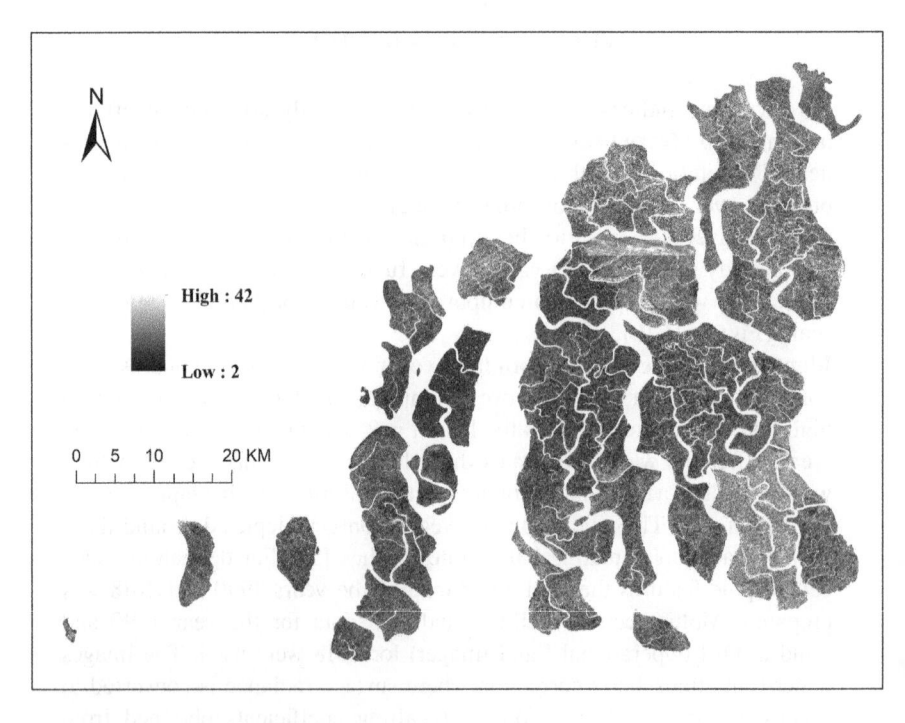

High : 42

Low : 2

0   5   10   20 KM

FIGURE 11.4   DEM of the study region.

11.3.5 NORMALIZATION OF COMPONENTS

An important phase in vulnerability assessment is the standardization or normalization of input maps. The components of mangrove vulnerability are usually measured on different scales; therefore, it is necessary to obtain comparable scales in raw data sets for which normalization of values is necessary. It involves defining the values of factors in [0, 1] using the appropriate mathematical formula to create a degree of membership of mangrove vulnerability map.

In this study, after creation of all component layers, the maximum and minimum values of these components were set and normalized as an index (as each of the components are measured on a different scale) using the formula

$$M_i^{\text{Index}} = \frac{M_i - M_{\min}}{M_{\max} - M_{\min}}$$

where M is the value of the major component i, and $M_{\min}$ and $M_{\max}$ are the minimum and maximum values, respectively, for each component (Table 11.2).

Now using the same formula each normalized component layer categorized into exposure, sensitivity and adaptive capacity was added and further normalized in the range (0–1) for respective islands to give them equal weightage and importance (Table 11.3). This second step of normalization finally provided individual Exposure, Sensitivity and Adaptive Capacity scores of each islands of the study site.

TABLE 11.2
Normalized Values of Components of Exposure, Sensitivity and Adaptive Capacity

Name	Erosion	Accretion	Saline Blank	Regeneration	Degeneration	Salinity	Mangrove Health	Earth Skin	Temperature	Wind Speed	Monsoon Rainfal	Elevation	Slope	SSH	SST	Current Speed	Tide
Lothian	0.01702	0.00000	0.06051	0.33423	0.21420	0.42855	0.87086	0.20577	0.14860	0.26640	0.62145	0.717948785	0.382993693	0.476654818	0.457532857	0.14192	0.32339
Bhangaduni North	0.72486	0.00000	0.89008	0.03120	0.90834	0.74913	0.39149	0.80256	0.73816	0.85829	0.13465	0.201751919	0.908592852	0.974271336	0.916518608	0.54429	0.58095
Dalhousie	0.06784	0.00000	0.29839	0.05608	0.67194	0.63347	0.49500	0.68607	0.62083	0.76417	0.18940	0.28332	0.74781	0.91956	0.85439	0.43669	0.56467
Bulchery	0.61307	0.00000	0.64485	0.04202	0.78306	0.75188	0.29899	0.84918	0.75622	0.86703	0.09619	0.201578	0.93278	0.97618	0.92269	0.57364	0.59147
Chulkati East	0.01543	0.00000	0.07478	0.31479	0.22798	0.42867	0.86196	0.21403	0.16000	0.27974	0.59712	0.71344	0.39893	0.50397	0.48200	0.14364	0.32728
Dulibhasani Central	0.02512	0.00000	0.08423	0.29283	0.33065	0.43023	0.86057	0.22035	0.16171	0.32317	0.57323	0.70283	0.40099	0.52787	0.50342	0.15199	0.33118
Chulkati North	0.00346	0.00000	0.03761	0.46450	0.01636	0.00001	1.00001	0.00003	0.00002	0.00036	0.99999	1.00007	7.75583	1.78816	0.03043	0.00692	0.00105
Dulibhasani South	0.05076	0.00000	0.25145	0.22352	0.49673	0.53174	0.60468	0.51129	0.40756	0.61615	0.27891	0.45112	0.49686	0.77947	0.72887	0.30632	0.51267
Ajmalmari North east	0.04961	0.00000	0.14706	0.16240	0.45967	0.53786	0.60081	0.51643	0.41595	0.64977	0.24839	0.43927	0.50700	0.81061	0.75677	0.31783	0.52252
Chotahaldi Central	0.36339	0.00000	0.98463	0.00002	0.76095	0.77254	0.28456	0.88363	0.80464	0.87669	0.07609	0.14210	0.93888	0.98079	0.939235051	0.63899	0.647885

TABLE 11.3
Normalized Values of Exposure, Sensitivity and Adaptive Capacity

NAME	Adaptive Capacity_N	Sensitivity_N	Exposure_N
Lothian	0.570200073	0.206042095	0.275182999
Bhangaduni North	0.183694122	0.905464555	0.804283987
Dalhousie	0.2461867	0.567965108	0.631350346
Bulchery	0.159332464	0.787266013	0.800773072
Chulkati East	0.560425564	0.22105882	0.283867976
Dulibhasani Central	0.550316747	0.260025423	0.296821332
Chulkati North	0.731380883	1.38784E-05	0.000308629
Dulibhasani South	0.378607947	0.40650047	0.495145405
Ajmalmari North east	0.355738952	0.361681526	0.508856178
Chotahaldi Central	0.124835245	0.897770202	0.79893344
Matla	0.345539474	0.35842122	0.518676667
Dhanchi	0.373574239	0.372475124	0.528592875
Herobhanga North	0.550941377	0.244664747	0.301385138
Netdhopani west	0.52497051	0.247386753	0.325351939
Gosaba south	0.099362979	0.809972679	0.866145294
Gosaba Chandraduania south	0.348839854	0.356922009	0.54110634
Gosaba Chandraduania north	0.519887665	0.29095311	0.361433092
Baraduania Khal west Gosaba	0.329026967	0.456844084	0.555402406
Khejurtala khal Gosaba	0.501869657	0.23248438	0.368009193
Bakultala khal central Gosaba	0.350227709	0.310203021	0.570052829
Baghmara khal Gosaba	0.238266575	0.582406366	0.65058744
N.Baikunthakhal Gosaba	0.324213493	0.386344345	0.58366889
Dakshin chamtakhal central	0.2316924	0.585829007	0.659838436
Netidhopani south	0.481125836	0.237324375	0.389672363
Netidhopani north	1.00000649	0.057216483	0.084656208
Katuajpuri west	0.479624798	0.269458376	0.407886384
Katuajpuri south	0.21694002	0.579699294	0.682646262
Saznekhali northwest	0.972265027	0.122027558	0.141530905
Saznekhali central	0.841303402	0.140538699	0.187101988
Saznekhali east	0.456162171	0.249124882	0.412248045
Saznekhali north	0.874245296	0.105318241	0.204307918
Jhilla east	0.303600902	0.451001433	0.598286878
Katuajpuri north	0.461651956	0.27428373	0.427791492
Katuajpuri	0.306091249	0.482043351	0.618490479
Jambudwip	0.062422974	0.999922113	0.968471174
Netidhopani east	0.438684684	0.278638484	0.428865631
Duania central	0.800533149	0.101651207	0.210533794
Gomdi khal Saznekhali	0.425695119	0.306325505	0.443412833
Hingalganj	0.711112563	0.111680821	0.218178224
Ajmalmari north	0.390135664	0.306348853	0.466131639
Ajmalmari south	0.974956972	0.131198159	0.236449616
Banka khal central Gosaba	1.46965E-09	0.81274862	1
Saznekhali	0.411239246	0.282613724	0.476725435
Dulibhasani east	0.738621245	0.189654339	0.262297271
Halliday island	0.213428835	0.58098516	0.703003348

11.3.6 Evaluation of Criteria

Once the scores of Exposure (*E*), Sensitivity (*S*) and Adaptive Capacity (AC) have been calculated, the three contributing factors were combined using the following equation to calculate VI for the 45 islands of Indian Sundarbans [102].

$$VI = (E - AC) \times S$$

where VI is the vulnerability Index, *E* is the calculated exposure score, AC is the calculated adaptive capacity score and *S* is the calculated sensitivity score. The VI value varies from 1 (very low vulnerable) to 1 (very highly vulnerable) (Table 11.4).

TABLE 11.4
The Vulnerability Index of Different
Islands of Indian Sundarbans

Name	Vulnerability
Lothian	2
Bhangaduni North	2
Dalhousie	5
Bulchery	4
Chulkati East	5
Dulibhasani Central	2
Chulkati North	2
Dulibhasani South	1
Ajmalmari North east	3
Chotahaldi Central	3
Matla	5
Dhanchi	3
Herobhanga North	3
Netdhopani west	2
Gosaba south	2
Gosaba Chandraduania south	5
Gosaba Chandraduania north	3
Baraduania Khal west Gosaba	2
Khejurtala khal Gosaba	3
Bakultala khal central Gosaba	2
Baghmara khal Gosaba	3
N.Baikunthakhal Gosaba	4
Dakshin chamtakhal central	3
Netidhopani south	4
Netidhopani north	2
Katuajpuri west	1
Katuajpuri south	2
Saznekhali northwest	4
Saznekhali central	1

(Continued)

TABLE 11.4 (*Continued*)
The Vulnerability Index of Different
Islands of Indian Sundarbans

Name	Vulnerability
Saznekhali east	1
Saznekhali north	2
Jhilla east	1
Katuajpuri North	3
Katuajpuri	2
Jambudwip	3
Netidhopani east	5
Duania Central	2
Gomdi khal Saznekhali	1
Hingalganj	2
Ajmalmari north	1
Ajmalmari south	2
Banka khal central Gosaba	1
Saznekhali	5
Dulibhasani East	2
Halliday Island	1
Halliday Island	4

11.4 RESULTS AND DISCUSSION

11.4.1 INFLUENCE OF SEA LEVEL RISE, TIDAL ACTION AND COASTAL EROSION ON MANGROVES

Response of mangrove habitats to sea level rise is strongly influenced by the geomorphological setting and tidal regimes of that region [103]. Tidal ranges have profound influences on wetland with respect to sea level rise, where areas witnessing smaller tidal ranges have greater exposure compared to areas with larger tidal ranges [104,105]. In the entire Bengal delta tidal range varies from 3.7 to 5 m. After analysis of tide gauge records of nearly two decades it is revealed that there is a 12 mm rise in sea level regionally. The satellite altimetry results of the period 1993–2014 further show that the sea level has registered a rise of 6–8 mm year^{-1} irrespective of the land movement. Such a rise is indicative of high rate of subsidence in the delta leading to enhanced vulnerability of mangroves. Moreover radar interferometry studies suggest a high rate of subsidence in the Sundarbans delta at the rate of 4 mm year^{-1}. The tidal range and sea surface height vary from 3.26 mm and 0.256 mm in Bhangaduani Island to 2.10 and 0.219 mm, respectively, in North Netidhopani. The shift of mangroves of Indian Sundarbans landwards is probably due to factors like sea level changes as well as coastal erosion and accretion [100]. The mangroves of entire island complex have registered three types of changes with respect to sea level fluctuations. With no change in sea level little-to-no change has been observed in mangrove position of the otherwise agricultural land dominated middle part of the island system. With rise in sea level mangroves in the sea ward margins of the southernmost islands like Jambudwip, Bulcherry, Bangaduani, Dalhousie have either retreated landwards or have been lost to coastal erosion.

A strong correlation exists between sea level rise and coastal erosion [51]. Coastal erosion and accretion are responsible for changing the configuration of shorelines as well as modifying the coastal areas. In the present study it is observed that the most erosion prone islands are those that are exposed to sea level rise. Thus the susceptibility of islands to coastal erosion increases towards the sea facing islands of Sundarbans region. Bhangaduni, south western Jambudwip, Bulchery, Chulkathi south, Dhanchi, and Baghmari South Dalhousie islands were observed to be most vulnerable to coastal erosion (Figure 11.5). The western flanks of the islands eroded more than their eastern counterparts due to tidal surges whereas Lohachara Bedford islands have completely lost. Phenomena of marginal accretion were observed predominantly along eastern and northern margin of islands, in the inner estuaries and along the sea facing islands of west Ajmalmari, northern parts of Dhulibasani, Herobhanga, Dhanchi, north eastern part of Jambudwip. Amount of land erosion and accretion over the past 28 years is estimated to be 122 and 48 km^2, respectively. The presence of thick mangroves and lesser anthropogenic activities has helped keeping the north eastern matured islands comparatively in a stable condition. Bhangaduani Island witnessed a recession of shoreline by approximately 1.07 km followed by Bulcherry and Dalhousie Island where various natural forcings like storm surge coupled with wave action and sea level rise facilitated in shoreline change.

According to [106] mangroves tends to adjust to a new sea level and higher frequency of inundation. This adjustment is always governed by geomorphic, hydrological settings, elevation or slope and sedimentation pattern of the deltaic coast. The islands with lower elevation and truncated sloping coastlines become more exposed

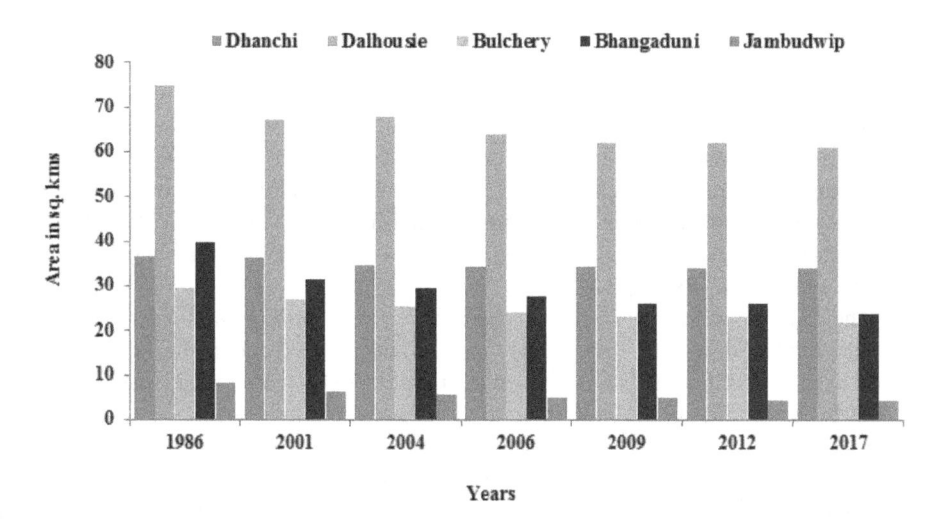

FIGURE 11.5 Area of land loss of different islands of Indian Sundarbans over the period (1986–2017).

to erosion and inundation and the mangroves of such regions tend to retreat landwards [109]. Tectonic activities and irregular rates of sediment deposition facilitated the eastern portion of the delta to subside more swiftly than the western part, resulting in an obvious west-to-east downward tilt of the deltaic complex [107,108]. This west to east tilt has shifted the entire river course eastward and Hooghly and Meghna Rivers being the only perennial sources of fresh water directly feeding the Sundarbans mangrove forest, the fresh water supply is extensively cut off in the western part [107]. Thus the islands with elevation of 3–8 m above sea level are significantly threatened by intense tidal activities as well as sea level rise induced inundation and coastal erosion. In spite of having the richest diversity and highest areal coverage (estimated as 1836.6 km^2 excluding creeks and water bodies), the mangroves of Sundarbans are facing the maximum threats. This may be attributed to construction of dams and canals leading to reduction of sediment supply by almost 68%. The rise in sea level, decline in sediment input from Ganges and its tributaries have forced the earlier dense mangrove patches of Bhangaduani, Bulcherry, Chotahaldi, south Gosaba, Katuajpuri, Chamtakhal and Bakultala islands to invade landwards.

11.4.2 INFLUENCE OF RAINFALL AND TEMPERATURE ON MANGROVES

The coastal regions of the Bengal delta are amongst the most vulnerable regions in the world experiencing salinity intrusion and other climate change impacts [110]. The biophysical characteristics of mangroves are largely determined by seasonal variability of temperature and rainfall along with potential fluctuations in water salinity and soil moisture [75]. The projected rise in SST of Indian Ocean is over 2°C within the year 2100. In the deltaic complex of Sundarbans regions, sea surface temperatures have increased at a rate of approximately 0.018°C/year. The average annual sea surface temperature was found to be lowest in the month of January (22.92°C) and highest in the month of June (30.91°C). Moreover the mean SST showed an increasing trend during the pre-monsoon season and increasing trend during monsoon and post-monsoon season. The surface air temperature anomaly data over the Sundarbans and adjacent parts of the Bay of Bengal were analyzed to find an annual increasing trend in the rise in temperature. Temperature increased at the rate of 0.019°C per year. Such climate change-induced temperature variations intend to affect the growth and development of growing stock of mangroves of Sundarbans [111]. Moreover, species diversity of phytoplankton in Sundarbans wetlands is higher in summer (March) and lower in winter (November), indicating its close correlation with ambient summer temperature (25°C–35°C) [112].

The Indian Sundarbans region receives an average annual rainfall of 1500 mm which steadily decreases from Sagar in the south to Kolkata in the north [16]. Moreover it was found that the amount of rainfall showed a steady decreasing trend from 1990 to 2018. While more than 80% of the rainfall occurs during June to September, the region receives some amount of downpour during pre- and post-monsoon seasons [68]. The average seasonal rainfall trends over Sundarbans region within the period 1990–2018 reveal increase in pre-monsoonal and monsoonal average from 16.65 to 147 mm and 274 to 283 mm over the years, respectively. During the months of December and January the average rainfall increased while there was a

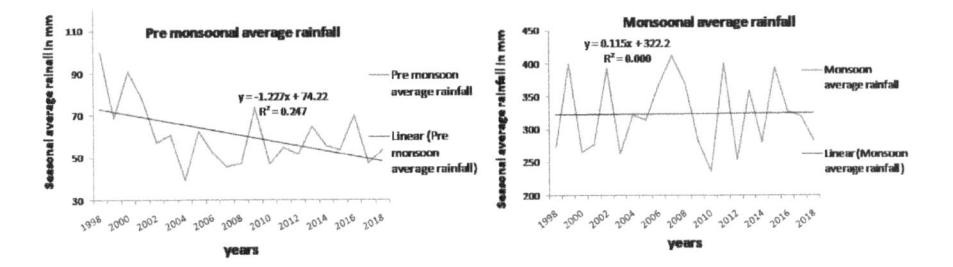

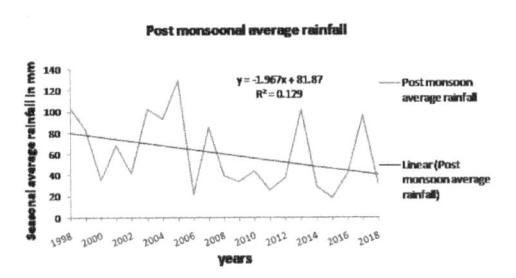

FIGURE 11.6   Rainfall regimes of Sundarbans region of the period 1998–2018.

sharp decrease in rainfall during the months of October and November (Figure 11.6). The increasing amounts of rainfall during the summer months were possibly due to ENSO phenomena. It is a well-established fact that increased riverine sediment load induced by augmented rainfall can contribute to higher primary production, increase in mangrove tree height and litterfall. But such increased runoff also mediates higher concentration of suspended sediments with higher turbidity. This in turn negatively affects the ecosystems [68]. Further in Indian Sundarbans region, mangrove species like Heritiera fomes and Nypa fruticans are gradually disappearing from the central region owing to siltation of Bidyadhari River. Moreover, the deteriorating mangrove health conditions in the six southern islands could also be attributed to the decreasing amount of rainfall received by these regions.

11.4.3   SALINITY REGIMES

Generally the water salinity in the Sundarbans region increases annually during the dry season and influences the seasonal variability of soil salinity. With peak salinity level (20–30 ppt in the downstream region) during the months of March and May and lowest during the monsoon seasons the soil salinity of the study site is significantly influenced by vagaries of monsoonal rains and river water discharge [68,69]. The global warming phenomenon has left its footprints in the Sundarbans deltaic complex in the form of increased salinity intrusion in the central sector of Matla and Thakuran estuaries due to rise in sea level [52]. Tectonic uplift in the west and subsidence by sediment compaction in the east [113] in combination with variation in freshwater inputs have created different salinity zones – hyposaline in the eastern

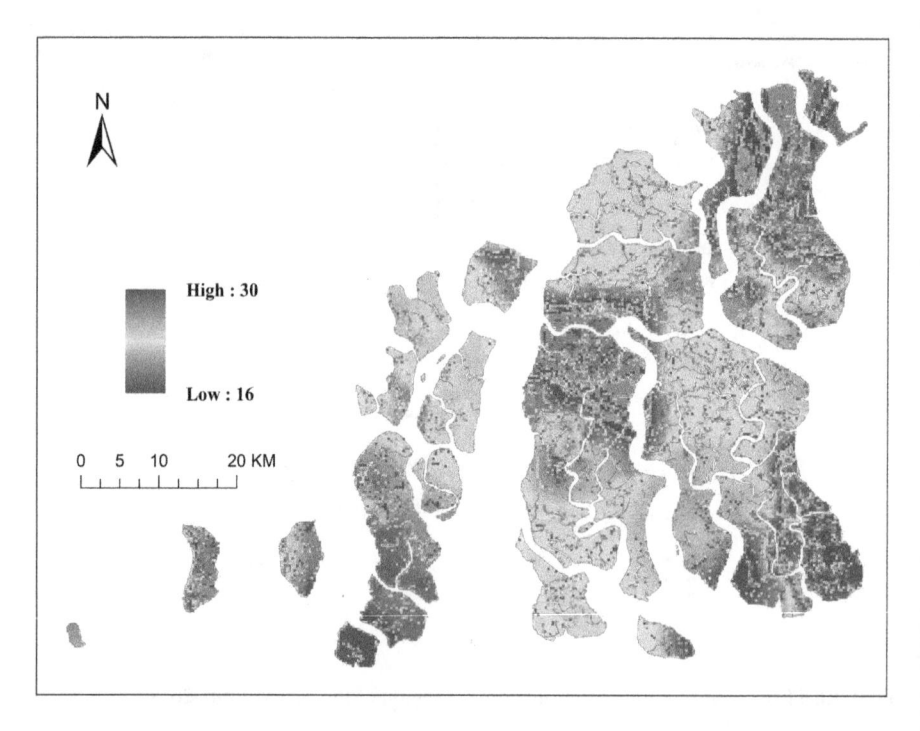

FIGURE 11.7 Soil salinity map of Sundarbans.

and western part, characterized by freshwater inputs by the rivers and hypersaline in the higher central part drained by the Matla, Bidyadhari or Harinbhanga rivers (Figure 11.7). Based upon changing soil salinity conditions the mangrove species assemblage also shows distinct patterns. *Avicennia marina*, *A. alba* and *Bruguiera cylindrica* are abundantly found in the lower coastal sections while *B. gymnorhiza*, *Ceriops decandra* and *Rhizophora mucronata* are more dominant in the upstreams. The more fresh water loving species like Sundari-tree Heritiera fomes, Excoecaria agallocha, and Sonneratia caseolaris dotting the river channels are commonly found in the east [15,114].Our observation showed that annual mean salinity in the surface waters of Matla Estuary was ~20 ppt (one of the principal estuaries of Sundarban). Such increased salinity due to siltation of Bidyadhari River has restricted the growth of fresh water species *Heritiera fomes*. On the contrary the species is thriving well in the western sector of the island complex due to dominance of fresh water that is being constantly added by the Farakka barrage. Soil salinity is coherently related to regional differences of carbon stock and dissolved nutrients in the Sundarbans [115] and with increasing salinity the mangrove plants adapt stunt growth. The observed salinity intrusion in the Sundarbans swamps may lead to decline in species richness and diversity. It has been already found that with increased soil salinity the mangrove canopy of various species decreased and important tree species like *Heritiera fomes* was infected with top dying disease.

11.4.4 DOMINANCE OF SALINE BLANKS

Soil samples were collected from the saline blank areas and analyzed. They showed high to very high salinity values. The analysis of land cover change data over the 28 years period showed an increase in area of saline blanks from 1990 to 2018 in the islands of Ajmalmari and Dhulibasani in the west Chotohaldi in the central part and Bakultala in the west (Figure 11.8). Apart from the existing patches of salt encrustations, some new patches are also developed in the regions of Suznekhali in the north and Katuajpuri in the north east. Although the presence of saline blanks is deterrent for the growth of mangroves, the salt tolerant mangrove species like *Ceriops decandra*, *Avicennia marina* and *Avicennia alba* prefer such saline environment in contrast to fresh water loving species like *Heritierra fomes* and *Xylocarpus granatum* [99].

11.4.5 LAND COVERS CHANGES

The impact of sea level rise and resultant erosion accretion processes significantly brought land cover changes along the tidal creeks and mangrove swamps (Table 11.5). The change of land cover over the time period of 28 years exhibited significant aggradation of land area along the eastern portions of the southern islands of Bulcherry, Dhulibasani, Chulkathi and western section of Gosaba, and Dalhousie Island. Land loss due to erosion in Jambudwip and Bhangaduani was also observed from the land cover maps (Figure 11.9). It could be observed that saline blanks over the period of

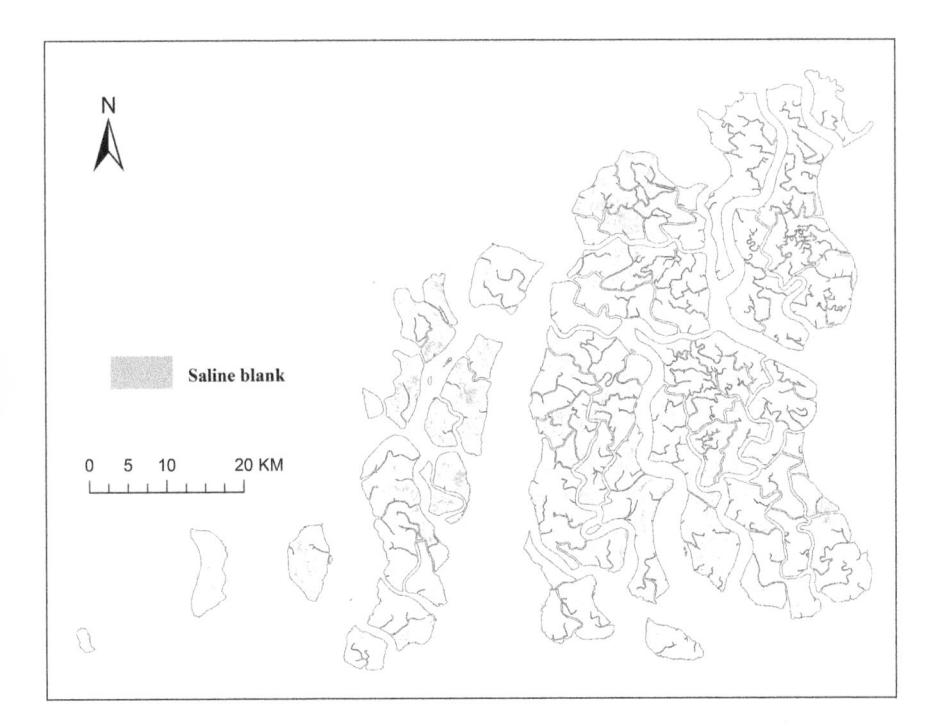

FIGURE 11.8 Saline blanks of the Sundarbans region.

TABLE 11.5
The Areal Change of Major Land Cover
Features between 1990 and 2018

Feature	1990	2018
Water	1782.3249	1826.3552
Mangrove	2174.5287	2120.5382
Saline blank	35.5239	46.6928
Beach and Mudflat	8.2593	7.0506

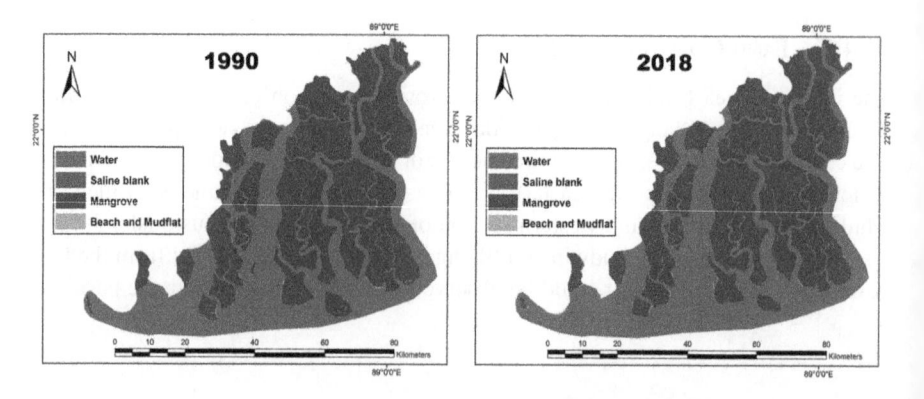

FIGURE 11.9   Land cover changes in Sundarbans between 1990 and 2018.

28 years have progressively increased by 31.54% whereas mangroves have decreased by 2.5% over the same period.

11.4.6   CYCLONES AND TIDAL BORE

A model developed by Unnikrishnan et al. [116] projected an increase in frequency of cyclones specifically in the late monsoon season as well as in the summer months between 2070 and 2100. The intensity of tidal bores and associated rise in tidal volumes has increased in the Sundarbans compared to the rest of the eastern coast of India [116]. These phenomena further trigger the destructive capabilities of cyclonic storms, escalated sea levels and periodic high tide events. Cyclones and tidal bores have been reported to be responsible for rapid land loss of Sagar [70] and Jambudwip [72] to erosion.

11.4.7   MANGROVE HEALTH

It was evident from the observations that when VCI valued 0 NDVI also decreased which signified severe vegetation stress and less green vegetation. And when VCI ranged 100, NDVI also increased which indicated healthy vegetation with

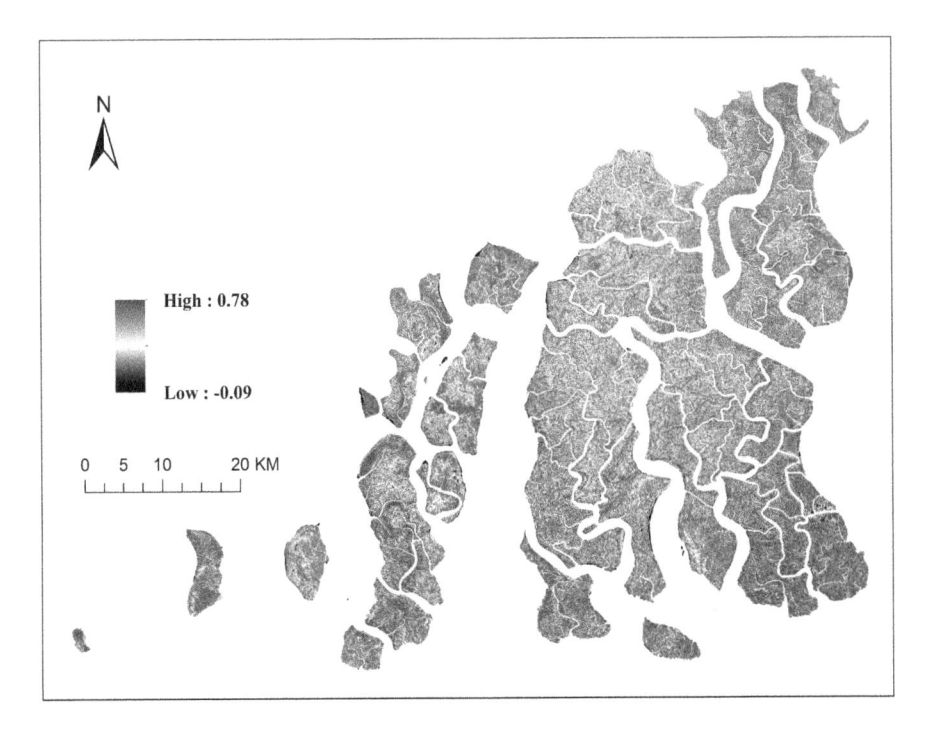

FIGURE 11.10   Mangrove health of Indian Sundarbans.

high pigment content. Further when TCI approached 0, BT increased indicating hotter weather and vice versa. From the temporal analysis (1990 and 2018) of vegetation heath, it could be observed that mangrove health decreased phenomenally (Figure 11.10). In the year 1990 almost the entire Indian Sundarbans mangroves were in a healthy condition and over the years the percentage of healthy vegetation were significantly replaced by percentage of stressed vegetation. Decrease in rainfall, increase in air temperature and soil salinity could be the potent causes behind the degrading health condition of these mangroves.

With vegetation health ranging between 0.2 and 0.4 six sea facing southern islands of Indian Sundarbans region exhibited stressed mangroves compared to rest of the eastern, western or northern island blocks where the mangrove vegetation health ranged between 0.5 and 0.75. The input of riverine fresh water which otherwise determines the health of mangroves were significantly reduced in these islands which increased the soil salinity and deterred the health of mangroves. The islands with stressed vegetation also observed decreasing trend of precipitation and increased air temperature which have a strong correlation with mangrove health.

11.4.8   ADAPTIVE CAPACITY OF MANGROVES

Mangroves have strong power of resistance and they have proved their ability through quaternary by withstanding large-scale sea-level fluctuations [103]. However, the ongoing climate changes in the recent times coupled with anthropogenic activities have put these halophytes in tremendous challenges to cope with the environmental impacts. The resistance powers of mangroves are determined by the mangroves ability to keep pace with sea level rise without any alterations in their process or structure. On the other hand the resilience referred to the mangroves ability to migrate landwards in response to change in sea level [117]. Now mangroves cannot keep pace with rise in sea level if the rate of sea level rise exceeds the rate of change in elevation of mangroves sediment surface. There are several processes that affect the elevation of mangroves sediment surface [104,118,119]. The beneficial usage of dredge spoils could augment mangrove sediment elevation [120]. Moreover there are various factors that allow the peat building process and in turn prompt the recovery of mangroves. Thus uninterrupted hydrological regimes and access to sediments and fresh water as well as supply of healthy propagules from adjacent mangrove areas, close proximity to neighbouring healthy mangroves, mangroves backed by salt flats, coastal plains can ensure appropriate habitat for landward colonization of mangroves as well as high success rate of mangrove health recovery.

All these factors altogether ensure mangrove regeneration and degeneration which have shown different results over all these islands. The rate of regeneration was very low in sea facing islands and high in western and northern islands.

11.4.9 Vulnerability Assessment

After assessing the exposure sensitivity and adaptive capacity components of each island of Sundarbans a Vulnerability map was prepared. The islands were grouped in the range 1–5 based on the classification output (Figure 11.11) where 1 represented low vulnerability and 5 represented high. Moreover ranks of 1–2 indicated mangrove areas that were currently resilient, but their resilience could be further enhanced by reducing any subtle impact of climate change. Mangrove islands with ranks 2–4 indicated some core vulnerability that required significant management options.

All the islands falling under category 5 and 4 ranking were found to have some innate vulnerability due to their extreme exposure to sea level rise and low tidal range. It showed that mangroves of Jambudwip, Gosaba south, Bakultala south, Bulchery, Chotahaldi east and Kultali followed closely by Shushnichara, Surendranagar, Katuajpuri southeast and south, Baghmara, Haliday and Dalhousie south have most exposure to climate change and sea level rise impacts. One of the most important sensitive components was also found to have deteriorated over the years in these islands. Mangroves in the southern islands like Bhangaduani, Bulcherry Jambudwip, and Chotahaldi were found in a relatively stressed condition probably due to dominance of saline blanks, high salinity intrusion and decreasing trends of pre- and post-monsoonal rainfall. These islands have further lost significant acreage of mangroves mostly to sea level rise, severe cyclonic impacts and associated storm surge and lower elevation above mean sea level. The rate of regeneration of mangroves in these islands was observed to be very low compared to other islands.

The mangrove islands falling under category 3, 2, and 1 were observed to have some resilience to climate change impacts with significant rate of accretion in these islands, where fine sediments carried in suspension with the coastal waters during

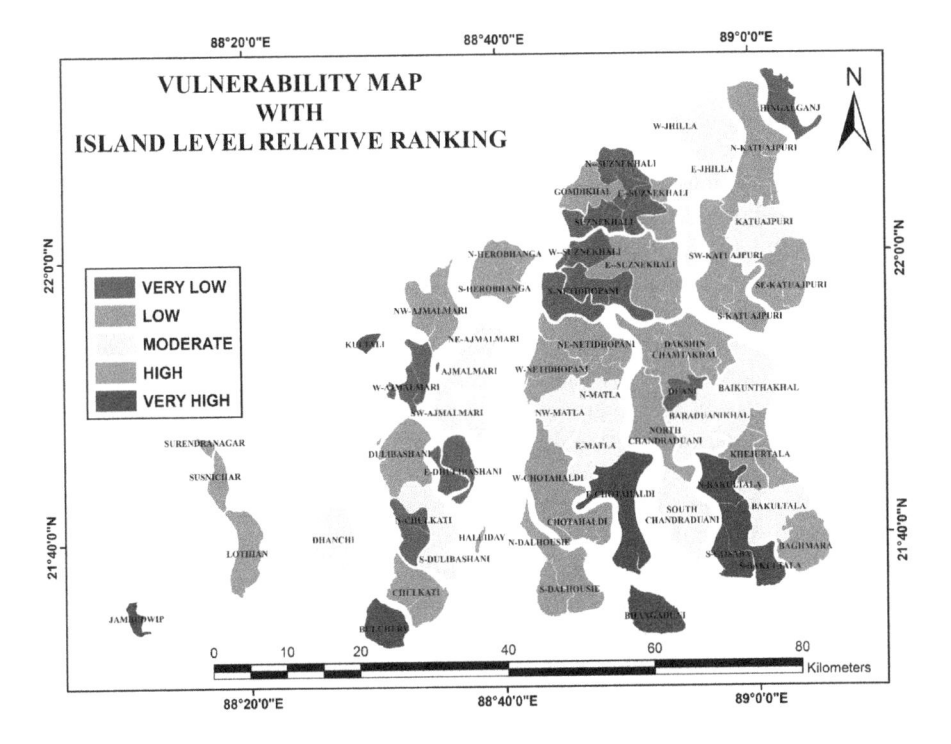

FIGURE 11.11 Bio-physical vulnerability map of Indian Sundarbans.

high tides finally settles in the forest cover with slowing down of tidal currents by vegetation cover [53]. The ability of particular mangrove species like *Avicennia* sp. and *Rhizophora* sp. in trapping the sediments has proven to be an important adaptive capacity of mangroves to rising sea level, where they can migrate landwards. The mangroves in these islands were found to be in healthy condition compared to their southern counterparts with the high potential of recovery from degeneration. Moreover every island in these 3 categories have experienced significant rate of erosion, sea level rise and scarce fresh water induced high salinity conditions as well as dominance of saline blanks.

Thus with highest and lowest rate of regeneration, Duania central and Chotahaldi central ranked 1 and 5, respectively, whereas in terms of mangrove health, Chulkathi north and Baikunthakhal central of Gosaba ranked highest and lowest, respectively. On the other hand with significant exposure to sea level rise, high rate of erosion and dominance of saline blanks, Jambudwip followed by Bhangaduani north, Gosaba south Bulcherry, Baikunthakhal and Chotahaldi were at maximum risk whereas Duania, Chulkathi, Saznekhali, Hingalganj, Ajmalmari, Dhulibasani, Nethidopani ranked one with highest rate of accretion and lowest values of salinity. It was further evident that all the six sea facing islands with steep slopes (0.04%) were at increased risk of erosion and loss of lands compared to those gently sloping islands with vulnerability ranking of 1 and 2.

11.5 CONCLUSION

The above study aimed at assessing the bio-physical vulnerability of mangroves of Indian Sundarbans region following the methodology laid down in IPCC AR4 framework. The results from the study indicated that the impacts of rising sea level coupled with low rate of sediment accretion proved to be detrimental for the health of mangroves and may further lead to habitat loss in the region. The rate of sea level rise was highly interrelated to land loss and coastal erosion in the islands like Jambudwip, Bhangaduani north, Gosaba south Bulcherry, Baikunthakhal, and Chotahaldi. It is the natural ability of mangroves to maintain its position during tidal fluctuations by migrating landwards in response to sea-level rise. However, in many islands of Sundarbans massive land conversions to agricultural fields, deforestation of mangroves in the name of socio-economic developments prevented the landward shift of these mangroves as a result of which vast tracts of mangroves were lost. Moreover, dominance of salt incrustations further deterred the growth of mangroves particularly in the sea facing island. The decreased fresh water supply particularly by river Matla, siltation of river Bidyadhari and sufficient monsoonal vagaries have resulted in increased salinization problem in the Sundarbans region. This has further resulted in significant shift of fresh water loving *Heritiera fomis* north ward.

The mangroves were found in stressed condition in those islands where the amount of precipitation was low and temperature was reportedly high coupled with other climatic stressors. Such islands were like Jambudwip, Bhangaduani north, Gosaba south Bulcherry, Baikunthakhal, and Chotahaldi. In contrast the mangroves were reportedly found to be in healthy condition in those islands which received ample amount of precipitation, experienced moderate temperatures, no significant cyclonic impacts less dominance of saline blanks and high rate of accretion.

Different exposure and sensitivity components were taken into account individually. The islandwise analysis revealed that for almost all the exposure (components like sea level rise, temperature, precipitation) and sensitivity components (mangrove health, erosion, salinity, saline blanks, slope) six southern islands, namely, Jambudwip, Bhangaduani north, Gosaba south Bulcherry, Baikunthakhal, and Chotahaldi ranked with very high vulnerability whereas Baghmarakhal, Katuajpuri south, Dalhousie, Chamtakhal south were ranked highly vulnerable. With high rate of accretion and regeneration rate, gentle intertidal slopes, low salinity, high precipitation and moderate temperature, islands like Duania, Chulkathi, Saznekhali, Hingalganj, Ajmalmari, Dhulibasani, Nethidopani ranked one.

Effective management options and legislative procedures can contribute to successful mangrove adaptation and protection strategies. Landward migration of mangroves by removing any kind of developmental barriers can be considered and planned in collaboration with local communities. Moreover appropriate areas that are compliant with sediment supply, tidal regimes and species diversity could be selected for artificial plantation of mangroves. In Lothian Island such initiatives have already been taken up. The protection of shoreline should also be considered as an important mitigation option which could facilitate the mangroves to keep pace with rising sea level by trapping suspended sediments and increasing surface accretion during transgressed period. From the above study it could be found that the rate of regeneration was found to be lowest in the sea facing islands and highest in the northern and western part of the regions which may be attributed to increasing rate of accretion. In the past few decades large tracts of mangrove swamps have been reclaimed or have to aquaculture or agricultural farming in various islands of Sundarbans. Such actions need to be checked through proper implementation of stringent laws and actions to protect the existing mangrove species from future extinction.

When overall vulnerability assessment of the islands are considered, Duania, Chulkathi, Saznekhali, Hingalganj, Ajmalmari, Dhulibasani, Nethidopani with healthy mangroves and insignificant impact of climate changes require least management options whereas stressed mangroves of Jambudwip, Bhangaduani north, Gosaba south Bulcherry, Baikunthakhal, and Chotahaldi followed by Haliday Baghmarakhal, Katuajpuri south, Dalhousie, Chamtakhal south require immediate adaptation strategies through Government interventions.

REFERENCES

1. Tomlinson, P.B., 1986. *The Botany of Mangroves*. Cambridge University Press, Cambridge, 413 pp.
2. Lin, P. and Wang, W.Q., 2001. Changes in the leaf composition, leaf mass and leaf area during leaf senescence in three species of mangroves. *Ecological Engineering*, 16(3), pp. 415–424.
3. Wang, Y., Bonynge, G., Nugranad, J., Traber, M., Ngusaru, A., Tobey, J., Hale, L., Bowen, R. and Makota, V., 2003. Remote sensing of mangrove change along the Tanzania coast. *Marine Geodesy*, 26(1–2), pp. 35–48.
4. Satyanarayana, B., Mohamad, K.A., Idris, I.F., Husain, M.L. and Dahdouh-Guebas, F., 2011. Assessment of mangrove vegetation based on remote sensing and ground-truth measurements at Tumpat, Kelantan Delta, East Coast of Peninsular Malaysia. *International Journal of Remote Sensing*, 32(6), pp. 1635–1650.
5. Paphavasit, N., Wattayakorn, G., Aksornkoae, S. and Clough, B.F., 1993. *Mangrove of Thailand: present status of conservation, use and managementThe Economic and environmental values of mangrove forests and their present state of conservation in the South-east Asia/Pacific region*. International Society for Mangrove Ecosystems, Okinawa (Japón) International Tropical Timber Organization, Yokohama (Japón) Japan International Association for Mangroves, Tokyo (Japón)
6. Dahdouh-Guebas, F., Mathenge, C., Kairo, J.G. and Koedam, N., 2000. Utilization of mangrove wood products around Mida Creek (Kenya) amongst subsistence and commercial users. *Economic Botany*, 54(4), pp. 513–527.
7. Dahdouh-Guebas, F., Jayatissa, L.P., Di Nitto, D., Bosire, J.O., Seen, D.L. and Koedam, N., 2005. How effective were mangroves as a defence against the recent tsunami? *Current Biology*, 15(12), pp. R443–R447.
8. Bahuguna, A., Nayak, S. and Roy, D., 2008. Impact of the tsunami and earthquake of 26th December 2004 on the vital coastal ecosystems of the Andaman and Nicobar Islands assessed using RESOURCESAT AWiFS data. *International Journal of Applied Earth Observation and Geoinformation*, 10(2), pp. 229–237.
9. Duarte, C.M., Middelburg, J.J. and Caraco, N., 2005. Major role of marine vegetation on the oceanic carbon cycle. *Biogeosciences*, 2(1), pp. 1–8.

10. Taillardat, P., Friess, D.A. and Lupascu, M., 2018. Mangrove blue carbon strategies for climate change mitigation are most effective at the national scale. *Biology Letters*, *14*(10), p. 20180251.

11. Kuenzer, C., Bluemel, A., Gebhardt, S., Quoc, T.V. and Dech, S., 2011. Remote sensing of mangrove ecosystems: A review. *Remote Sensing*, *3*(5), pp. 878–928.

12. Mukhopadhyay, A., Mondal, P., Barik, J., Chowdhury, S.M., Ghosh, T. and Hazra, S., 2015. Changes in mangrove species assemblages and future prediction of the Bangladesh Sundarbans using Markov chain model and cellular automata. *Environmental Science: Processes & Impacts*, *17*(6), pp. 1111–1117.

13. Borges, A.V., Djenidi, S., Lacroix, G., Theate, J., Delille, B., and Frankignoulle, M., 2003. Atmospheric CO_2 flux from mangrove surrounding waters. *Geophysical Research Letters*, *30* (11), p. 1558.

14. Mukhopadhyay, A., Payo, A., Chanda, A., Ghosh, T., Chowdhury, S.M. and Hazra, S., 2018. Dynamics of the Sundarbans mangroves in Bangladesh under climate change. In *Ecosystem Services for Well-Being in Deltas* (pp. 489–503). Palgrave Macmillan, Cham.

15. Ghosh, A., Schmidt, S., Fickert, T. and Nüsser, M., 2015. The Indian Sundarban mangrove forests: History, utilization, conservation strategies and local perception. *Diversity*, *7*(2), pp. 149–169.

16. Gopal, B. and Chauhan, M., 2006. Biodiversity and its conservation in the Sundarban mangrove ecosystem. *Aquatic Sciences*, *68*(3), pp. 338–354.

17. Giri, C., Ochieng, E., Tieszen, L.L., Zhu, Z., Singh, A., Loveland, T., Masek, J. and Duke, N., 2011. Status and distribution of mangrove forests of the world using earth observation satellite data. *Global Ecology and Biogeography*, *20*(1), pp. 154–159.

18. DasGupta, R. and Shaw, R., 2013. Changing perspectives of mangrove management in India–an analytical overview. *Ocean & Coastal Management*, *80*, pp. 107–118.

19. Selvam, V., 2003. Environmental classification of mangrove wetlands of India. *Current Science*, *84*(6), pp. 757–765.

20. Mandal, R.N. and Naskar, K.R., 2008. Diversity and classification of Indian mangroves: A review. *Tropical Ecology*, *49*(2), pp. 131–146.

21. Upadhyay, V.P., Ranjan, R. and Singh, J.S., 2002. Human-mangrove conflicts: The way out. *Current Science*, *83*(11), pp. 1328–1336.

22. Mukhopadhyay, A., Mondal, P., Barik, J., Chowdhury, S.M., Ghosh, T. and Hazra, S., 2015. Changes in mangrove species assemblages and future prediction of the Bangladesh Sundarbans using Markov chain model and cellular automata. *Environmental Science: Processes & Impacts*, *17*(6), pp. 1111–1117.

23. Valiela, I., Bowen, J.L. and York, J.K., 2001. Mangrove Forests: One of the World's Threatened Major Tropical Environments: At least 35% of the area of mangrove forests has been lost in the past two decades, losses that exceed those for tropical rain forests and coral reefs, two other well-known threatened environments. *Bioscience*, *51*(10), pp. 807–815.

24. Iftekhar, M.S. and Saenger, P., 2008. Vegetation dynamics in the Bangladesh Sundarbans mangroves: A review of forest inventories. *Wetlands Ecology and Management*, *16*(4), pp. 291–312.

25. Duke, N.C., Meynecke, J.O., Dittmann, S., Ellison, A.M., Anger, K., Berger, U., Cannicci, S., Diele, K., Ewel, K.C., Field, C.D. and Koedam, N., 2007. A world without mangroves? *Science*, *317*(5834), pp. 41–42.

26. Gilman, E.L., Ellison, J., Duke, N.C. and Field, C., 2008. Threats to mangroves from climate change and adaptation options: A review. *Aquatic Botany*, *89*(2), pp. 237–250.

27. Gilman, E., Ellison, J., Sauni, I. and Tuaumu, S., 2007. Trends in surface elevations of American Samoa mangroves. *Wetlands Ecology and Management*, *15*(5), pp. 391–404.

28. Alatorre, L.C., Sánchez-Carrillo, S., Miramontes-Beltrán, S., Medina, R.J., Torres-Olave, M.E., Bravo, L.C., Wiebe, L.C., Granados, A., Adams, D.K., Sánchez, E. and Uc, M., 2016. Temporal changes of NDVI for qualitative environmental assessment of mangroves: shrimp farming impact on the health decline of the arid mangroves in the Gulf of California (1990–2010). *Journal of Arid Environments*, *125*, pp. 98–109.

29. Kathiresan, K. and Rajendran, N., 2005. Coastal mangrove forests mitigated tsunami. *Estuarine, Coastal and Shelf Science*, *65*(3), pp. 601–606.

30. McCarthy, J.J., Canziani, O.F., Leary, N.A., Dokken, D.J., White, K.S. (Eds.), 2001. *Climate Change 2001: Impacts, Adaptation and Vulnerability*. Cambridge University Press, Cambridge.

31. Adger, W.N., 2006. Vulnerability. *Global Environmental Change*, *16*(3), pp. 268–281.

32. Polsky, C., Neff, R. and Yarnal, B., 2007. Building comparable global change vulnerability assessments: The vulnerability scoping diagram. *Global Environmental Change*, *17*(3–4), pp. 472–485.

33. Ellison, J.C., 2015. Vulnerability assessment of mangroves to climate change and sea-level rise impacts. *Wetlands Ecology and Management*, *23*(2), pp. 115–137.

34. Field, C.D., 1995. Impact of expected climate change on mangroves. In *Asia-Pacific Symposium on Mangrove Ecosystems* (pp. 75–81). Springer, Dordrecht.

35. Füssel, H.M. and Klein, R.J., 2006. Climate change vulnerability assessments: An evolution of conceptual thinking. *Climatic Change*, *75*, pp. 301–329.

36. Alongi, D.M., 2008. Mangrove forests: Resilience, protection from tsunamis, and responses to global climate change. *Estuarine, Coastal and Shelf Science*, *76*(1), pp. 1–13.

37. Waycott, M., McKenzie, L.J., Mellors, J.E., Ellison, J.C., Sheaves, M.T., Collier, C. and Schwarz, A.M., 2011. Vulnerability of mangroves, seagrasses and intertidal flats in the tropical Pacific to climate change.

38. Ebi, K.L., Kovats, R.S. and Menne, B. 2006. An approach for assessing human health vulnerability and public health interventions to adapt to climate change. *Environmental Health Perspectives*, *114*, pp. 1930–1934.

39. Lugo, A.E. and Snedaker, S.C., 1974. The ecology of mangroves. *Annual Review of Ecology and Systematics*, *5*(1), pp. 39–64.

40. Saenger, P. and Snedaker, S.C., 1993. Pantropical trends in mangrove above-ground biomass and annual litterfall. *Oecologia*, *96*(3), pp. 293–299.

41. Faraco, L.F., Andriguetto-Filho, J.M. and Lana, P.C., 2010. A methodology for assessing the vulnerability of mangroves and fisherfolk to climate change. *Pan-American Journal of Aquatic Sciences*, *5*(2), pp. 205–223.

42. Omo-Irabor, O.O., Olobaniyi, S.B., Akunna, J., Venus, V., Maina, J.M. and Paradzayi, C., 2011. Mangrove vulnerability modelling in parts of Western Niger Delta, Nigeria using satellite images, GIS techniques and Spatial Multi-Criteria Analysis (SMCA). *Environmental Monitoring and Assessment*, *178*(1–4), pp. 39–51.

43. Naskar, K. and Mandal, R., 1999. *Ecology and Biodiversity of Indian Mangroves* (Vol. 1). Daya Books, New Delhi.

44. Nandy, S. and Kushwaha, S.P.S., 2011. Study on the utility of IRS 1D LISS-III data and the classification techniques for mapping of Sunderban mangroves. *Journal of Coastal Conservation*, *15*(1), pp. 123–137.

45. Barik, J. and Chowdhury, S., 2014. True mangrove species of Sundarbans delta, West Bengal, eastern India. *Check List*, *10*(2), pp. 329–334.

46. Giri, S., Mukhopadhyay, A., Hazra, S., Mukherjee, S., Roy, D., Ghosh, S., Ghosh, T. and Mitra, D., 2014. A study on abundance and distribution of mangrove species in Indian Sundarban using remote sensing technique. *Journal of Coastal Conservation*, *18*(4), pp. 359–367.

47. Mukhopadhyay, S.K., Biswas, H.D.T.K., De, T.K. and Jana, T.K., 2006. Fluxes of nutrients from the tropical River Hooghly at the land–ocean boundary of Sundarbans, NE Coast of Bay of Bengal, India. *Journal of Marine Systems*, 62(1–2), pp. 9–21.

48. Mandal, R.N., Das, C.S., Naskar, K.R., 2010. Dwindling Indian Sundarban mangrove: The way out. *Science & Culture*, 76 (7e8), p. 275e282.

49. Houghton, J.T., Ding, Y.D.J.G., Griggs, D.J., Noguer, M., van der Linden, P.J., Dai, X., Maskell, K. and Johnson, C.A., 2001. *Climate Chage 2001*.

50. Mitra, A., 2013. Impact of climate change on mangroves. In *Sensitivity of Mangrove Ecosystem to Changing Climate* (pp. 131–159). Springer, New Delhi.

51. Hazra, S., Ghosh, T., DasGupta, R. and Sen, G., 2002. Sea level and associated changes in the Sundarbans. *Science and Culture*, 68(9/12), pp. 309–321.

52. Mitra, A., Gangopadhyay, A., Dube, A., Schmidt, A.C. and Banerjee, K., 2009. Observed changes in water mass properties in the Indian Sundarbans (northwestern Bay of Bengal) during 1980–2007. *Current Science*, 97, pp. 1445–1452.

53. UNEP (United Nations Environment Programme). 1994. Assessment and Monitoring of Climate Change Impacts on Mangrove Ecosystems. UNEP Regional Seas Reports and Studies No. 154. Nairobi: UNEP.

54. Ellison, J.C., 2000. How South Pacific mangroves may respond to predicted climate change and sea-level rise. In *Climate Change in the South Pacific: Impacts and Responses in Australia, New Zealand, and Small Island States* (pp. 289–300). Springer, Dordrecht.

55. Church, J., Gregory, J., Huybrechts, P., Kuhn, M., Lambeck, K., Nhuan, M., Qin, D., Woodworth, P., 2001. Changes in sea level. In: Houghton, J., Ding, Y., Griggs, D., Noguer, M., van der Linden, P., Dai, X., Maskell, K., Johnson, C. (Eds.), *Climate Change 2001: The Scientific Basis (Publishedfor the Intergovernmental Panel on Climate Change)* (pp. 639–693). Cambridge University Press, Cambridge, UK; New York, USA (Chapter 11),

56. McLeod, E. and Salm, R.V., 2006. *Managing Mangroves for Resilience to Climate Change*. World Conservation Union (IUCN), Gland.

57. Lange, G.-M., Dasgupta, S., Thomas, T., Murray, S., Blankespoor, B., Sander, K., and Essam. T., 2010. Economics of Adaptation to Climate Change-Ecosystem Services. World Bank Discussion Paper No. 7. World Bank, Washington, DC.

58. Mukhopadhyay, A., Wheeler, D., Dasgupta, S., Dey, A. and Sobhan, I., 2018. *Aquatic salinization and mangrove species in a changing climate: Impact in the Indian Sundarbans*. The World Bank.

59. Dasgupta, S., Huq, M., Khan, Z. H., Ahmed, M. M. Z., Mukherjee, N., Khan, M. F., and Pandey, K., 2014. Vulnerability of Bangladesh to cyclones in a changing climate: Potential damages and adaptation cost. *Climate and Development*, 6, pp. 96–110.

60. Dasgupta, S., Kamal, F.A., Khan, Z.H., Choudhury, S., and Nishat, A., 2015a. River salinity and climate change: Evidence from coastal Bangladesh. In J. Whalley and J. Pan (eds.), *Asia and the World Economy: Actions on Climate Change by Asian Countries* (pp. 205–242). World Scientific Press, Singapore.

61. Dasgupta, S., Hossain, Md. M., Huq, M., and Wheeler, D. 2015b. Climate change and soil salinity: The case of coastal Bangladesh. *Ambio*, 44(8), pp. 815–826.

62. Komiyama, A., Ong, J.E., and Poungparn, S., 2008. Allometry, biomass, and productivity of mangrove forests: A review. *Aquatic Botany*, 89(2), pp. 128–137.

63. Dasgupta, S., Sobhan, I., and Wheeler, D., 2016. *Impact of climate change and aquatic salinization on mangrove species and poor communities in the Bangladesh Sundarbans*. The World Bank.

64. Banerjee, K., Gatti, R.C., and Mitra, A., 2017. Climate change-induced salinity variation impacts on a stenoecious mangrove species in the Indian Sundarbans. *Ambio*, *46*(4), pp.492–499.

65. Balaguru, K., Taraphdar, S., Leung, L.R., and Foltz, G.R., 2014. Increase in the intensity of postmonsoon Bay of Bengal tropical cyclones. *Geophysical Research Letters*, *41*(10), pp. 3594–3601.

66. Elsner, J. B., Kossin, J. P., and Jagger, T. H., 2008. The increasing intensity of the strongest tropical cyclones. *Nature*, 455, pp. 92–95. doi:10.1038/nature07234.

67. Islam, M.T., 2014. Vegetation changes of Sundarbans based on landsat imagery analysis between 1975 and 2006. *Acta Geographica Debrecina Landscape & Environment*, *8*(1), pp. 1–9.

68. Neogi, S.B., Dey, M., Lutful Kabir, S.M., Masum, S.J.H., Kopprio, G.A., Yamasaki, S., and Lara, R.J., 2016. Sundarban mangroves: diversity, ecosystem services and climate change impacts. *Asian Journal of Medical and Biological Research*, 2, pp. 488–507.

69. Lara, R.J., Neogi, S.B., Islam, M.S., Mahmud, Z.H., Islam, S., Paul, D., Demoz, B.B., Yamasaki, S., Nair, G.B., and Kattner, G., 2011. Vibrio choleraein waters of the Sunderban mangrove: Relationship with biogeochemical parameters and chitin in seston size fractions. *Wetlands Ecology and Management*, *19*, pp. 109–119.

70. Gopinath, G. and Seralathan, P., 2005. Rapid erosion of the coast of Sagar island, West Bengal-India. *Environmental Geology*, *48*(8), pp. 1058–1067.

71. Hazra, S., Samanta, K., Mukhopadhyay, A., and Akhand, A., 2010. Temporalchange detection (2001–2008) study of Sundarban (final report).School of Oceanographic Studies, Jadavpur University, Jadavpur.

72. Das, M., 2014. Deformation of the Jambudwip island of Sundarban region, Eastern India. *International Journal of Geomatics and Geosciences*, 5, pp. 9–18.

73. Ghosh, T., Hajra, R., Mukhopadhyay, A., 2014. Island erosion and afflicted population: Crisis and policies to handle climate change. In Filho, W.L., Alves, F., Caeiro, S., Azeiteiro, U.M., (eds.), *International Perspectives on Climate Change* (pp. 217–225). Springer, Berlin.

74. Shrestha, S., Miranda, I., Kumar, A., Pardo, M.L.E., Dahal, S., Rashid, T., Remillard, C. and Mishra, D.R., 2019. Identifying and forecasting potential biophysical risk areas within a tropical mangrove ecosystem using multi-sensor data. *International Journal of Applied Earth Observation and Geoinformation*, *74*, pp. 281–294.

75. Kumar, D., Gautam, A.K., Palmate, S.S., Pandey, A., Suryavanshi, S., Rathore, N. and Sharma, N., 2017. Evaluation of TRMM multi-satellite precipitation analysis (TMPA) against terrestrial measurement over a humid sub-tropical basin, India. *Theoretical and Applied Climatology*, *129*(3–4), pp. 783–799.

76. Clough, B. and Sim, R.G., 1989. Changes in gas exchange characteristics and water use efficiency of mangroves in response to salinity and vapour pressure deficit. *Oecologia*, *79*(1), pp. 38–44.

77. Cheeseman, J.M., Clough, B.F., Carter, D.R., Lovelock, C.E., Eong, O.J. and Sim, R.G., 1991. The analysis of photosynthetic performance in leaves under field conditions: A case study using Bruguiera mangroves. *Photosynthesis Research*, *29*(1), pp. 11–22.

78. Cheeseman, J.M., Herendeen, L.B., Cheeseman, A.T. and Clough, B.F., 1997. Photosynthesis and photoprotection in mangroves under field conditions. *Plant, Cell & Environment*, *20*(5), pp. 579–588.

79. Buyadi, S.N.A., Mohd, W.M.N.W., and Misni, A., 2013. Impact of land use changes on the surface tem-perature distribution of area surrounding the National Botanic Garden, Shah Alam. *Procedia Social Behaviour Science*, *101*, pp. 516–525.

80. Islam, M. S., and Islam, K. S., 2013. Application of thermal infrared remote sensing to explore the relation-ship between land use–land cover changes and urban heat Island effect: A case study of Khulna City. *Journal of Bangladesh Institute of Planners*, 6, pp. 49–60.

81. Thakur, S., Maity, D., Mondal, I., Basumatary, G., Ghosh, P.B., Das, P. and De, T.K., 2020. Assessment of changes in land use, land cover, and land surface temperature in the mangrove forest of Sundarbans, northeast coast of India. *Environment, Development and Sustainability*, 23, pp. 1–27.

82. Cahoon, D.R. 2006. A review of major storm impacts on coastalwetland elevations. *Estuar Coast*, 29, p. 889.

83. Faraco, L.F.D., Andriguetto-Filho, J.M., and Lana, P.C. 2010. A method-ology for assessing the vulnerability of mangroves andfisherfolk to climate change. *Pan-American Journal of Aquatic Sciences*, 5, pp. 205–223.

84. Fiu, M., Areki, F., Rounds, I. and Ellison, J.C., 2010. Vulnerability Assessment of Coastal Mangroves to Impacts of Climate Change: case studies from Fiji.

85. Ghosh, T. and Mukhopadhyay, A., 2014. *Natural Hazard Zonation of Bihar (India) using Geoinformatics: A Schematic Approach*. Springer Science & Business Media, Berlin.

86. Ellison, J.C. and Zouh, I., 2012. Vulnerability to climate change of mangroves: Assessment from Cameroon, Central Africa. *Biology*, *1*(3), pp. 617–638.

87. Ellison, J. and Strickland, P., 2015. Establishing relative sea level trends where a coast lacks a long term tide gauge. *Mitigation and Adaptation Strategies for Global Change*, *20*(7), pp. 1211–1227.

88. Punwong, P., Marchant, R. and Selby, K., 2013. Holocene mangrove dynamics and environmental change in the Rufiji Delta, Tanzania. *Vegetation History and Archaeobotany*, *22*(5), pp. 381–396.

89. Marimuthu, N., Wilson, J.J., Vinithkumar, N.V. and Kirubagaran, R., 2013. Coral reef recovery status in South Andaman Islands after the bleaching event 2010. *Journal of Ocean University of China*, *12*(1), pp. 91–96.

90. Szlafsztein, C. and Sterr, H., 2007. A GIS-based vulnerability assessment of coastal natural hazards, state of Pará, Brazil. *Journal of Coastal Conservation*, *11*(1), pp. 53–66.

91. Shepherd, I. 1991. Information integration and GIS. In: Maguire, D., Goodchild, M., Rhind, D. (eds.) *GIS–Principles and Applications* (pp. 337–360). Longman, London.

92. Nayak, S., 2002. Use of satellite data in coastal mapping. *Indian Cartographer*, 22, pp. 147–157.

93. Mujabar, S., and Chanrasekhar, N., 2013. Shoreline change analysis along the coast between Kanyakumari and Tuticorin of India using remote sensing and GIS. *Arabian Journal of Geosciences*, 6, pp. 647–664.

94. Singh, A., 1986. Change detection in the tropical forest environment of northeastern Indiausing Landsat data. In Edenand, M.J., and Parry, J.T. (eds.), *Remote Sensing and Tropical Land Management* (pp. 237–254). Wiley, London.

95. https://power.larc.nasa.gov/data-access-viewer/.

96. Kogan, F.N., 1995. Application of vegetation index and brightness temperature for drought detection. *Advances in Space Research*, *15*(11), pp.91–100.

97. Rahman, A., Roytman, L., Krakauer, N.Y., Nizamuddin, M. and Goldberg, M., 2009. Use of vegetation health data for estimation of Aus rice yield in Bangladesh. *Sensors*, *9*(4), pp. 2968–2975.

98. Kogan, F.N., 1997. Global drought watch from space. *Bulletin of the American Meteorological Society*, *78*(4), pp. 621–636.

99. Chakravortty, S. and Ghosh, D., 2018. Development of a model for detection of saline blanks amongst mangrove species on hyperspectral image data. *Current Science*, *115*(3), p. 541.

100. Pramanik, M.K., 2015. Assessment of the impacts of sea level rise on mangrove dynamics in the Indian part of Sundarbans using geospatial techniques. *Journal of Biodiversity, Bioprospecting and Development*, *3*(155), pp.2376–0214.

101. Joshi, H. and Ghose, M., 2003. Forest structure and species distribution along soil salinity and pH gradient in mangrove swamps of the Sundarbans. *Tropical Ecology*, *44*(2), pp. 195–204.

102. Jiao, X. and Moinuddin, H., 2016. Operationalizing analysis of micro-level climate change vulnerability and adaptive capacity. *Climate and Development*, *8*(1), pp. 45–57.

103. Woodroffe, C., Robertson, A. and Alongi, D., 1992. Mangrove sediments and geomorphology. *Tropical mangrove ecosystems. Coastal and estuarine studies*, *41*. doi: 10.1029/CE041p0007.

104. Woodroffe, R., Macdonald, D.W. and Da Silva, J., 1995. Dispersal and philopatry in the European badger, Meles meles. *Journal of Zoology*, *237*(2), pp. 227–239.

105. Semeniuk, V., 1994. Predicting the effect of sea-level rise on mangroves in northwestern Australia. *Journal of Coastal Research*, *10*, pp. 1050–1076.

106. Limaye, R.B. and Kumaran, K.P.N., 2012. Mangrove vegetation responses to Holocene climate change along Konkan coast of south-western India. *Quaternary International*, *263*, pp. 114–128.

107. Ellison, A.M., Mukherjee, B.B. and Karim, A., 2000. Testing patterns of zonation in mangroves: Scale dependence and environmental correlates in the Sundarbans of Bangladesh. *Journal of Ecology*, *88*(5), pp. 813–824.

108. Deb, S.C., 1956. PalaeoclimatologyandgeophysicsoftheGangeticDelta. *Geological Reviews of India*, *18*, pp. 11–18.

109. Gilman, E., Ellison, J. and Coleman, R., 2007. Assessment of mangrove response to projected relative sea-level rise and recent historical reconstruction of shoreline position. *Environmental Monitoring and Assessment*, *124*(1–3), pp.105–130.

110. IPCC, 2007. Climate Change 2007), Synthesis report. In: Fourth Assessment Report of the Intergovernmental Panel on Climate Change. Cambridge University Press, Cambridge, United Kingdom and New York, NY, USA.

111. Rahman, M.R. and Asaduzzaman, M., 2010. Ecology of sundarban, bangladesh. *Journal of Science Foundation*, *8*(1–2), pp. 35–47.

112. Manna, S., Chaudhuri, K., Bhattacharyya, S. and Bhattacharyya, M., 2010. Dynamics of Sundarban estuarine ecosystem: eutrophication induced threat to mangroves. *Saline Systems*, *6*(1), p. 8.

113. Hanebuth, T.J., Kudrass, H.R., Linstädter, J., Islam, B. and Zander, A.M., 2013. Rapid coastal subsidence in the central Ganges-Brahmaputra Delta (Bangladesh) since the 17th century deduced from submerged salt-producing kilns. *Geology*, *41*(9), pp. 987–990.

114. Spalding, M., Kainuma, M., Collins, L., 2010. *World Atlas of Mangroves*. Earthscan, London, p. 319.

115. Wahid, S.M., Babel, M.S. and Bhuiyan, A.R., 2007. Hydrologic monitoring and analysis in the Sundarbans mangrove ecosystem, Bangladesh. *Journal of Hydrology*, *332*(3–4), pp. 381–395.

116. Unnikrishnan, A.S., Kumar, M.R.R., and Sindhu, B., 2011. Tropical cyclones in the Bay of Bengal and extreme sea level projections along the east coast of India in a future climate scenario. *Current Science India*, 101, pp. 327–331. van der Valk, A.G. and P.M. Attiwill, 1984. Acetylene reduction in an *Avicennia marina* community in Southern Australia. *Australian Journal of Botany*, 32, pp. 157–164.

117. Bennett, E.M., Cumming, G.S., Peterson, G.D., 2005. A systems modelapproach to determining resilience surrogates for case studies. *Ecosystems*, 8, pp. 945–957.

118. Ellison, J.C. and Stoddart, D.R., 1991. Mangrove ecosystem collapse during predicted sea-level rise: Holocene analogues and implications. *Journal of Coastal Research*, 7, pp. 151–165.

119. Woodroffe, C.D., 2002. *Coasts: Form, Process and Evolution*. Cambridge University Press, Cambridge.

120. Lewis, J., 1990. The vulnerability of small island states to sea level rise: The need for holistic strategies. *Disasters*, *14*(3), pp. 241–249.

12 Monitoring the Effects of the Tropical Cyclone 'Amphan' on the Indian Sundarbans Using Microwave Remote Sensing

Rituparna Acharyya
Central University of Karnataka

Niloy Pramanick
Jadavpur University

Kaushik Gupta
Jadavpur University
University of Manitoba

Atreya Basu
University of Manitoba

Tuhin Ghosh
Jadavpur University

Anirban Mukhopadhyay
Jadavpur University
University of Manitoba

CONTENTS

12.1   INTRODUCTION

In the context of growing concern over the global depletion of mangroves, their role in India has become increasingly significant. The mangrove share in India is 0.67% of the total forest cover, of which 60% are on the eastern coast of the Indian Sundarbans (DasGupta and Shaw, 2013). Sundarbans are the largest patch of tidal halophytic forests in India and Bangladesh (Islam, 2014). They are vast, elongated, dense and mature mangrove forests, situated along the delta formed by the convergence of three rivers, Ganga, Brahmaputra and Meghna (Osti et al., 2009; Mondal and Saha, 2018). They save millions from flooding, cyclones, protect the coast by trapping nutrients, sediments and provide fuel, timber, medicines, food and building material, to deter coastal erosion (Osti et al., 2009; Ghosh et al., 2016). Sundarban mangroves are considered essential because they are high in productivity and important to the people of the Ganges delta as well as to the local and global ecosystems (Rahman et al., 2020). They retain and sequester carbon in wetlands and sediments by decomposition of litter and absorb other water contaminants (Ghosh et al., 2016). Mangroves, with annual global loss of 1%–2%, are heavily affected by both the climate and anthropogenic factors; if this rate persists, these forests may not be visible by the end of this century (Ghosh et al., 2015).

a. Anthropogenic Factors: Sundarbans support over 300,000 people including woodcutters, fishermen, honey-pickers with few alternatives and their growing population is damaging mangroves (Ghosh et al., 2016). Since 1995, large areas of Indian mangroves have been deforested for crop production (DasGupta and Shaw, 2013). Tourism increases at the expense of mangroves and it becomes polluted by sewage disposal in nearby ports and market areas next to tidal creeks (Paul et al., 2017). The fishing industry has expanded rapidly, including dry fish processing, development of aquaculture ponds and, fishermen's colonies, causing damage to the mangrove areas (Rahman et al. 2010).

b. Climatic Factors: Climate change threatens the biodiversity of Sundarbans, increasing flooded areas and salinity in coastal areas, the sea level may rise to 0.98 m by 2100 and net subsidence can occur at 2.5 mm year^{-1} (Payo et al., 2016). Erosion was 5.9% between 1975 and 2006 and is the major cause of the damage endured by Sundarban mangroves rather than by flooding (Hazra et al., 2002; Islam, 2014). The 12 southern sea-facing islands are the predominant erosion zones from Sagar (west) to Bhangaduni (east), with a total erosion of 162.88 km^2 over 30 years, and their western banks are more vulnerable to erosion than the east, indicating the effects of tidal waves (Payo et al., 2016). The tidal waves inundate the mangroves and undergo a long period of evaporation, which increases salinity, affecting their growth, recovery, and hypersaline patches forms, with encrusted algal surfaces on the shore (Paul et al., 2017). The forest cover of 47% has changed from the deforested to the degraded regions to the saline banks in succession from 2000–2012 (Samanta and Hazra, 2017).

The magnitude and severity of cyclones in the Bay of Bengal have increased since 1970 and has had an immediate effect on the level of coastal flooding, erosion and salinity (Hazra et al., 2002). The strong tropical cyclones degraded Sundarbans mangroves by 19.3% in 1977, 1988, and it took 25 years for dense forests to recover (Islam, 2014). There has been a rise in rainfall between 2000 and 2008 at a rate of 0.0041 m hour^{-1} and the relative mean sea level has risen at a rate of 17.8 mm year^{-1} (Hazra et al., 2010). Sundarbans receive 75% of rainfall in monsoon (June - September) and cyclones like 'Aila' and frequent floods cause severe coastal damage (Mondal and Saha, 2018). Mangroves got ravaged on November 1991, 1998, 2007 and April 2009 by tropical cyclones along the coast of Bangladesh and West Bengal (India), and the catastrophe left the mangrove stumps and sediments were stripped from the forest floor, exposing the mangrove root system (Paul et al., 2017). The scale and intensity of the damage to floristic diversity caused by the tropical cyclone 'Sidr' (15 November 2007) have devastated three major floristic taxa – *Heritiera fomes* (Sundari), *Excoecaria agallocha* (Gewa) and *Sonneratia apetala* (Kewra) (Bhowmik and Cabral, 2013). Cyclone 'Aila' (25 May 2009) has left vast areas of land unproductive and largely strained the socio-economic status of the coastal communities, shifting their reliance from land to forest resources, putting more stress on already stressed mangroves. (Rahman et al., 2017).

The Tropical cyclone 'Amphan' made landfall on the coast of West Bengal, Odisha and Bangladesh on 20th May 2020 (Ranganathan, 2020). 'Amphan' struck around 2.30 p.m. South 24 Parganas district in West Bengal, near the Sagar Islands of Sundarbans (Singh and Barik, 2020; Das, 2020). It was the most recent one, and its landfall in the region was much more severe (240 km hour^{-1}) than its predecessors, Aila (120 km hour^{-1}) in 2009, Bulbul (155 km hour^{-1}), in 2019 (Chakraborty and Thakur, 2020). It has shattered the embankments in the Sundarbans Delta and ruined many coastal huts due to a huge storm surge over 5 m (Singh and Barik, 2020) Seawater even infiltrated wells, ponds, depriving thousands of people from their access to water, and winds swept saltwater through trees: guava and banana, particularly mangroves, burnt with brackish water, leaves yellow and red (Thomas, 2020).

Remote sensing has been identified as an effective technique for assessing the degradation of mangrove forests induced by cyclones as well as for monitoring their initial recovery from the degraded state (Mondal and Saha, 2018; Bhowmik and Cabral, 2013; Elmore et al., 2000). The development in satellite data analysis applications and hardware has resulted in further use of remote sensing (Giri et al., 2007). Climate data (e.g. sea surface temperature, rainfall, and wind) is extracted using high-resolution (HR) ocean satellite datasets (Aqua MODIS Lev.3, TRMM, and AMSRE) (Hazra et al., 2010). The normalized difference vegetation index is the tool used to assess actual forest cover, to distinguish between land cover types, the minimum and the maximum value for the open and closed forest (Samanta and Hazra, 2017; Paling et al., 2008; Bhowmik and Cabral, 2013). The supervised classification method with multi-temporal satellite data is used to examine the nature of deforestation, size and composition of the mangrove species of Sundarbans (Giri et al., 2007; Paling et al., 2008). The Object-Based Image Analysis used to evaluate the extent of mangrove cover over a 40-year span (1975–2015) which shows that 4% of the mangrove has become water (Mondal and Saha, 2018). Landsat 5 and OLI 8 data are used to map the Spatio-temporal distribution of carbon-nitrogen (C:N) ratio of senescent leaves to assess primary productivity and nutrient balance in mangroves (Rahman et al., 2020).

Synthetic aperture radar (SAR), contrary to optical imagery, can be used efficiently to assess habitat depletion in the Sundarbans (Cornforth et al., 2013). The advent of freely accessible Sentinel-1 C-band SAR imagery provides the advantage of land cover classification in extreme cloud-covered conditions (Fonteh et al., 2016). SAR data are used to derive the segregation parameters for mangrove ecosystem classes mapping the extent of the mangroves (Abdel-Hamid et al., 2018). Radarsat-2 C-band SAR data are used to improve mangrove forest classification by analyzing the spectral and backscattering signatures of forests (Zhen et al., 2018). Here in this study, the SAR imageries from Sentinel-1 C band are used to assess the damage caused to the Sundarban mangroves by Cyclone 'Amphan'. In other words, the status of the mangrove coverage of Indian Sundarbans before and after the landfall of the 'Amphan' cyclone has been assessed by using the application of SAR.

12.2 AIM AND OBJECTIVES

This study aims to obtain information regarding the areal extent of mangroves before and after the 'Amphan' disaster by using the Sentinel-1 SAR imageries for assessing the damage caused by this tropical super cyclone to mangroves in the Indian Sundarbans.

The main objectives of this study are to assess the following:

i. To estimate the total area covered by the mangroves in the Indian Sundarbans.
ii. To delineate the exact areal extent of mangroves got damaged by the cyclone 'Amphan'.
iii. To evaluate the usefulness of SAR data in extracting information about cyclone damage.

12.3 STUDY AREA

Sundarbans, located within 21°32′N to 22°40′N and 88°05′E to 89°51′E, is the largest mangrove forest of the world, covering nearly 10,000 km², 62% of which is located in Bangladesh and 38% in India (Dahdouh-Guebas, 2011; Ghosh et al., 2015). Sundarbans was declared as a 'World Heritage Site' in 1987 and 'Biosphere Reserve' in 1989 by UNESCO. The area of the current study is based on Indian Sundarbans (21°322′N to 22°402′N and 88°852′E to 89°002′E) covering low-lying coastal plains in the state of West Bengal, including the districts, North and South 24 Parganas. It comprises an area of 6312.768 km² (2012) out of which mangroves covered 2246.839 km² in 1986, declined to 2201.41 km² in 1996, to 2168.91, and 2122.421 km² in 2012 (Samanta and Hazra, 2017). The current study over the world heritage site Indian Sundarbans is to estimate the actual coverage of mangrove in 2020, and how much of the mangrove area have got ravaged by the catastrophe of recently happened Tropical Cyclone 'Amphan' (Figure 12.1).

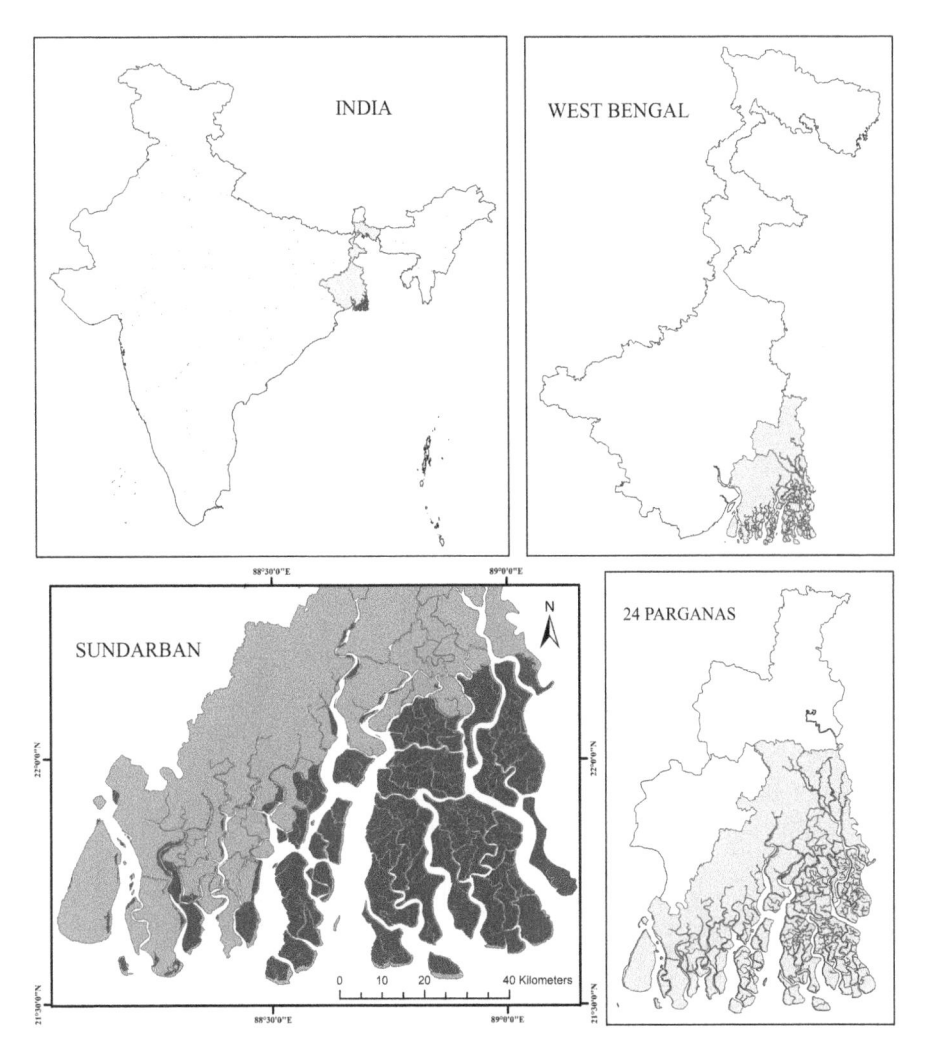

FIGURE 12.1   Location map of the study area (Indian Sundarbans).

12.4   METHODOLOGY

The study used images of ESA (European Space Agency), Sentinel-1-SAR data from Copernicus Open Access Hub. The Sentinel-1 C-band images were acquired on the dates before and after the cyclone for the study shown in Table 12.1. They were collected in IW (Interferometric Wide Swath) sensor mode with a 250 km swath width. The images having polarisations, VH (Vertical transmit–Horizontal receive) and VV (Vertical transmit–Vertical receive), and high-resolution Level-1 GRD (Ground Range Detected) processing level.

The main discussion in this chapter is regarding the steps used in data processing for the study. Figure 12.2 shows the flow chart comprising an outline of all the steps used in data for this study. The steps have been performed in the application of SNAP

TABLE 12.1

List of the Data Sets (Sentinel 1 Imagery) Used for the Study

Mission	Name	Cell Size (m)	Acquisition Date	Instrument	Product	Sensor Mode	Polarization
Sentinel-1A	S1A_IW_GRDH_1SDV_20200507T121204_20200507T121229_ 032459_03C246_B374.SAFE	10	07.05. 2020	SAR-C	Level 1-GRD-HR	IW	VV/VH
Sentinel-1A	S1A_IW_GRDH_1SDV_20200507T121229_20200507T121254_ 032459_03C246_4BAF.SAFE	10	07.05.2020	SAR-C	Level 1-GRD-HR	IW	VV/VH
Sentinel-1A	S1A_IW_GRDH_1SDV_20200531T121206_20200531T121231_ 032809_03CCE0_1D31.SAFE	10	31.05.2020	SAR-C	Level 1-GRD-HR	IW	VV/VH
Sentinel-1A	S1A_IW_GRDH_1SDV_20200531T121206_20200531T121231_ 032809_03CCE0_1D31.SAFE	10	31.05. 2020	SAR-C	Level 1-GRD-HR	IW	VV/VH

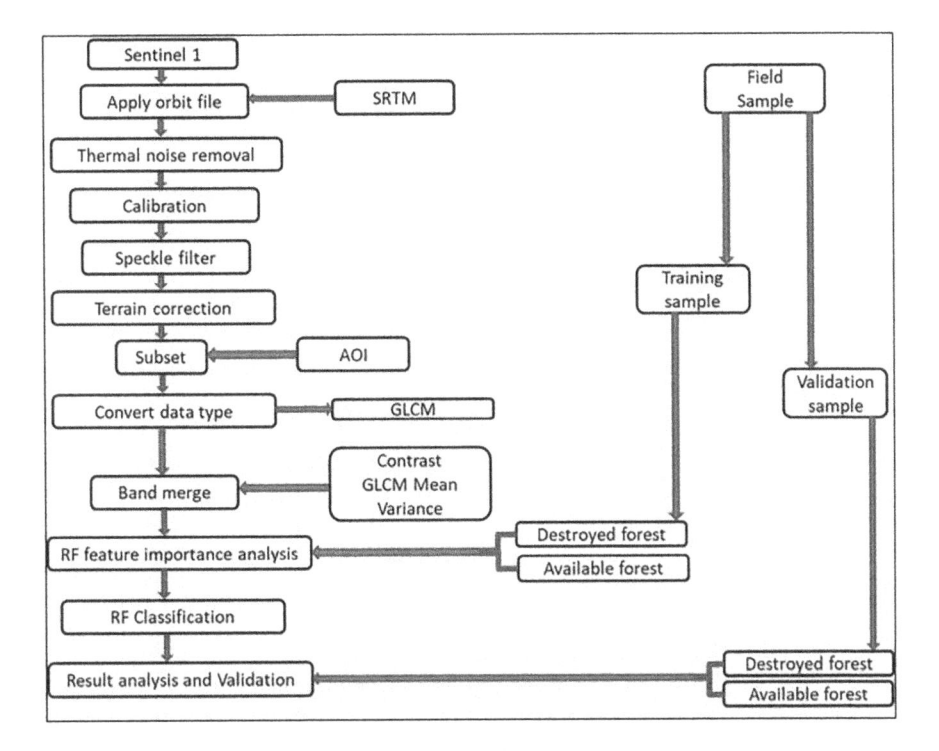

FIGURE 12.2 Flow chart showing the methodology of the current study.

(Sentinel's Application Platform), version 7.0 of the ESA (European Space Agency), for processing the Sentinel-1 images covering the study area.

- **Applying Orbit File:** The Orbital State vectors in the metadata of SAR are typically not accurate and can be refined in days-to-weeks after product development. The application of specific orbit files available in SNAP allows the download and the update of state orbit vectors into the product metadata of each SAR scene and provides reliable details of satellite position and velocity (ESA, 2020).
- **Thermal Noise Removal:** In specific, thermal noise reduction prevents the noise effects in the inter-sub-swath texture. It has been applied to Sentinel-1 Level-1 GRD products which have not already been corrected. In SNAP, where thermal noise reduction operator is available, data from Sentinel-1 can also trigger a noise signal that may have been removed during level-1 generation and update the product annotations to enable the re-application of the correction (ESA, 2020).
- **Calibration:** Image calibration has been applied to SAR imagery using Sentinel-1. It is the process that to the SAR images in such a way that the pixel values of the SAR images actually represent the radar backscatter of the reflecting object. The details required to apply the calibration equation are included in the Sentinel-1 GRD product, precisely (Filipponi, 2019).

The calibration vector used as an appendix in the software enables easy conversion of the image intensity values into sigma nought values. Calibration inverts the scaling factor implemented during the level-1 product generation and imposes a constant offset and a range-dependent boost, including an absolute calibration constant (ESA, 2020).

- **Speckle Filtering:** The SAR images include implicit salt and pepper, like textures called 'speckles', which reduce the image quality and render it more complex to interpret the objects. The occurrence of fluctuations in the intensity of the speckle in complex SAR images is an inevitable consequence of the consistent form of the image. Each cell contains several individuals scatters, each of which contributes to the total signal returned from the resolution cell. Since the radar wavelength is usually much shorter than the resolution cell, the phase obtained from each scatters is random (ESA, 2020). The filters available in the SNAP single product filter operator are: 'Boxcar', 'Median', 'Frost', 'Gamma Map', 'Lee', 'Refined Lee', 'Lee Sigma' 'IDAN' (Filipponi, 2019). Among these filters, the 'Lee Sigma' filter being used for the study, which assumes Gaussian noise distribution, filters the middle pixel in the sliding window with the average pixel within the two-sigma range (ESA, 2020).
- **Terrain Correction:** The SAR images are generally sensed at different angles above 0°, usually results in images that are related to side-looking geometry with some distortion. Terrain corrections are oriented to compensating for these variations to ensure the image is geometrically as close to the actual world as possible (Filipponi, 2019). The process in terrain correction downloads the file of DEM, externally for proper terrain correction. The STRM v.4 (3″ tiles) from the FTP Joint Research Centre (xftp.jrc.it) will be automatically downloaded in tiles for ortho-rectification of the image covering the study area. Different types of Digital Elevation models can also be used (ACE, GETASSE30, ASTER, SRTM 3Sec GeoTiff).
- **Subset:** The operator is being utilized to generate either the spatial and/or spectral subsets of the data object. The spatial subset can be specified by pixel position or geographical polygon (ESA, 2020).
- **Convert Data Type:** The Convert Datatype Operator converts the data and formats it in several forms for the different data groups. This operator can interpret, internally extract the full data and the metadata in many compatible data objects, and use the plug-in writer modules to create data in a specific file format. It can convert between the following data types: 32-bit, 32-bit, 32-bit, Integer matrix integral (16-bit + 16-bit) integer, 32-bit integer (32-bit + 32-bit).
- **GLCM (Grey Level Co-occurrence Matrix):** The GLCM is a calculation of how pixel combinations vary from one another. The grey levels (brightness values) are shown in the image. The GLCM texture recognizes the relation between two pixels at a time called the reference and the neighbouring pixels.
  - The GLCM mean is not only the sum of all the initial pixel values in the picture frame.

$$\mu_i = \sum_{i,j=0}^{N-1} i\left(P_{i,j}\right) \qquad u_j = \sum_{i,j=0}^{N-1} i\left(P_{i,j}\right)$$

The left-hand approach is used to calculate the mean based on the reference digit, μ_i. It is also essential to measure the average using the nearby pixels, μ_j as in the right-hand equation. The symmetric GLCM, where each pixel in the window is counted as a reference, and once as a relative, the two values are the same.

- The GLCM variance uses the GLCM values it deals with specific combinations of references and a neighbour pixel. In this case, it differs from the variance of the grey levels in the original image.

$$v_i^2 = \sum_{i,j=0}^{N-1} P_{i,j}\left(x+a\right)^n \qquad v_i^2 = \sum_{k=0}^{n} P_{i,j}\left(i-\mu_i\right)^n$$

$$u_j = \sqrt{v_i^2} \qquad u_j = \sqrt{v_i^2}$$

Variance (v) is calculated using the grey levels i or j gives the same result as the GLCM is symmetrical (Dutta et al., 2008).

- GLCM contrast is referred to as the sum of squares variance. This calculates the amount of variation in the image and it is the opposite of homogeneity.

$$\text{Contrast} = \sum_{i,j=0}^{N-1} P_{i,j}\left(i-j\right)^2$$

When i and j are equal, the cell is diagonal and $(i-j)=0$, indicates pixels that are exactly similar to their neighbours, thus given a weight of 0. When i and j differ by 1, there is a small resemblance, and the weight is 1. If they differ by 2, the contrast is rising and the weight is 4. The weight continues to rise exponentially (Dutta et al., 2008).

- **RF Classification:** Random Forest (RF) is an algorithm with several data sets (Breiman, 2001). As the bootstrapping technique is used, the simple algorithm flow of the RF classifiers is two-thirds of this data collected by training samples (in-bag data) and the remaining samples are validation samples (OOB) that could be used to calculate the internal error (Gislason et al., 2006). For each training sample collection, classification and regression tree is built to create a RF of N trees. As in the creation phase of the growing tree, m is chosen randomly from all the features M (usually $m=$pM). The optimum segmentation function in m is selected according to the Gini Coefficient.

$$\text{Gini} = 1 - \sum_{C} p^2\left(C/N\right)$$

In RF, errOOB1$_i^f$ is computed utilizing OOB data of attribute f; then noise intervention is placed arbitrarily to f of OOB1$_i^f$ and errOOB2$_i^f$ is then re-calculated.

$$FI^f = 1/N \sum_C p^2 \left(errOOB2\ _i^f - errOOB1\ _i^f \right)$$

- **Field Sample Collection:** As specified in the methodology, field samples are obtained to verify the findings of the study. The main purpose of this field survey is to verify the findings of a remote sensing-based analysis utilizing SAR Sentinel 1 data sets. The field survey has been conducted over the location that has been hugely damaged the tropical cyclone 'Amphan' as well the area that have no got damaged by the cyclone. Forty field samples collected from the study area. Figure 12.3 shows the location points where the field survey has been conducted. The exact locations where the field survey has been conducted are Sagar, Jambudwip, Mousuni, Lothian, Patharpratima, Kultali, Hiranmalpur Basanti and Ajmalmari.

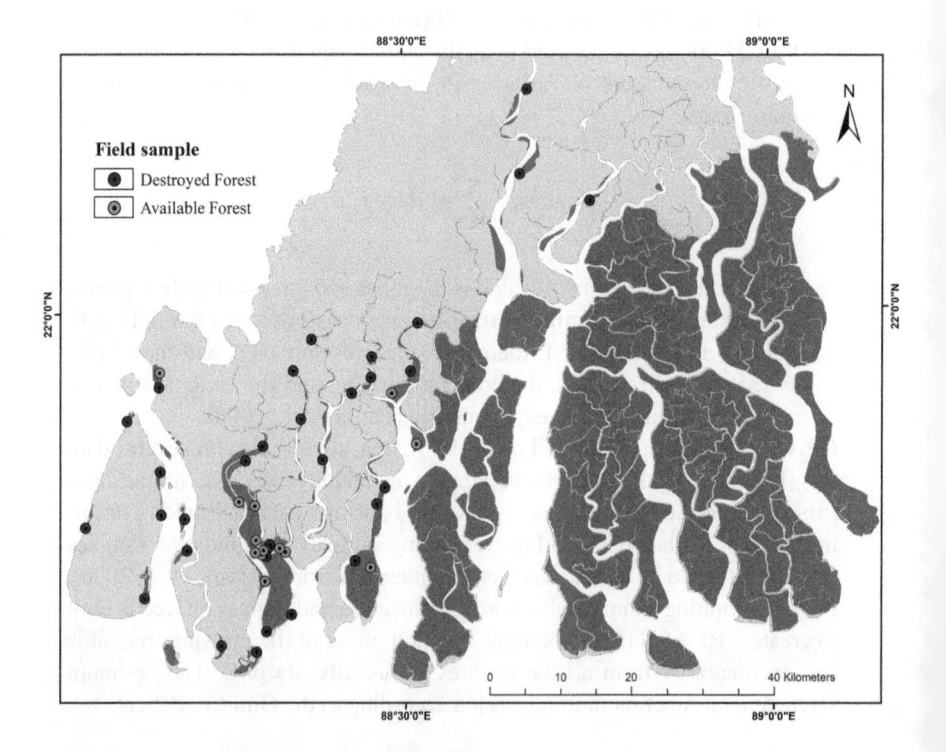

FIGURE 12.3 Field sample points acquired from the study area.

12.5 RESULTS

This study focuses on determining immediate changes in the health of mangroves within a very short period when the 'Amphan' cyclone was the main cause of destabilization in Sundarbans. The study has been undertaken to analyze the expected degradation of the cyclone, particularly in the Indian Sundarbans. The results of the analysis supported this expectation. This session illustrates the results of delineating the extent of the areal damage done by the cyclone. Figures 12.4 and 12.5 show the classification results of the study area within the time frame. The maps in the figures were processed from SAR imageries comprising consecutive pairs of Sentinel-1 C band GRD data, acquired before (7th May 2020) and the after (31st May 2020) the cyclone passage. The images were classified by the method of supervised classification using the algorithm of RF Classifier because it is prominent owing to its ability to classify large quantities of data with high accuracy. GLCM textural parameters are required to enhance the use of RF classifiers in mapping, as the GLCM texture of the mean, correlation, contrast and entropy of size nine windows has been stacked with enhanced Lee filters (Fonteh et al., 2016). The results are generated by applying the methods mentioned above are based on the areal coverage of the mangrove forests of the Indian Sundarbans.

The changes in the areal coverage or extent of the mangrove forests have taken place due to the damage inflicted by the cyclone. Those changes reveal the regions, where the expansion and shrinkage of the mangrove cover can be identified. Thus, the total area covered by Indian Sundarbans mangroves is 2,109 km^2, as per the assessment done by Forest Survey of India in 2017 (Forest Survey of India, 2017). Figure 12.4 shows the classified image of the study area before the 'Amphan' cyclone arrived and struck the shores of the mangrove delta. The map indicates that the actual extent of the Indian mangroves is approximately 2210.58 km^2, comprising the main mangroves (2099.34 km^2) and the marginal mangroves (111.24 km^2) based on the results obtained from the classified radar image of the study area, acquired before the 'Amphan' disaster. The image has been classified into two distinct classes, 'remained forest' and 'destroyed forest'.

The light black patches display the remaining mangrove forests (2210.58 km2) in the Indian Sundarbans, and the deep black patches are the regions where mangroves have been destroyed before the arrival of 'Amphan' which accounts to 92.79 km^2. The dense mangrove patches are concentrated mostly on the eastern side of the Sundarbans Delta, stretching from Dhulibasani, Chulkati to Katuajhuri, and Gosaba. Whereas the western side consists of comparatively small patches, i.e. Jambudwip, Lothian, Susnicahra, Dhanchi as well as discontinuous mangrove patches in Sagar, Mousuni, Patharpratima, Namkhana. The spots and patches in deep black clearly show the destroyed portions of mangroves which are also mostly clustered on North and Central Saznekhali, Gorankhal South Gosaba, Bakultala Khal East Gosaba and North Matla. The devastation of these mangrove patches can be attributed to anthropogenic and climatic factors.

The total area under by the mangroves in the Indian Sundarbans has got severely altered by the effect of 'Amphan'. The total area estimated from the classified imagery (Figure 12.4) of the study area acquired after the 'Amphan' is 1,834 km^2.

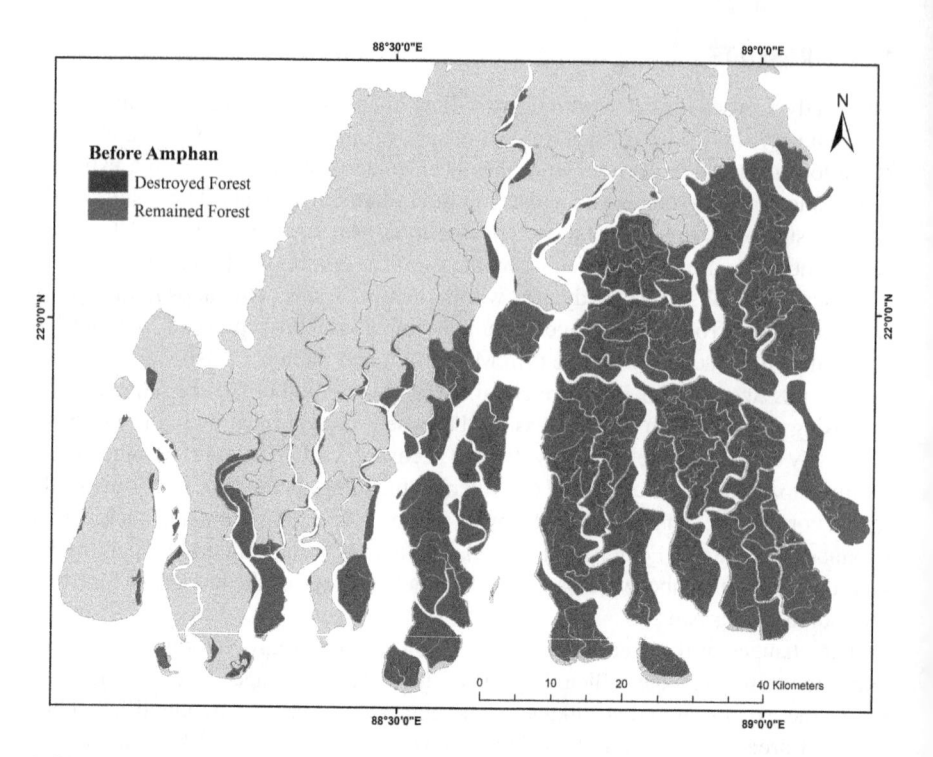

FIGURE 12.4   Indian Sundarbans before 'Amphan' cyclone.

The area occupied by mangrove forests has abruptly decreased by 375 km2, including regions previously degraded by climate and anthropogenic factors (92.79 km²) and adverse impacts of 'Amphan' (283.16 km²). A significant change in the area of mangroves after the cyclone shows an estimated change in the area has occurred by 13.37%. In other words, 'Amphan' has destroyed approximately 13.4% of mangroves of Indian Sundarbans.

Similarly, like the previous figure, Figure 12.5 shows the classified image with two distinct groups, 'remained forest' and 'destroyed forest'. The light black patches denote the remaining mangrove forests and the deep black patches show the mangrove areas got destroyed before and after the 'Amphan' landfall. Figure 12.5 shows the way of the cyclone 'Amphan' which was a Category II storm with winds over 100 miles hour[-1] (Bloch, 2020). Since the path of the cyclone was inclined towards the west of the Sundarbans, the mangroves were ravaged on that portion.

Aside from the mangroves that were devastated before the cyclone (92.79 km²), patches that got had been affected only by this cyclone (283.16 km²) need to be distinctly identified to assess the mangrove destruction induced by 'Amphan' alone. Figure 12.6 shows the mangroves after 'Amphan' and highlights those certain regions of the mangroves which have been affected only by the intensity of this cyclone. The deep black spots and patches in Figure 12.6 are denoting the regions that have undergone destruction induced by 'Amphan'. As mentioned above, due to the inclination of the path of 'Amphan' to the west, the deep black spots and patches are

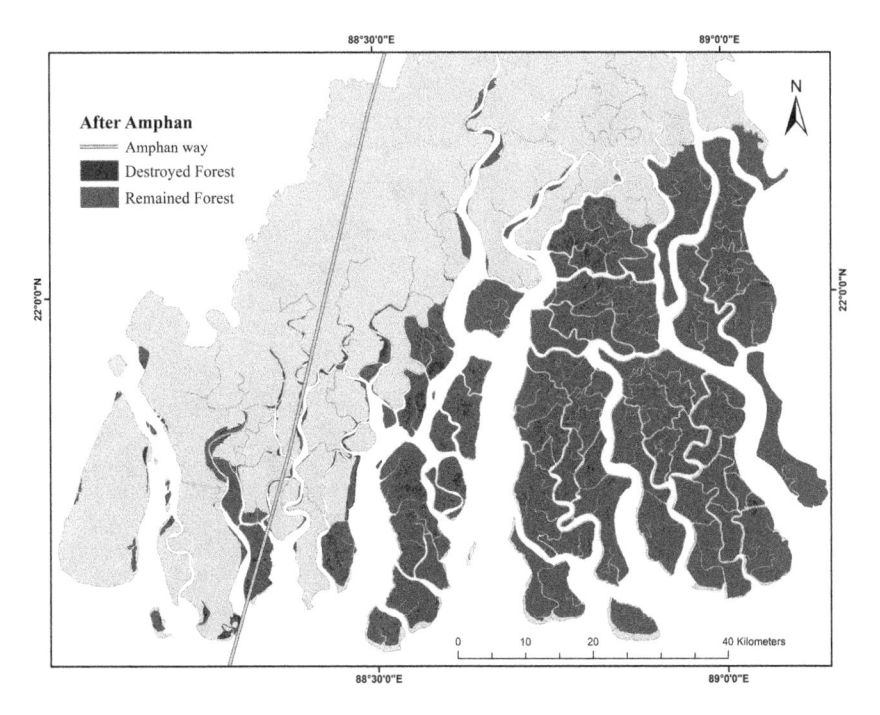

FIGURE 12.5   Indian Sundarbans after 'Amphan' cyclone.

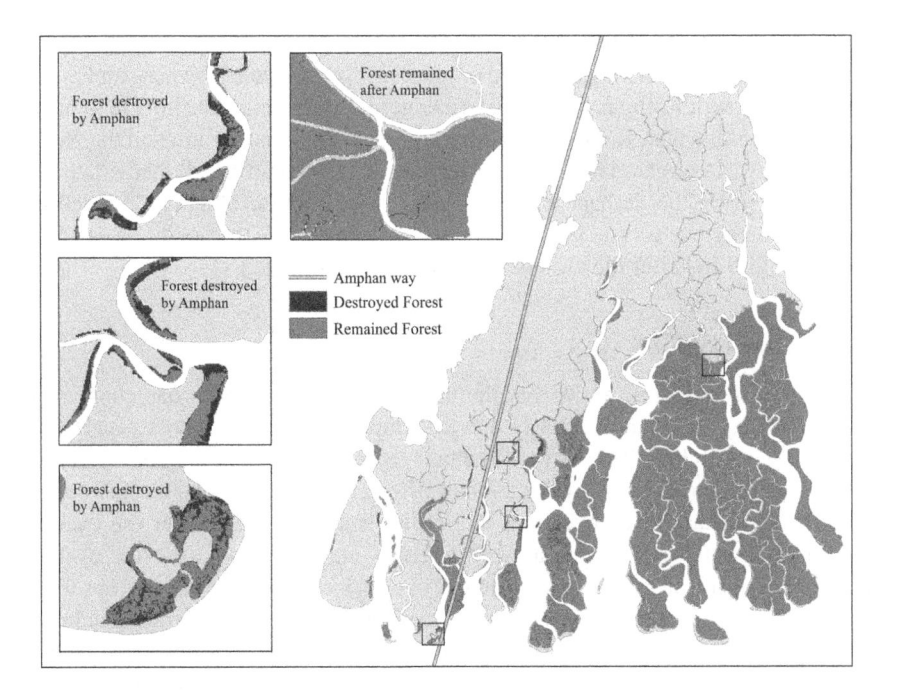

FIGURE 12.6   Mangrove patches in Indian Sundarbans damaged by 'Amphan' cyclone.

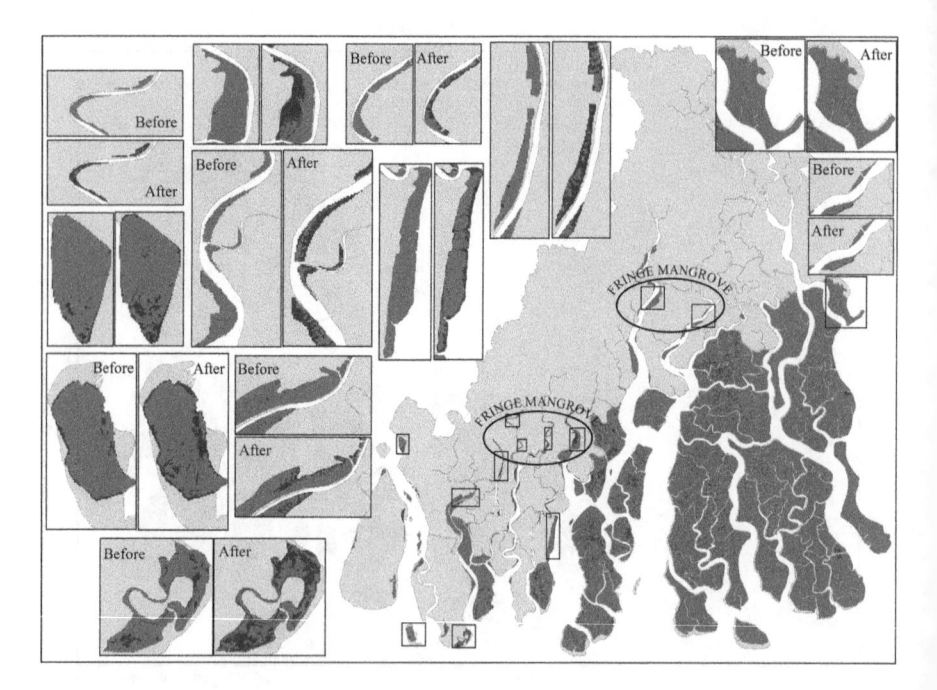

FIGURE 12.7   Mangrove patches in Indian Sundarbans shown Before and After 'Amphan' cyclone.

mainly concentrated on the western side of the study area. These spots and patches in the deep black are mainly clustered on the coastline region of Western and South-western portion of Lothian, South-eastern edges of Mainland, Dakshin Gopalnagar, Maheshpur, Shibnagar, Rakhalpur (Patharpratima), Namkhana main and the North-eastern fringe of Sagar. These are the regions of the Indian Sundarbans that have been affected by the carnage of the tropical super cyclone 'Amphan', which has destroyed mangrove forests in these regions. In Figure 12.7 the condition of theses patches before and after the cyclone has also been highlighted.

12.6   DISCUSSION

The results of the study show that the extent of the Sundarbans has changed by 283.16 km^2 (approximately 13.4%) due to 'Amphan'. As stated above, apart from the effect of 'Amphan' 92.79 km^2 of the mangrove cover had already been affected before the incidence of this cyclone. Thus, after a comparison is made between the percentage of mangrove cover loss that took place before and after the cyclone, it is found that there has been a 67.23% rise in mangrove loss after the cyclone. This increase in the percentage of mangrove loss can be assessed by observing Figure 12.7, which focuses on the condition of particular mangrove patches in Sundarbans before and after 'Amphan'. Figure 12.7 shows the mangrove patches of Mainland, Patharpratima, Namkhana and Sagar that underwent the damage induced by this cyclone and their condition before the cyclone. The cyclone has devastated mostly the fringe mangroves (Figure 12.6) in the above mentioned regions of Sundarbans.

The mangrove ecosystem of Sundarbans continues to shift on both the spatial and temporal scales due to numerous natural causes like this cyclone 'Amphan', hydrological changes, sea-level rise as well as anthropogenic causes like land use changes, clear-cutting, industrial pollution, etc. There are several scientific reasons behind every natural disaster and, despite having the explanations for them, many believe that behind these cyclonic disasters there is a supernatural power, but that is by no means real and scientific. The tropical cyclones typically develop along the equator over warm ocean waters. The hot and humid air continuously rising above the ocean creates a lower air pressure region below it, which in turn enables cold air to swell in and this process continues to cause a cyclonic storm. 'Amphan' struck the deltaic shores of Sundarbans in the Bay of Bengal was a natural disaster and also a super-cyclone. It took 40 hours for 'Amphan' to evolve into a super-cyclone, induced by high sea surface temperatures between 32°C and 34°C at Bay of Bengal (Sangomla, 2020). According to the scientists, long-term heating has been the main factor behind the strong intensification of Amphan, from 140 km hour^{-1} up to a maximum wind speed of 260 km hour^{-1} in 18 hours (Sangomla, 2020). The intensity of these tropical cyclones striking the adjoining areas of the Bay of Bengal every year during the pre-monsoon and monsoon cycle has increased in recent times. This time 'Amphan', particularly in certain sections of the mangrove zone, has caused tremendous destruction. Nevertheless, the key factor behind minimized disruption to the adjacent area is the presence of mangrove trees. Regardless of the government's desperate efforts to cope with the disaster, the mangrove has always assumed the responsibility for coping with this kind of disasters from time to time. As a consequence, the mangroves endured the wrath of the cyclone and unprecedented damage, but were able to decrease the intensity of the cyclone that prevailed during the landfall. In addition, the tidal surge caused by the cyclone in the mangrove delta caused extensive saltwater flooding. The dual effects of cyclones and flooding have impacted mangroves and the surrounding area. The local economy has been affected and indigenous livelihood became disrupted. The tidal surges have breached the banks of the river, and the cultivable lands were submerged under saline water for a long time, leading to huge losses in the agricultural sectors. The paddy fields were flooded by saline water following the breach of 32 embankments which left the farmers utterly helpless (Chakraborty and Thakur, 2020). The coastal regions were flooded by Amphan, which pushed up to 15 km of water as it struck the Ganges Delta. The wind speed recorded by IMD (Indian Meteorological Department) was around 150–160 km hour^{-1}, but the cyclone induced quite less damage because of the mangroves and later the wind speed became reduced to 112 km hour^{-1}, recorded in Kolkata. (Singh and Barik, 2020; Ranganathan, 2020). The mangroves are even more effective when disasters become severe and their roots form a complex interweaving with tree trunks, serving as pace breakers to delay or avoid tides. Unfortunately, the extent and health of Sundarbans have been diminished by deforestation, land-use changes, aquaculture and tourism. It must have played a significant part in making Amphan as dangerous as it was. Hence it must be kept in mind that if certain responsibilities and measures are not taken to restore the mangroves, then 1 day the region may be destroyed completely.

12.7 CONCLUSION

According to the aforementioned findings, the famous mangrove forests of Sundarbans have been degraded by 13.4% as a consequence of the latest tropical cyclone ('Amphan'). Numerous cyclones have damaged the impeccable existence of mangrove forests in recent times. The 'Amphan' cyclone struck at a speed of 240 km hour^{-1} when the mangrove trees were demolished by the storm, which in effect was able to substantially decrease the wind speed to a great extent. Therefore, the cyclone was not able to do as much damage as it was supposed to do in the adjacent rural localities and nearby urban areas. Yet the damage inflicted by this cyclone upon the mangroves and the coastal communities is unprecedented. The region's socio-economic infrastructure has been disrupted and many people are struggling to live a life of poverty and hunger. They are deprived of basic necessities such as food, housing, medical care, and education. The only way out of this crisis of public life is by massive support from the state and central government system and several NGOs. It should not be forgotten that if we want to defend ourselves from this repeated assault of the tropical cyclones, we need to reform the mangrove cover in Sundarbans. It will take a long time for mangroves to naturally regenerate and recover from this damage. No one knows beforehand that we will not be the victims of another cyclonic disaster, so the possibility of it cannot be ruled out. Therefore, we cannot afford to say that we will not fall into the grip of this kind of catastrophe in the near future. In other words, this devastating natural disaster is inevitable whether it is today or tomorrow. In order to survive from this calamity, we need to think of manual reconstruction of the mangrove region by planting mangrove trees and the concerned authorities have to play a leading role in this.

REFERENCES

Abdel-Hamid, A., et al. (2018) 'Mapping mangroves extents on the red sea coastline in Egypt using polarimetric SAR and high resolution optical remote sensing data', *Sustainability (Switzerland)*, 10(3), pp. 1–22. doi: 10.3390/su10030646.

Bhowmik, A. K. and Cabral, P. (2013) 'Cyclone sidr impacts on the sundarbans floristic diversity', *Earth Science Research*, 2(2), pp. 62–79. doi: 10.5539/esr.v2n2p62.

Bloch, M. (2020) 'Live cyclone amphan map: Tracking the storm's path', *The NewYork Times*, 20 May, p. 1.

Breiman, L. (2001) 'Random forests', *Machine Learning*, 45. doi: 10.1023/A:1010933404324.

Chakraborty, T. and Thakur, B. (2020) 'Sundarbans needs immediate help', *The Statesman*, 15 July. Available at: https://www.thestatesman.com/opinion/sundarbans-needs-immediate-help-1502905854.html.

Cornforth, W. A., et al. (2013) 'Advanced land observing satellite phased array type L-Band SAR (ALOS PALSAR) to inform the conservation of mangroves: Sundarbans as a case study', *Remote Sensing*, 5(1), pp. 224–237. doi: 10.3390/rs5010224.

Dahdouh-Guebas, F. (2011) 'World Atlas of Mangroves: Mark Spalding, Mami Kainuma and Lorna Collins (eds)', *Human Ecology*, 39(1), pp. 107–109. doi: 10.1007/s10745-010-9366-7.

Das, S. (2020) 'Extremely severe cyclonic storm Amphan makes landfall', *Deccan Herald*, 20 May, pp. 1–2. doi: 10.1017/CBO9781107415324.004.

DasGupta, R. and Shaw, R. (2013) 'Changing perspectives of mangrove management in India - An analytical overview', *Ocean and Coastal Management*, 80, pp. 107–118. Elsevier Ltd. doi: 10.1016/j.ocecoaman.2013.04.010.

Dutta, R., Stein, A. and Patel, N. R. (2008) 'Delineation of diseased tea patches using Mxl and texture based classification', *The International Archives of the Photogrammetry, Remote Sensing and Spatial Information Sciences*, (January), pp. 1693–1700.

Elmore, A. J., et al. (2000) 'Quantifying vegetation change in semiarid environments', *Remote Sensing of Environment*, 73(1), pp. 87–102. doi: 10.1016/s0034-4257(00)00100-0.

ESA (2020) *SNAP Software, Help Document*. Available at: https://step.esa.int/main/toolboxes/snap/ (Accessed: 18 July 2020).

Filipponi, F. (2019) 'Sentinel-1 GRD preprocessing workflow'. In *Proceedings*, p. 11. doi:10.3390/ecrs-3-06201.

Fonteh, M. L., et al. (2016) 'Assessing the utility of sentinel-1 C band synthetic aperture radar imagery for land use land cover classification in a tropical coastal systems when compared with landsat 8', *Journal of Geographic Information System*, 8(4), pp. 495–505. doi: 10.4236/jgis.2016.84041.

Forest Survey of India (2017) *Mangrove Cover - Forest Survey of India*. Dehradun. Available at: https://fsi.nic.in/isfr2017/isfr-mangrove-cover-2017.pdf.

Ghosh, A., et al. (2015) 'The Indian Sundarban Mangrove Forests: History, utilization, conservation strategies and local perception', *Diversity*, 7(2), pp. 149–169. doi: 10.3390/d7020149.

Ghosh, M. K., Kumar, L. and Roy, C. (2016) 'Mapping long-term changes in Mangrove species composition and distribution in the Sundarbans', *Forests*, 7(12), pp. 1–17. doi:10.3390/f7120305.

Giri, C., et al. (2007) 'Monitoring mangrove forest dynamics of the Sundarbans in Bangladesh and India using multi-temporal satellite data from 1973 to 2000', *Estuarine, Coastal and Shelf Science*, 73(1–2), pp. 91–100. doi: 10.1016/j.ecss.2006.12.019.

Gislason, P. O., Benediktsson, J. A. and Sveinsson, J. R. (2006) 'Random forests for land cover classification', *Pattern Recognition Letters*, 27(4), pp. 294–300. doi: 10.1016/j.patrec.2005.08.011.

Hazra, S., et al. (2002) 'Sea level and associated changes in the Sundarbans', *Science and Culture*, 68(9), pp. 309–321.

Hazra, S., et al. (2010) *Temporal Change Detection (2001–2008) Study of Sundarban*, World Wildlife Fund, Jadavpur University. Kolkata.

Islam, M. T. (2014) 'Vegetation changes of Sundarbans based on landsat imagery analysis between 1975 and 2006', *Landscape & Environment*, 8(1), pp. 1–9.

Mondal, B. and Saha, A. K. (2018) 'Spatio-temporal analysis of Mangrove loss in vulnerable islands of Sundarban World Heritage Site, India', In *Lecture Notes in Geoinformation and Cartography*, pp. 93–109. doi: 10.1007/978-3-319-78208-9_5.

Osti, R., Tanaka, S. and Tokioka, T. (2009) 'The importance of Mangrove forest in Tsunami disaster mitigation', *Disasters*, 33(2), pp. 203–213. doi: 10.1111/j.1467-7717.2008.01070.x.

Paling, E. I., Kobryn, H. T. and Humphreys, G. (2008) 'Assessing the extent of mangrove change caused by Cyclone Vance in the eastern Exmouth Gulf, northwestern Australia', *Estuarine, Coastal and Shelf Science*, 77(4), pp. 603–613. doi: 10.1016/j.ecss.2007.10.019.

Paul, A. K., et al. (2017) 'Mangrove degradation in the Sundarban', In Finkl, C. and Makowski, C. (eds) *Coastal Wetlands: Alteration and Remediation*. Springer International Publishing, Berlin, pp. 357–392. doi: 10.1007/978-3-319-56179-0.

Payo, A., et al. (2016) 'Projected changes in Area of the Sundarban Mangrove Forest in Bangladesh due to SLR by 2100', *Climatic Change*, 139(2), pp. 279–291. doi: 10.1007/s10584-016-1769-z.

Rahman, M. M., et al. (2020) 'Remote sensing-based mapping of senescent leaf C:N ratio in the sundarbans reserved forest using machine learning techniques', *Remote Sensing*, 12(9), pp. 5–8. doi: 10.3390/RS12091375.

Rahman, M. M., Rahman, M. M. and Islam, K. S. (2010) 'The causes of deterioration of Sundarban mangrove forest ecosystem of Bangladesh: Conservation and sustainable management issues', *AACL Bioflux*, 3(2), pp. 77–90.

Rahman, S., et al. (2017) 'The impact of cyclone aila on the sundarban forest ecosystem 7KH ', *International Journal of Ecology & Development*, 32, pp. 1–11.

Ranganathan, P. (2020) 'Amphan in the Sundarbans: How Mangroves protect the coast from tropical storms', *The WIRE*, 28 May, pp. 1–4. Available at: https://science.thewire.in/environment/cyclone-amphan-sundarbans-delta-mangrove-trees-climate-change/.

Samanta, K. and Hazra, S. (2017) 'Mangrove forest cover changes in Indian Sundarban (1986–2012) using remote sensing and GIS', *Environment and Earth Observation*, pp. 97–108. doi: 10.1007/978-3-319-46010-9_7.

Sangomla, A. (2020) 'Less atmospheric aerosol may have intensified Cyclone Amphan', *Down To Earth*, 21 May, pp. 1–3. Available at: https://www.downtoearth.org.in/news/climate-change/-less-atmospheric-aerosol-may-have-intensified-cyclone-amphan--71280#:~:text=Amphan took 40 hours after, rapid intensification%2C according to experts.

Singh, S. S. and Barik, S. (2020) 'Cyclone Amphan batters West Bengal, Odisha', *The HINDU*, May, pp. 1–5. Available at: https://www.thehindu.com/news/national/other-states/cyclone-amphan-batters-west-bengal-odisha/article31634110.ece.

Thomas, S. A. (2020) 'After Amphan', *London Review of Books*, 23 June, p. 1. Available at: https://www.lrb.co.uk/blog/2020/june/after-ampha.

Zhen, J., Liao, J. and Shen, G. (2018) 'Mapping Mangrove Forests of Dongzhaigang Nature Reserve in china using landsat 8 and radarsat-2 Polarimetric SAR data', *Sensors (Switzerland)*, 18(11). doi: 10.3390/s18114012.

13 Satellite Remote Sensing to Evaluate the Social Consequences of Bio-Physical Damage: A Case Study of Cyclone Amphan in Sundarbans Adjoining Areas in Bangladesh

Sheikh Tawhidul Islam
Jahangirnagar University

Indrajit Pal
Asian Institute of Technology

Krishna Prosad Mondal
Jahangirnagar University

CONTENTS

13.1 INTRODUCTION

The world's largest mangrove forest, the Sundarbans, has evolved and sustained with its neighbouring human population through a symbiotic unison for thousands of years. The ecosystem services of different kinds such as supporting, provisioning, regulatory and cultural services provided by this forest attracted people to make settlements and live close to the forests. The harvest of forest resources by the local inhabitants remained at a sustainable level in the past. Commercial exploitation based on the sole aim to generate revenue from the forest resources was introduced by the British (East India Company) administration since the beginning of eighteen hundred centuries and formally started with the establishment of Zamindari system in the Indian sub-continent through the enactment of Permanent Settlement Act of 1793. The British administration undertook a thorough topographic survey to produce large-scale (1:63360) topographic maps (conducted between 1908 and 1942) and medium-scale topographic maps (1924) which were revised through planimetry survey using air-borne remote sensing (aerial photographic survey conducted from 1943 to 1945). These survey activities facilitated to retrieve information about the distribution of forest resources and map out canals, creeks, rivers and land tracts in the Sundarbans. The map products produced at various scales assisted extraction of forest resources on a commercial scale, that is, harvesting large wood logs and non-timber forest products by the government agencies including private entrepreneurs. This indiscriminate extraction process exceeds sustainability threshold levels at some points and degradation of the forest started to happen. This unsustainable extraction of forest resources was further aggravated by the impacts of disasters in the area, especially with the cyclonic disturbances. The disasters directly caused damage to the forest resources and at the same time ravaged human communities making them poorer who in turn create more pressure on Sundarbans resources to offset their loss. At this point, forest survey activities conducted by the state agencies with a different aim to monitor the changes that happened in the Sundarbans either as results of disaster impacts or through over extraction of forest resources. The aim of survey activities also included to devise appropriate strategies for improving the conditions of forests through scientific management. Assessment of forest quality in terms of forest biophysical properties through forest inventory had also been the purpose of the survey activities. The application of remote sensing was found to be the cardinal method to facilitate the survey process. It suggests that the application of remote sensing in forest tract mapping especially in the Sundarbans mangrove forests should be seen from the contextual institutional, physical, and social backgrounds. This may inform about the rationale in choosing a satellite sensor as a data source from a wide array of choices (in terms of spatial, temporal and spectral resolution, cost-effectiveness, revisit interval for change detection purposes, etc.). This chapter, in these contexts, aimed to show how satellite remote sensing played roles in supplying data for mapping Sundarbans forest resources in the contexts of the man–nature interface. And, secondly, how remote sensing data were used in Bangladesh parts of Sundarbans by different authors in their exercises.

This chapter is primarily based on a literature review that gave an understanding of the objectives and purposes of different research activities where satellite remote sensing data were used. Secondly, Sentinel-1 radar data were used through the Google Earth Engine (GEE) to assess the impacts of cyclone Amphan of Sundarbans forest areas. The purpose of doing the exercise was to know the impacts of disaster that caused further pauperization of the communities leading to, in a passive way, create more pressure on Sundarbans resources. At the outset, a discussion is given on the attributes of Sundarbans located in the Bangladesh part followed by an examination about the man–nature interface in the forest areas. The discussion and arguments were supported with data provided by remote sensing sources. A section is also provided that summarizes the major works on Sundarbans where authors used remote sensing data. The necessity of using the state-of-the-art remote sensing data and methods for sustainable forest management purposes were discussed in the latter part of the paper.

13.2 SUNDARBANS MANGROVE FOREST: BANGLADESH PART

The single largest mangrove forest tract of the world, spread in the mouth of Ganges tidal plains, is known as the Sundarbans. Deposition of billions of tonnes of sediments over thousands of years from upstream Himalayan areas, the supply of huge amounts of water from large watersheds primarily caused by monsoon climatic processes created living conditions for hundreds of floral and faunal species in this mangrove forests. The ecosystem in the Sundarbans is characterized as a brackish water environment resulted from the mix of salt and freshwater. The water ecosystem is rich in nutrients and hence creates spawning grounds of a variety of fish species. The land areas of the forests are also characterized by rich and productive plant and animal biodiversity. The Sundarbans ecosystem thus becomes the suppliers of a variety of food, fibre, fish, medicinal plants and wood for the people living in forest nearby areas (Figure 13.1).

It also harbours world famous Royal Bengal Tiger, Gangetic Dolphins and Deer as to uphold its strong and distinctive ecological identity and diversity. The Sundarbans forest acts as shields to protect human settlements from impacts of cyclonic disturbances such as storm surge and gusty winds. Thus, the mangrove forests provide a number of provisioning, regulatory, supportive and spiritual services (known as ecosystem services) and attracted people to live in forest adjoining areas for centuries.

The interaction between man and nature in the forest areas gradually increased from subsistence to commercial levels and eventually creates a number of threats to the ecological integrity of the forest ecosystems. People living in nearby areas altered land use unsustainably (e.g. changed large areas from subsistence agriculture into shrimp farms), established industries such as wood saw mills, brick making industry where fuels are generally supplied from forest woods, constructed port activities and urban centres, performed unsustainable tourism activities and thus exposed the Sundarbans forests to a number of risks including biodiversity loss, environmental pollution from upstream urban wastes and industrial effluent, intentional or unintentional oil leakage and discharge from ships and vessels,

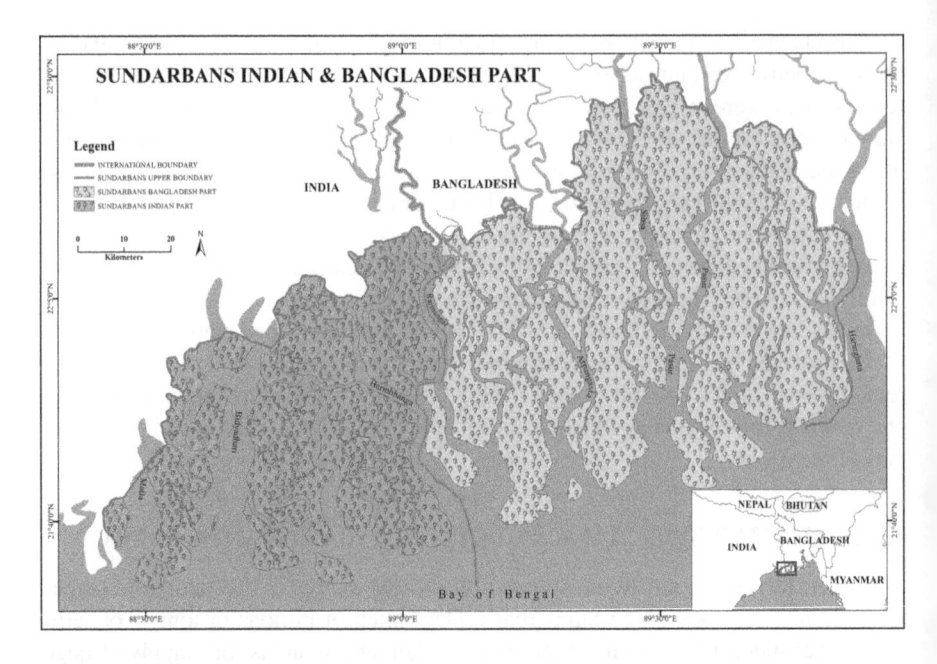

FIGURE 13.1 Sundarbans mangrove forests with river systems.

emission of hazardous gas from industries, pressure from the unsustainable number of tourists. In addition, the increase in the frequency and magnitude of cyclones and repeated impacts on forests, an increase of salinity in water and soil and shortage of fresh water supply because of climate change are also posing huge threats to the Sundarbans (Pal and Gosh 2018). It suggests that the ecosystem in this mangrove forests is complex, resources are diversified, interaction with man is complicated, disturbance from natural disasters and subtle changes happening as results of climate change difficult to assess and apprehend and all these in a combined fashion characterize the Sundarbans mangrove forest especially that occurs in Bangladesh.

13.3 MAN–NATURE INTERFACE IN SUNDARBANS: OPPORTUNITIES AND CHALLENGES TOWARDS SOCIAL WELL-BEING

It is estimated that about one million people live in the Sundarbans bordering villages (Laskar & Rahman 2016) and 20% of them (about 200,000 in numbers) directly interact with Sundarbans almost every day for making their livelihoods. People (male, female and children) collect both aquatic and land-based resources from the Sundarbans forests. Golpata (*Nipa Palm*), honey and wood are harvested from the land-based resources and fish, shrimp and mud-crab are collected from the rivers, creeks and canals of forests. Land-based product collection activities are mostly commercially organized by big merchants (especially for the *golpata*, honey collection) that needs big investments to operate. In contrast, collection of fish (including shrimp fry collection), shrimp and mud crab are mainly individual or household-level activities. About two-thirds of the people who are engaged in water-based resource collection engage in fish-shrimp harvesting and one-third in mud-crab collection. In Sundarbans, people sometimes change their profession with the change of the seasons as quality and quantity of resources vary with seasons and thus remain in the resource collection process throughout the year. Thousands of women are engaged in (adult) shrimp and shrimp fry collections from the Sundarbans rivers and canals. These women generally live in the river/canal adjoining areas and their activities are governed by the rules of tides of the rivers. They generally go twice a day in the river during ebb/low tides for harvesting shrimp fries. They push the nets for miles along the river banks to catch shrimp fries.

This suggests that people living in Sundarbans adjoining areas are heavily dependent on the forest products and the thriving ecosystem services provided by the mangrove forests make the local environment a context of fusion for man and nature. The Sundarbans forests have always been seen as a single unit where both the components (man and nature) coexist and the survey activities and mapping exercises took place by taking this phenomenon into account.

The harmony between the man and nature indicated above sometimes breaks down with the impacts of natural disasters especially the cyclones. In recent times super cyclone Amphan (made landfall on 20th May 2020) ravaged Sundarbans and forest adjoining human settlements and villages and cause huge damage (Figures 13.2 and 13.3). Change detection and cyclone impact assessment exercise were conducted in this research using Sentinet-1 radar data of the European Space Agency with the help of the Sentinel-1 toolbox and GEE (GEE 2020).

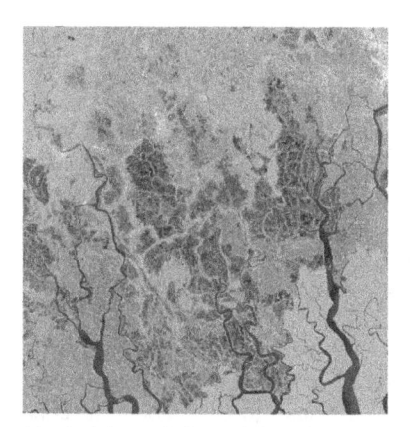
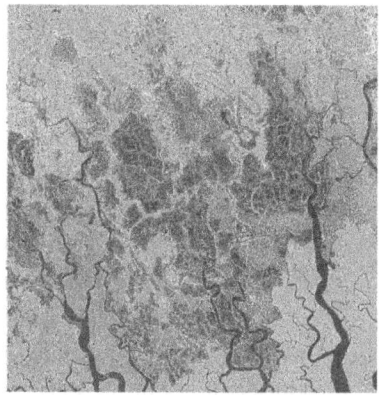

Sentinel-1 mosaic image before the flood of the study area in Google Earth Engine (16th May 2020).

Sentinel-1 mosaic image after the flood of the study area in Google Earth Engine (22th May 2020)

FIGURE 13.2 Impacts of cyclone Amphan in Sundarbans forests and adjoining areas detected by using the Sentinel-1 radar image.

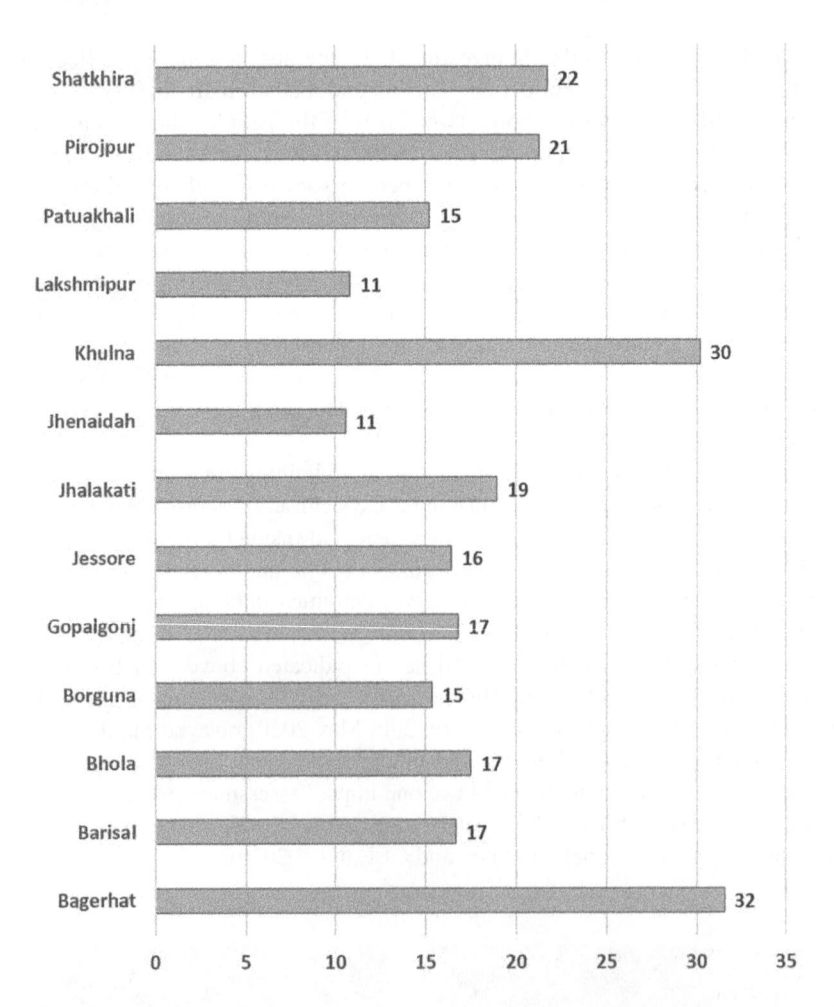

FIGURE 13.3 Percentage of lands inundated due to cyclone Amphan 2020 in different coastal districts of Bangladesh (using Sentinel-1 radar image). (Modified after ICIMOD 2020 (Uddin & Matin 2020).)

13.4 SATELLITE REMOTE SENSING IN RESOURCE ASSESSMENT IN SUNDARBANS

This section summarizes the use of remote sensing data and techniques by different authors in the Bangladesh part of Sundarbans. The results of these applications are widely used by a range of institutions and professionals to know about the condition of the forests (i.e. the biophysical health) and the degree of loss that happened as results of natural and human-induced stresses and also to devise and deploy strategies towards effective forest management. For example, Bangladesh Government has declared 10-km forest adjoining buffer areas as Ecologically Critical Area (ECA) and imposed a set of rules to follow by different entities to safeguard the forests. In doing that Landsat image has been used to generate maps of the mangrove forests and to identify designated areas to be declared as ECA (Figure 13.4). The Bangladesh Forest Department and other nongovernment agencies use this map and information to co-manage the forests by involving local people.

Remote sensing techniques have been used to research and mapping exercises in the Sundarbans forest areas since 1980s (Cornforth et al. 2013) although the initial application was carried out in 1943–1945 using air-borne aerial photography. Imhoff et al. (1986) used space-borne radar data to detect the forest canopy characterization of Sundarbans. Optical and radar images have been used mostly since then to detect the edge, habitat fragmentation, change detection, land cover mapping, vegetation categorization, forest characterization and wetland mapping. However, images from optical sensors in earlier times were used most of the time to analyze the change detection of the area.

Dutta et al. (2015) assessed the ecological disturbance of three major cyclones using MODIS data; henceforth, a number of research activities have been done to evaluate the impact of cyclone Sidr occurred in 2007 using Landsat MSS, SPOT 5, ASTER, MODIS and ALOS – PALSAR data (Mandal and Hosaka 2020; Cornforth et al. 2013). No analysis has been done using SENTINEL 1 to assess the cyclone's impact and change detection up until now (Table 13.1).

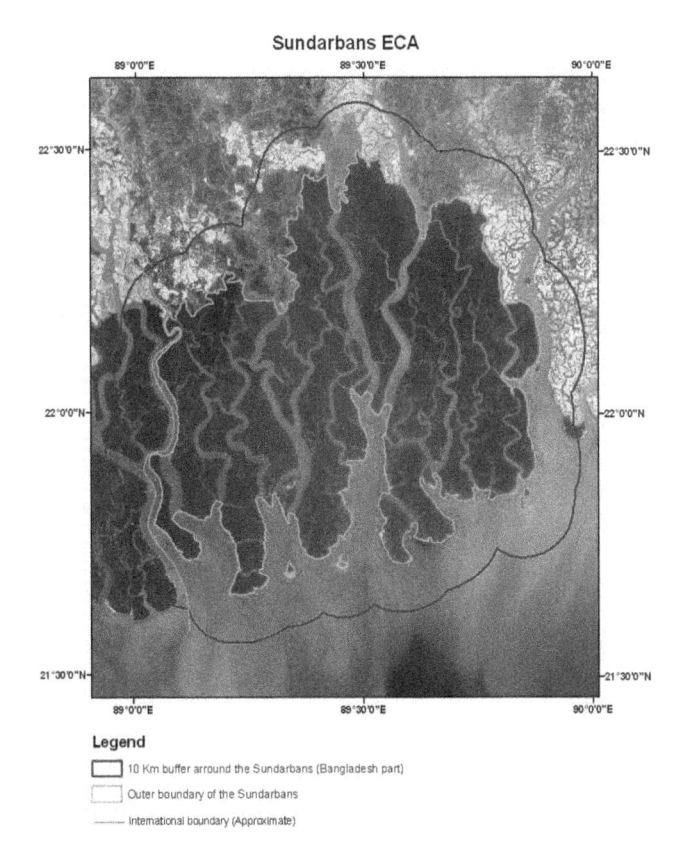

FIGURE 13.4   Sundarbans ECAs. (Landsat and DoE.)

TABLE 13.1
Geospatial Applications on Sundarbans Mangrove Forest (Bangladesh Part).

Application	Data Used	Method	References
Assessing cyclone disturbances (1988–2016) in the Sundarbans mangrove forests	Landsat and GEE	NDVI, Supervised classification, Kappa coefficient	Mandal and Hosaka (2020)
Forest vulnerability assessment	Landsat-5 and Landsat-8	Geo-statistical hot spot ($G * i$) model, NDVI and FDI.	Hussain and Islam (2020)
Senescent Leaf C:N Ratio mapping in the Sundarbans	LandsatTM5 and Landsat 8	Machine learning models	Rahman et al. (2020)
Climate Change Adaptations in the Mangrove Ecosystem	CRU TS v3.21, GPCC v6, WorldClim Data	Climate model simulations, Fuzzy Cognitive Mapping	Singh et al. (2019)
To assess and monitor the land use changes associated with deforestation in Bangladesh	Landsat TM, OLI	Change Detection, NDVI	Shimu et al. (2019)
Impact of cyclone Aila on the Sundarbans	Landsat TM	Change Detection	Rahman et al. (2017)
To assess the relationship between temperature, rainfall pattern and mangrove species in the Sundarbans	Landsat TM/ETM/OLI, SPOT, CBERS, SIR, ASTER, IKONOS and Quick Bird	Supervised classification, Change Detection	Ghosh et al. (2017)
Tropical cyclone impacts	SPOT 5, Quickbird-2, Worldview-1	Change Detection	Hoque et al. (2016)
To assess the physical factors of cyclonic storm surge Aila and impacts in Khulna and adjacent area	Field Data, Google Earth and SRTM DEM	Participatory Geographic Information System (PGIS)	Kabir (2016)
Assessing the Bangladesh Sundarbans present status	Landsat TM and Landsat ETM+	Change Detection	Aziz and Paul (2015)
Ecological disturbances in Sundarbans from 2001 to 2011	MODIS, Landsat-8, v03r05 data	MGDI, EVI, LST	Dutta et al. (2015)
Floristic Diversity of Sundarbans affected by Sidr 2007	Landsat 7 ETM+	Unsupervised classification, NDVI	Bhowmik and Cabral (2013)
Conservation of mangroves	ALOS-PALSAR	Change Detection	Cornforth et al. (2013)
Status and distribution of mangrove forests	Landsat MSS, TM, QuickBird and IKONOS	Change Detection, Image classification	Giri et al. (2011)
Damage estimation by cyclone	ASTER	Supervised classification	Akhter et al. (2008)
Impact assessment	MODIS	Normalized Difference Infrared Index (NDII)	GOB (2008)

13.5 SOCIAL CONSEQUENCES OF FOREST BIOPHYSICAL DAMAGE OF CYCLONE AMPHAN

The discussions given above indicate that the interaction between man and nature is strong and intricate in Sundarbans forest areas. Majority of the people are poor (refer to Section 13.3) and interacted with the mangrove ecosystems for their subsistence while a small part are big merchants who commercially organize their activates. Cyclone Amphan destroyed the habitats of people, local infrastructure, agricultural farms and fishery, and it happened in a time when the communities are suffering from COVID-19 crisis including lockdown conditions. This damaged conditions and breakdown of provisions for livelihoods worsened the conditions of the forest dependent poor people. It was reported by the Department of Disaster Management of Bangladesh Government that the strong winds caused by cyclone Amphan destroyed 217,667 house structure and inundated 149,000 ha of agricultural lands and breached embankments in 80 places in the Sundarbans adjoining areas. Although the number of deaths of human population was less (only 26 and 2.6 million people affected) as government evacuated about 2.4 million people in 14,636 permanent and temporary cyclone shelters. The road communications were seriously damaged and affected the transportation of the local produce such as fish and crabs to regional markets. These conditions made the local poor people further poor as people did not able to receive usual financial returns from the catch due to the fall of market price of the produce. On top of that, they had to invest money for house repair and meeting other costs including COVID-19-related health expenditures. Discussions with local people suggest that local people started going inside the forests once the cyclone is over and situation becomes normal and have been putting extra pressure on the Sundarbans forests by extracting more resources to offset the loss caused by cyclone Amphan. It was also informed that some local fishermen poison waters of small creeks inside the forests for getting the best possible bulk of fish catch. In addition, activities of wildlife (especially the Royal Bengal Tiger and Deer) poachers have increased in the forests after the cyclone Amphan impacts in the area. Satellite remote sensing, in this regard, could play vital roles in assessing the forest biophysical damage and also the damage happened in the human settlement areas located in forest adjoining areas. This scientific data-driven knowledge may help to estimate the direct and indirect loss caused by cyclone Amphan impacts and may play useful roles in disaster loss and damage recovery planning (of the reforests and also forest dependent communities) and related program towards build-back-better of the systems.

13.6 CONCLUSION

The world's largest mangrove forests have been facing several internal and external challenges. The internal challenges include rapid changes in the topographical and hydrological properties leading to an alteration in the composition of floral species and ecosystem characteristics. The external challenges are generally originated from human-induced actions such as over-extraction of resources, conversion of forest lands into other land uses (Figure 13.5), disturbance in freshwater supply from upstream catchments areas into the forests by the construction of polders, dams. Natural

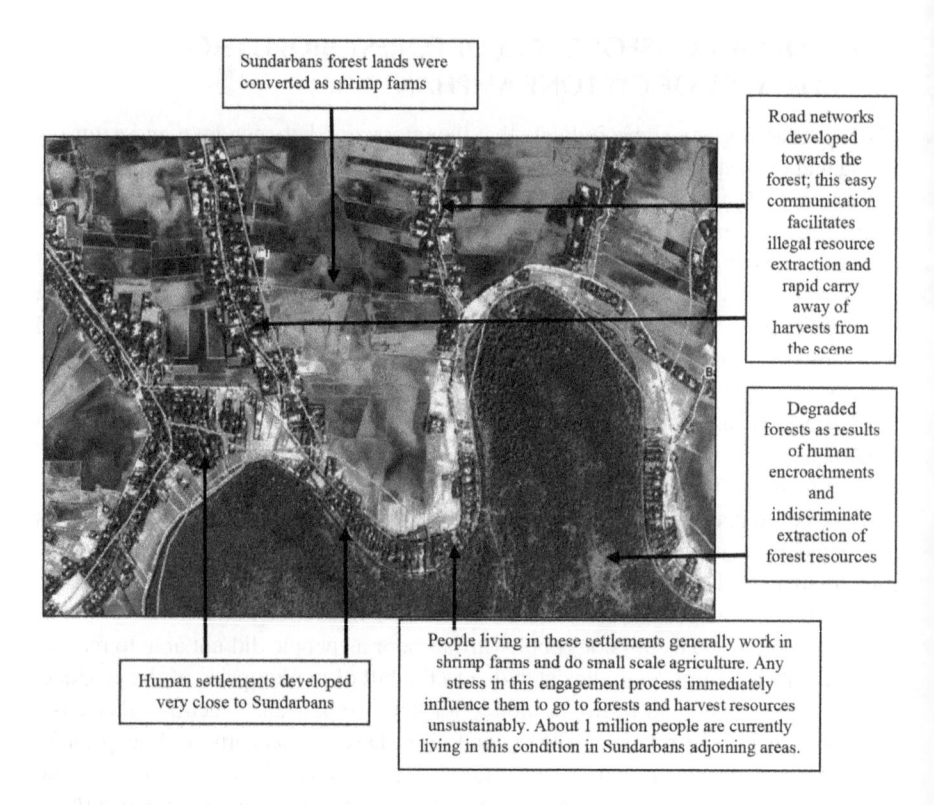

FIGURE 13.5   Google Earth image (January 2012) used to show the man–nature interface in Sundarbans adjoin areas.

disasters like cyclonic disturbances also create challenges in both direct and indirect ways (as discussed previously) on the Sundarbans. Forest mapping exercises are taking place using remote sensing data and methods in Sundarbans areas (e.g. recently conducted forest inventory) primarily to identify the distributions of species in terms of biophysical properties, composition, assessing the spatial extent of forest cover, etc. In some of these exercises, in-situ measurement of biophysical variables from sample plots in association with optical satellite imageries was used. Studies are also conducted to ascertain the impacts of natural disasters on Sundarbans and adjoining areas where forest-dependent human communities live. However, assessment of different environmental factors and processes such as (i) change in the salinity regimes as results of the reduction of fresh water supply from the upstream rivers, (ii) change in the land topography inside the forest areas due to sediment deposition, (iii) impacts of change in the climatic variables on disease prevalence and ecosystem properties remained to be less examined and as a result knowledge about the underlying reasons for the change in forest compositions/ecosystems and related threats also remained poor. This information gaps, in many instances, do not allow policymakers and forest managers/planners to take appropriate policies in regard to forest management in a rapidly changing condition of the environmental, hydrological, geomorphological and climatic processes. Co-ordination in complex social systems depends on the extent and effectiveness of information and communication processes operating within the inter-organizational system (Pal et al. 2017). It is indicated earlier that survey exercises and application of remote sensing in Bangladesh parts of the Sundarbans were conducted based on firstly to exploit forest resources and secondly to develop better management strategies. But in recent times, satellite and air-borne remote sensing techniques become more sophisticated in terms of technological improvements of the sensor systems, improvements in allied technologies such as LiDAR (Light Detection and Ranging), the cost-effectiveness of the data (in some cases data can be obtained free of cost), rapid image processing facilities using GEE platform collectively create opportunities to generate knowledge about different aspects of the Sundarbans more precisely and almost real-time. It could be said in this backdrop that more research aided with different types of remote sensing technologies (optical, radar LiDAR) needs to be done at an appropriate spatial and temporal scale so that old knowledge could be improved towards safeguarding the Sundarbans resources more effectively.

REFERENCES

Akhter, M., Iqbal, Z. and Chowdhury, R.M. (2008) 'ASTER imagery of forest areas of Sundarban damaged by cyclone Sidr', *ISME/GLOMIS Electronic Journal* 6(1), pp. 1–3.

Aziz, A. and Paul, A., (2015) 'Bangladesh Sundarbans: Present status of the environment and biota', *Diversity*, 7(3), pp. 242–269. Available at doi: 10.3390/d7030242.

Bhowmik, A. K. and Cabral, P. (2013) 'Cyclone Sidr impacts on the Sundarbans floristic diversity', *Earth Science Resources*, 2(2), p. 62.

Cornforth, W. A., et al. (2013) 'Advanced land observing satellite phased array type L-Band SAR (ALOS PALSAR) to inform the conservation of Mangroves: Sundarbans as a case study', *Remote Sensing*, 5(1), pp. 224–237. doi: 10.3390/rs5010224.

Dutta, D., et al. (2015) 'Assessment of ecological disturbance in the mangrove forest of Sundarbans caused by cyclones using MODIS time-series data (2001–2011)', *Natural Hazards*, 79(2), pp. 775–790. doi: 10.1007/s11069-015-1872-x.

GEE (2020). Sentinel-1 Algorithms. Google Earth Engine. https://developers.google.com/earth-engine/guides/sentinel1(accessed June 15, 2020).

Ghosh, M., Kumar, L. and Roy, C. (2017) 'Climate variability and Mangrove cover dynamics at species level in the Sundarbans, Bangladesh', *Sustainability*, 9(5), p. 805. Available at doi: 10.3390/su9050805.

Giri, C., E. Ochieng, L. L. Tieszen, Z. Zhu, A. Singh, T. Loveland, J. Masek, and N. Duke. (2011). "Status and distribution of mangrove forests of the world using earth observation satellite data." *Global Ecology and Biogeography* 20 (1):154–159. doi: 10.1111/j.1466-8238.2010.00584.x.

GOB (2008) 'Super Cyclone Sidr 2007: Impacts and strategies for interventions', *Ministry of Food and Disaster Management. Bangladesh Secretariat*, Dhaka, Bangladesh. Available at: https://www.prevention web.net/engli sh/profe ssion al/polic ies/v. php?id=9470. (Accessed on 02 September 2020).

Hoque, M. A.-A., et al. (2016) 'Assessing tropical cyclone impacts using object-based moderate spatial resolution image analysis: a case study in Bangladesh', *International Journal of Remote Sensing*, 37(22), pp. 5320–5343. doi: 10.1080/01431161.2016.1239286.

Hussain, N. and Islam, M. N. (2020) 'Hot spot (G_i^*) model for forest vulnerability assessment: A remote sensing-based geo-statistical investigation of the Sundarbans mangrove forest, Bangladesh', *Modeling Earth Systems and Environment*, 6(4), pp. 2141–2151. doi: 10.1007/s40808-020-00828-4.

Imhoff, M., et al. (1986) 'Forest canopy characterization and vegetation penetration assessment with space-borne radar', *IEEE Transactions on Geoscience and Remote Sensing*, GE-24(4), pp. 535–542. doi: 10.1109/TGRS.1986.289668.

Kabir, T. (2016). Assessment of biophysical factors for storm surge hazard and their implications for food security. MSc Thesis. Bangladesh University of Engineering and Technology. Dhaka. http://lib.buet.ac.bd:8080/xmlui/handle/123456789/4424 (accessed June 20, 2020).

Laskar, M. and Rahman, L. (2016) 'Development initiatives of the Sundarban of Bangladesh', *Malaysian Forester*, 79, pp. 77–88.

Mandal, M. S. H., and Hosaka, T. 2020. 'Assessing cyclone disturbances (1988–2016) in the Sundarbans mangrove forests using Landsat and Google Earth Engine'. Natural Hazards 102 (1):133-150. doi: 10.1007/s11069-020-03914-z.

Pal, I. and Ghosh, T. (2018) 'Risk governance measures and actions in Sundarbans delta (India): A holistic analysis of post-disaster situations of cyclone Aila', in Pal, I. and Shaw, R. (eds.) *Disaster Risk Governance in India and Cross Cutting Issues*. Singapore: Springer, pp. 225–243. doi: 10.1007/978-981-10-3310-0_12.

Pal, I., Ghosh, T. and Ghosh, C. (2017) 'Institutional framework and administrative systems for effective disaster risk governance – Perspectives of 2013 Cyclone Phailin in India', *International Journal of Disaster Risk Reduction*, 21, pp. 350–359. doi: 10.1016/j.ijdrr.2017.01.002.

Rahman, S., et al. (2017) 'The impact of cyclone Aila on the sundarban forest ecosystem', *International Journal of Ecology & Development*, 32, pp. 87–97.

Rahman, M.M., et al. (2020) 'Remote sensing-based mapping of senescent leaf C:N ratio in the Sundarbans reserved forest using machine learning techniques', *Remote Sensing*, 12(9), p. 1375. Available at doi: 10.3390/rs12091375.

Shimu, S. A., et al. (2019) 'NDVI based change detection in Sundarban Mangrove forest using remote sensing data', In *2019 4th International Conference on Electrical Information and Communication Technology (EICT)*. IEEE, pp. 1–5. doi: 10.1109/EICT48899.2019.9068819.

Singh, P.K., et al, (2019) 'Evaluating the effectiveness of climate change adaptations in the world's largest mangrove ecosystem', *Sustainability*, 11(23), p. 6655. Available at doi: 10.3390/su11236655.

Uddin, K. and Matin, M. A., 2020 *Mapping floods in Bangladesh caused by Cyclone Amphan to support humanitarian response*. ICIMOD, Available at: https://www.icimod.org/article/mapping-floods-in-bangladesh-caused-by-cyclone-amphan-to-support-humanitarian-response/ (Accessed on 02 September 2020).

14 Vulnerability to Saltwater Intrusion Along Coastal Bangladesh Using GIS and Hydrogeological Data

Md Ashraful Islam
University of Dhaka

D. Mitra
Indian Institute of Remote Sensing

CONTENTS

14.1 INTRODUCTION

Saltwater intrusion is a serious environmental and sociological problem in coastal Bangladesh (Sarker et al., 2018). The main reason for this is the rapid sea-level rise due to global warming, together with various anthropogenic coastal activities in the coastal regions, such as the pumping of excessive groundwater from coastal aquifers and various construction works that have resulted in the cessation of river discharge (Brammer, 2014). Climate change and sea-level rise are expected to exacerbate the salinity problem that has already endangered the use of drinking and irrigation water by the marginalized coastal communities in the coastal areas of Bangladesh (Abedin et al., 2014, Khan et al., 2011).

Both the surface and ground water in many places on the coastal plains of Bangladesh are currently experiencing high salinity, for example, in the Jessore, Satkhira, Khulna, Narail, Bagherhat, and Gopalganj districts in the southwest of Bangladesh; the high salinity has jeopardized the health of people (Rakib et al., 2020, Sultana et al., 2015). Agricultural activities have also been disrupted and crop production has declined significantly, as farmers often struggle to grow multiple crops in 1 year (Khanom, 2016). In addition, the rising sea level is heading towards semi-permanent inundation and a subsequent upsurge in salinity in and near the coastal areas (Zahid et al., 2009).

The invasion of salinity is more acute in the dry season, when river flows are severely reduced; as a consequence, saltwater goes up to 240 km inland, extends well to the north (Rahman et al., 2000) and covers around 40% of the coastal area. Saltwater intrusion is almost non-existent in the rainy season (June–October), due to extreme freshwater flows, covering at most 10% of the coastal area; it gradually increases in the dry season, especially in winter (Shammi et al., 2017).

Bangladesh has one of the world's most dynamic coastal systems which, because of the intricate nature of sediment distribution and the aquifer system, has a complex hydro-geology (Shamsudduha et al., 2009). The aquifer-aquitard sediments change noticeably within a short distance, making it difficult to delineate proper aquifer boundaries in this region (Uddin and Abdullah, 2003). The level of groundwater salinity in the coastal plain varies significantly due to the variation in the vertical distribution of aquifer materials (Rahman et al., 2018). This sort of variation makes it difficult to measure the precise amount of salinity and hinders the process of getting rid of it.

To ensure a safe drinking- and irrigation-water supply to the large marginalized coastal community, it is necessary to consider salinity in this complex aquifer system. Mathematical modeling and numerical simulations may predict the variations, but the use of these advanced techniques can be significantly hampered in a resource-scare country such as Bangladesh by the lack of both expertise and relevant data. The use of GIS-based index modeling to map vulnerability to saltwater intrusion may help overcome this problem; GALDIT index-based GIS modeling could be the best solution. This study has adopted the GALDIT method to determine the spatial variation in the potential for saltwater intrusion across Bangladesh's coastal region, the first time this has been attempted.

14.2   STUDY AREA

The present study area, 20°34′–26°38′N, 88°01′–91°41′E, comprises 13 districts with a total area of 31,610 km², about 21% of the total area of the country (Figure 14.1). However, it contains about 66% of the total coastal region of Bangladesh and around 290 km of the shoreline. The population is approximately 120 million out of the total population of the country of around 144 million; the average population density is 630 persons per km² (BBS, 2011).

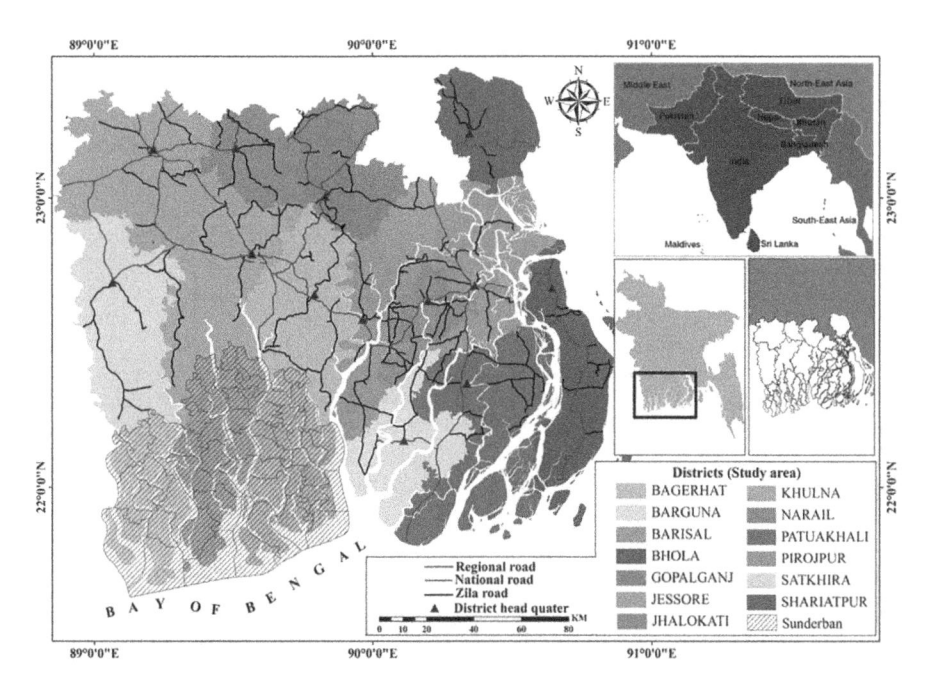

FIGURE 14.1   Location map of the study area.

The study area is characterized by varied land use, mostly agriculture and settlement areas. Cultivation of salt and shrimp is carried out in the southwestern part. A few areas are water-logged and contain other water bodies such as rivers, streams and beels. A major part is occupied by the world's largest mangrove forest, the Sunderbans.

Hydrogeologically, the Bangladesh coast is complex in nature due to frequent changes in aquifer sediments. This coast is a part of the dynamic Ganges-Brahmaputra-Meghna delta, the cause of the complexity in the entire aquifer system. Based on isotopic studies, Aggarwal et al. (2000) proposed a three-tier division of coastal aquifers in Bangladesh. DPHE-BGS (1999) and DPHE-BGS (2001), with slight modifications to the BWDB-UNDP (1982) study, also established a three-tier classification. According to their classifications, the upper shallow aquifer, the Upper Holocene aquifer, extends down to from 50 m to over 100 m, and is composed of fine sand with clay lenses over a thick clay and silt layer. The second aquifer, considered to be the main water-bearing zone, extends down to 250–350 m, and is predominantly fine-to-medium sand with inter-bedded clay layers (in some places as lenses); this particular layer is occasionally underlain and overlain by silty clay layers in many localities (Figure 14.2).

The third and deepest aquifer occurs at depths of 300–350 m, and is predominantly composed of fine-to-medium sand, with occasional course sand and, in several places, alternating thin silty clay or clay lenses. There is also a fourth aquifer in some locations.

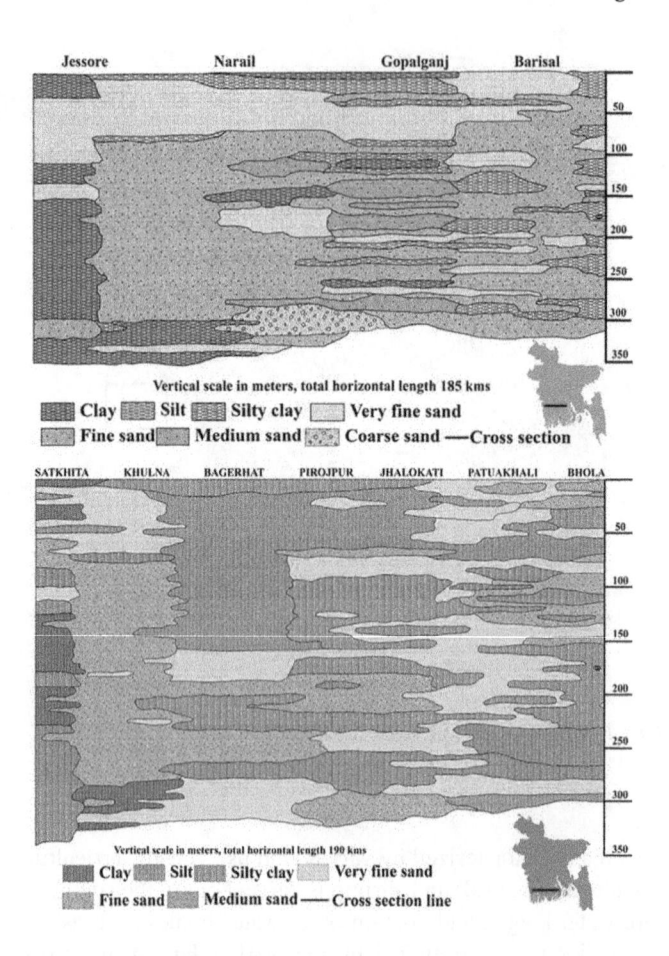

FIGURE 14.2 Lithological cross-section showing the aquifer distribution on the Ganges deltaic coast.

14.3 SALTWATER INTRUSION

By definition, saltwater intrusion is a process in which the fresh water in coastal aquifers is replaced by saline water under the influence of a hydraulic gradient. This phenomenon usually happens near the coast but its effect may be seen far inland. Saltwater intrusion can be a natural or human-induced process. Natural drivers include cyclones, storm surges, tsunamis, sea-level rise, and land subsidence in coastal areas (Barlow, 2003). Anthropogenic drivers include human-induced climate change, groundwater pumping near the coast, reduction in river discharges by upstream dam construction and land use/landform changes (Rapti-Caputo, 2010). On 26th December, 2004, a tsunami caused widespread destruction and contamination of coastal aquifers. For instance, over 40,000 wells (drinking-water source) were either destroyed or contaminated in Sri Lanka (Illangasekare et al., 2006). In any coastal area, to what degree saltwater intrusion will occur depends on the intrinsic characteristics of the aquifers, and is mostly beyond human control.

Rivers and estuaries allow sea-water inflow through backwater from the sea which, in turn, makes the surface water saline (Oude Essink, 2001). Various geological factors, such as structural elements, the geomorphology and the lithology, are important in controlling saltwater intrusion. Hydraulic gradients, pumping rates, and replenishment govern the extent to which saltwater replaces fresh water (Ammar et al., 2016, Barlow, 2003). Buried paleochannels forming freshwater zones near coastal regions are also affected by saltwater (Sharma et al., 2016). A study by Mulligan et al. (2007) that simulated the coastal environment confirmed that paleochannels permit both seawater inflow and outflow.

Sea-level rise due to the climate change is the most important factor that affects saltwater intrusion (Pustry and Farook, 2020). The risk of saltwater intrusion remains higher in the dry season than in the wet season. In summer, the decrease in rainfall and increase in evapotranspiration results in a lowering of the groundwater level, increasing the probability of saltwater ingress (Rapti-Caputo, 2010). Increases in global temperature have caused glacial melting and thermal expansion of sea water which, in turn, causes arise in sea level. The rising sea level contributes to an increase in the head of seawater at the ocean boundary, and a resulting migration of the freshwater-saltwater mixing zone towards the land, thus enhancing saltwater intrusion (Custodio, 1997).

Groundwater has been overexploited in recent years with the rise in water demand. Another significant cause of saltwater intrusion in many parts of the world is known to be excessive groundwater withdrawal in coastal regions. Such excessive withdrawal even causes land subsidence, with the consequent destruction of buildings and other coastal structures (Minderhoud et al., 2017). Excessive diversion of upstream river discharge for agricultural purposes in many coastal areas may also lower the groundwater level and facilitate saltwater intrusion (Barlow, 2003).

It is important to understand which areas are more vulnerable to saltwater intrusion for management purposes. Towards this, various models using sub-surface hydro-geological data have been used. In general, the adoption of an index has the benefit of removing or minimising subjectivity in the risk ranking phase in different areas. Such a standardized index is currently used in the US, Canada, and South Africa. The DRASTIC index, developed by Aller et al. (1987) for the US EPA, is one such index having simple and useful features for modeling saltwater intrusion. In the present study, the GALDIT model was used; it is same semi-empirical additive model as DRASTIC but with the addition of several new parameters which are important for the study area here (Chachadi, 2005).

14.4 MODEL FOR VULNERABILITY MAPPING OF SALTWATER INTRUSION

The physical characteristics that influence the potential for saltwater intrusion are inherent in each hydro-geologic setting. The most crucial factors that can be mapped are as follows:

1. Groundwater occurrence (aquifer type: unconfined; confined; or leaky confined)

2. Aquifer hydraulic conductivity
3. Depth to groundwater Level above sea level
4. Distance from the shore (distance inland perpendicular to the shoreline)
5. Impact of existing status of saltwater intrusion in the area
6. Thickness of the aquifer which is being mapped

This gives the acronym GALDIT.

These factors include the basic requirements for assessing the potential for saltwater intrusion in each hydro-geological setting. The GALDIT factors are measurable parameters; these variables are often obtainable from diverse sources without the need for detailed reconnaissance.

GALDIT – a numerical ranking system to assess the saltwater-intrusion in different hydro-geological settings contains three significant parts: weights, ranges and ratings. The relative importance of each GALDIT factor must be evaluated. Each factor here has been assigned a relative weight, ranging from 1 to 4, with the most significant factors having a weight of 4:

	GALDIT Factor	Weight
1.	Groundwater occurrence (aquifer type)	1
2.	Aquifer hydraulic conductivity	3
3.	Depth to groundwater level above sea level	4
4.	Distance from the shore	2
5.	Impact of existing status of saltwater intrusion	1
6.	Thickness of aquifer being mapped	2

A value, the importance rating, between 1 and 10 is assigned to each parameter, depending on local conditions; high values correspond to high vulnerability, with the values generally obtained from tables. The minimum value of the GALDIT index is therefore 13, the maximum value 130.

14.5   DATA PREPARATION FOR GALDIT

Collection of data used in the GALDIT method is complex and time-consuming. The method uses borehole information for the subsurface hydrological data. Borehole data are rare, and collecting information from boreholes is expensive as well. The present analysis was carried out with support from the Bangladesh Water Development Board, which provided data from boreholes in the coastal areas (BWDB, 2013). Data were provided from around 600 boreholes on aquifer type, thickness, hydraulic conductivity, and water table in various localities; 132 boreholes were selected for the GALDIT model.

14.5.1   GROUNDWATER OCCURRENCE (AQUIFER TYPE)

In general, aquifers can be categorized into three classes: confined aquifer, unconfined aquifer and leaky aquifer. Different aquifer types have different susceptibility to saltwater intrusion. For example, confined aquifers have greater vulnerability than the other two types because of their larger cones of depression and immediate discharge of water to wells during pumping. Moreover, once a confined aquifer is contaminated with saltwater, there is very little chance of a drop in salt concentration. This is because confined aquifers have very limited opportunities to be recharged with fresh rainwater, as they are overlain by impermeable layers, clay, or silty clay. In this context, unconfined and leaky aquifers are relatively less exposed. Several cross-sections as well as lithologs were prepared using Rockwork software for the identification of aquifer type. To be sure of aquifer type, both local and regional cross-sections were prepared using representative borelogs.

14.5.2 Hydraulic Conductivity

It is an intrinsic property of aquifers that they transmit water because of their porosity, with interconnected pores in sediments as well as fractures in consolidated rocks. In terms of aquifer vulnerability, a higher hydraulic conductivity poses a higher risk because, if the interconnectivity between grains is smooth, a saltwater front can move easily inland. A pumping test is one way to obtain information on the hydraulic conductivity of different aquifers. In the present analysis, the hydraulic conductivity was determined for only of 86 of the 132 boreholes in the GALDIT calculation. To resolve this problem, an interpolated raster was generated from the hydraulic-conductivity data using the IDW technique. All boreholes used for the GALDIT calculation were assigned a hydraulic-conductivity value from the raster using the extract values to points tool in ArcGIS10.

14.5.3 Depth to Groundwater

The water table is the depth of the groundwater below the ground surface. It can be measured in permanent piezometric holes or by a geophysical technique such as a resistivity survey. The groundwater level with respect to mean sea level (MSL) is an important factor in assessing potential saltwater intrusion in an area because it determines the hydraulic pressure that will resist the saltwater front. Ghyben-Herzberg relationship indicates that every meter of freshwater stored above MSL results in 40 m of freshwater down to the interface (Fetter, 1994). For the present purpose, water-table data were collected and converted to water level with respect to MSL by subtracting the water table from the reduced level.

14.5.4 Distance from the Shore

The closer to the coast, the greater the influence of sea water, while it normally decreases as one advances inland at right angles to the shore. The shoreline was taken from a recent satellite image, and the perpendicular distance from the shoreline measured of all boreholes used in the GALDIT calculation.

14.5.5 Impact of Existing Status of Saltwater Intrusion

This particular parameter gives an idea of the amount of saltwater at any location. As far as the salt concentration is concerned, freshwater is dominated by bi-carbonate and carbonate ions, sea water by chloride ions. Thus, the ratio of Cl⁻ to (CO_3^- + HCO_3^-) in an observation well gives the current status of saltwater intrusion; this was determined in a laboratory.

14.5.6 THICKNESS OF THE AQUIFER BEING MAPPED

The degree and extent of seawater invasion in coastal areas mainly depends on the thickness of the aquifer. Greater aquifer thickness reduces the likelihood of penetration of sea water and vice versa. Thickness was measured from the collected borelogs and incorporated into the borehole database that was for final GALDIT calculation.

14.6 GALDIT CALCULATIONS

In the present study, the vulnerability evaluation and ranking was done according to the GALDIT method (Chachadi and Lobo Ferreira, 2001), in which each of the six parameters has a predetermined fixed relative weight that reflects its relative importance to vulnerability (Table 14.1). With the GALDIT technique employed, the index of vulnerability to saltwater intrusion was obtained by the following expression:

$$GVI = \sum_{i=1}^{6} \left\{ (W_i) R_i \right\} / \sum_{i=1}^{6} W_i$$

where W_i is the weight and R_i is the importance rating of the ith indicator. Once the GALDIT index has been calculated, it is then possible to classify the coastal areas into various categories of saltwater-intrusion vulnerability. Figure 14.3 includes a brief flow chart of the GALDIT assessment method.

Chachadi and Lobo Ferreira (2001) stated that GALDIT is an open-ended model, permitting addition or subtraction of variables from the final calculation; individual classes could be modified if required. For example, in *Distance to shoreline*, more than 1,000 m is given high rating of 10. In the study area, 95% of the boreholes were greater than 1,000 m from the shoreline; a high rating was given for boreholes more than 40,000 m from the shoreline. Another justification of this particular adjustment is that, in the coastal area, tidal action is experienced up to 40–45 km inland. As a consequence, a distance less than 40,000 m was given a rating of 10. Additionally, an aquifer thickness greater than 20 m was given rank 10, a rank given to aquifers of thickness greater than 10 m in the original GALDIT model. The main reason behind this is that most of the aquifers used for drinking or irrigation purposes in the study area were more than 10 m thick.

14.7 MODEL OUTCOME

In the current study, each of the six GALDIT input parameters was individually rated, multiplied by the proposed GALDIT weights (Chachadi, 2005) and all summed to give the final GALDIT vulnerability maps of the Ganges deltaic

TABLE 14.1
Hydro-Geological Parameters and Their Weights for the GALDIT Model
(Chachadi and Ferreira, 2001; Chachadi et al. 2002; Chachadi 2005)

GALDIT Model

	Indicator	Weight	Indicator VARIABLE	Importance RATING
G	Groundwater occurrence/Groundwater type	1	Confirmed aquifer	10
			Unconfirmed aquifer	7.5
			Leaky confirmed aquifer	5
			Bounded aquifer	2.5
A	Aquifer hydraulic conductivity (m day^{-1})	3	>40 m day^{-1}	10
			10–40 m day^{-1}	7.5
			5–10 m day^{-1}	5
			<5 m day^{-1}	2.5
L	Height of ground water level above MSL	4	<1.0 m	10
			1.0–1.5 m	7.5
			1.5–2.0 m	5
			1.5–2.0	2.5
D	Distance from shore/high tide (m)	4	<500 m	10
			500–750 m	7.5
			750–1,000 m	5
			>1,000 m	2.5
I	Impact status of existing seawater intrusion	1	>2	10
			1.5–2.0	7.5
			1–1.5	5
			<1	2.5
T	Aquifer thickness (saturated) in meters	2	>10 m	10
			7.5–10 m	7.5
			5–7.5 m	5
			<5 m	2.5

coastal areas covering 13 major districts. Vulnerability maps were created for the dry season, the pre-monsoon, and the wet season, post-monsoon, for the three major aquifer systems in the study area. The index obtained ranged from 2.8 to 8.3 and 3.7 to 8.3 for the wet and dry season, respectively, for the first aquifer, from 2.8 to 8 and 4.7 to 8.7, respectively, for the second aquifer and from 3.7 to 8.3 and 4.7 to 9, respectively, for the third aquifer. An equal-area classification technique was adopted to classify output maps for the two seasons, with classifications from very low vulnerability to high vulnerability, to display the spatial distributions of various vulnerability classes across the study area. Vulnerability statistics were calculated for the overall study area (Table 14.2), in the three physiographic settings (Table 14.3) and for the individual districts (Table 14.4); these are described in the following sections.

The GALDIT model was executed over around 26,300 km^2, covering the three major aquifers in the Ganges deltaic coast of Bangladesh.

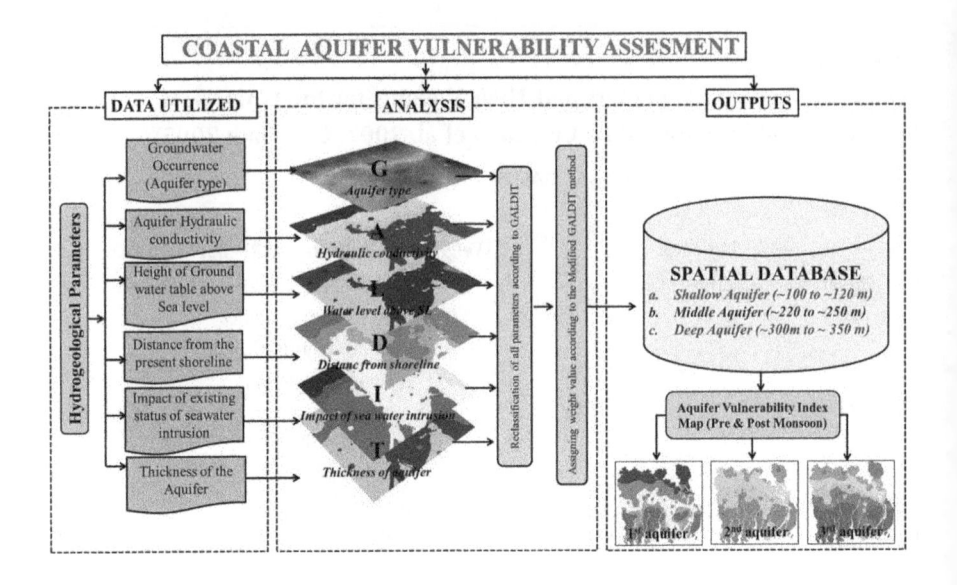

FIGURE 14.3   Flow chart for the GALDIT method.

The results (Table 14.4; Figures 14.4 and 14.5) show that in the first aquifer 38% of its total area, 10,000 km², lay in the high-vulnerability category in the wet season (post-monsoon), covering the south-eastern and southwestern coastal areas. The high-vulnerability area increased to 41%, 10,866 km², during the dry season (pre-monsoon), covering mostly of the southern coastal areas and extending up to around 60 km inland.

32% of the total area, 8,382 km², was moderately vulnerable during the wet season, covering mostly the middle and mid-southern parts of the study area. During the dry season, 42%, 11,010 km², of the area was moderately vulnerable, confined to the central coastal areas of the study area.

The low-vulnerability class in the wet season was 24%, 6,346 km², of the total area, in the extreme northwest, middle-north and extreme northeast of the study area. Surprisingly, it decreased to 16%, 4,374 km², during the dry season, mostly in the north-western and extreme eastern parts. Only 6%, 1,596 km², was in the very low vulnerability class during the wet season, in the extreme north-eastern part of the study area. This fell to only 1%, 74 km², of the total area during the dry season as two isolated clusters in the northern periphery of the study area.

The high-vulnerability class was only 8%, 2,131 km², of the total area of the second aquifer during the wet season, concentrated in the south-eastern peripheral areas and extending up to around 25 km inland. There was only a slight change during the dry season, with the high-vulnerability class 9%, 2,486 km², of the total area. The moderate-vulnerability class occupied about 29% of the total area in both seasons, concentrated in the middle-to-extreme northern periphery of the study area.

The low-vulnerability class was 35%, 9,244 km², of the total area during the wet season and 55%, 14,362 km², during the dry season, mostly concentrated in

TABLE 14.2

Area (km²) and Percent Area of the GALDIT Vulnerability Categories in the Study Area

	Vulnerability Intensity															
	High				Moderate				Less				Very Less			
	Wet		Dry		Wet		Dry		Wet		Dry		Wet		Dry	
Aquifer Division	Area	%	Area	%	Area	%	Area	%	Area	%	Area	%	Area	%	Area	%
First	10,000	38	10,866	41	8,382	32	11,010	42	6,346	24	4,374	16	1,596	6	74	1
Second	2,131	8	2,486	9	7,644	29	7,598	29	9,244	35	14,362	55	7,306	28	1,878	7
Third	5,64	2	471	4	10,105	38	9,688	36	11,517	44	15,793	59	4,138	16	371	2

Area in km² (total area 26,324 km²).

TABLE 14.3

GALDIT Vulnerability Statistics in the Three Major Physiographic Settings

GALDIT Vulnerability

| | | High | | | | Moderate | | | | Less | | | | Very less | | | |
| | | Wet | | Dry | | Wet | | Dry | | Wet | | Dry | | Wet | | Dry | |
Aquifer Division		Area	%	Area	%	Area	%	Area	%	Area	%	Area	%	Area	%	Area	%
Tidal delta	First	9,989	54	10,875	59	6,950	38	7,241	40	1,426	8	-	-	-	-	-	-
	Second	2,132	12	2,500	14	7,608	41	7,610	41	7,017	38	7,992	43	1,612	9	316	2
	Third	490	3	450	2	9,814	53	9,439	52	7,752	42	8,085	44	278	2	371	2
Active delta	First	3	0.1	-	-	415	9.9	1,343	32	2,402	57	2,820	67	1,418	33	74	2
	Second	-	-	-	-	-	-	-	-	1,198	28	4,235	100	3,037	72	-	-
	Third	-	-	-	-	73	2	90	2	2,807	66	4,145	98	1,355	32	-	-
Inactive delta	First	7	0.2	-	-	1,015	27	2,427	65	2,519	67.8	1,292	35	178	5	-	-
	Second	-	-	-	-	37	1	-	-	1,028	28	2,154	58	2,656	71	1,567	42
	Third	-	-	-	-	147	4	159	4	887	24	3,562	96	2,687	72	-	-

TABLE 14.4
District GALDIT Vulnerability Statistics

			High				Moderate				Less				Very less			
			Wet		Dry		Wet		Dry		Wet		Dry		Wet		Dry	
	Aquifer division		Area	%	Area	%	Area	%	Area	%	Area	%	Area	%	Area	%	Area	%
TIDAL DELTA	Satkhira	First	1,827	54	1,672	49	1,464	43	1,488	44	100	3	235	7	–	–	–	–
		Second	–	–	–	–	1,200	35	728	22	2,062	60	2,549	75	189	5	175	5
		Third	–	–	–	–	1,779	52	1,553	46	1,600	46	1,899	54	73	2	–	–
	Bagerhat	First	2,322	65	2,167	61	1,187	33	1,378	38.5	43	2	10	0.5	–	–	–	–
		Second	–	–	280	8	1,408	39	1,640	46	2,082	59	1,635	46	65	2	–	–
		Third	80	2	–	–	1,821	51	1,719	48	1,563	44	1,786	50	91	3	50	2
	Patuakhali	First	1,556	66	2,303	97	806	34	56	2.5	–	–	2	0.5	–	–	–	–
		Second	1,275	53.5	1,248	53	1,100	46	1,129	47	2	0.5	–	–	–	–	–	–
		Third	144	6	212	9	1,819	77	1,677	71	414	17	488	20	–	–	–	–
	Jhalokathi	First	23	3	–	–	656	92	712	100	33	5	–	–	–	–	–	–
		Second	–	–	–	–	171	24	300	42	252	35	337	47	289	41	75	11
		Third	–	–	–	–	252	35	228	32	441	62	369	52	19	3	115	16
	Bhola	First	1,106	68	1,174	72	379	23	456	28	147	9	–	–	–	–	–	–
		Second	548	33	482	28	813	49	1,070	62	301	18	1,158	10	–	–	–	–
		Third	–	–	–	–	973	60	1,200	73	653	40	436	27	–	–	–	–
ACTIVE DELTA	Khaulna	First	2,092	57	1,726	47	1,471	40	1,771	48	87	3	156	5	–	–	–	–
		Second	–	–	–	–	1,338	35.5	1,045	28	2,318	63.5	2,329	64	5	0.5	287	8
		Third	–	–	–	–	608	52	572	49	555	48	574	48.8	503	43	2	0.2
	Pirojpur	First	402	34	386	33	751	65	767	66	8	1	8	1	–	–	–	–
		Second	–	–	–	–	438	38	591	49	555	48	574	49	–	–	12	2
		Third	–	–	–	–	6–8	52	572	49	555	48	574	49	–	–	12	2

GALIDIT VULNNERABILITY

(*Continued*)

TABLE 14.4 (*Continued*)
District GALDIT Vulnerability Statistics

			GALIDIT VULNNERABILITY														
			High				Moderate				Less				Very less			
		Aquifer	Wet		Dry		Wet		Dry		Wet		Dry		Wet		Dry	
		division	Area	%	Area	%	Area	%	Area	%	Area	%	Area	%	Area	%	Area	%
	Barisal	First	132	8	173	11	399	26	1,382	89	1,024	66	–	–	–	–	–	–
		Second	10	1	–	–	174	11	276	18	649	41	1,174	75	734	47	117	7
		Third	–	–	–	–	39	0.5	379	25	1,385	90	1,261	80	143	9.5	189	13
	Gopalgannj	First	–	–	–	–	38	3	379	25	1,377	93	1,066	72	63	4	33	3
		Second	–	–	–	–	–	–	–	–	296	20	1,480	100	1,184	80	–	–
		Third	–	–	–	–	–	–	–	–	645	46	1,480	100	835	54	–	–
	Satriatpur	First	–	–	–	–	–	–	252	21	49	4	943	79	1,146	96	–	–
		Second	–	–	–	–	–	–	–	–	1,197	100	1,197	100	–	–	–	–
		Third	–	–	–	–	–	–	–	–	1,197	100	1,197	100	–	–	–	–
	Natail	First	–	–	–	–	25	3	150	15	776	77	815	81	205	20	42	4
		Second	–	–	–	–	–	–	–	–	244	24	1,007	100	763	76	–	–
		Third	–	–	–	–	–	–	–	–	205	20	1,007	100	802	80	–	–
	Barisal	First	–	–	–	–	–	–	555	94	552	94	35	6	38	6	–	–
		Second	–	–	–	–	–	–	–	–	–	–	591	100	591	100	–	–
		Third	–	–	–	–	–	–	19	3	582	98	572	97	9	2	–	–
INACTIVE	Jessore	First	–	–	–	–	296	11	1,504	58	2,105	82	1,076	42	179	7	–	–
DELTA		Second	–	–	–	–	–	–	–	–	353	14	1,349	48	2,230	86	1,431	52
		Third	–	–	–	–	–	–	–	–	252	10	2,583	100	2,331	90	–	–

Barisal* *includes in the tidal deltaic setting &* **Barisal'** *includes in the active deltaic settings (Muladi upazila)*

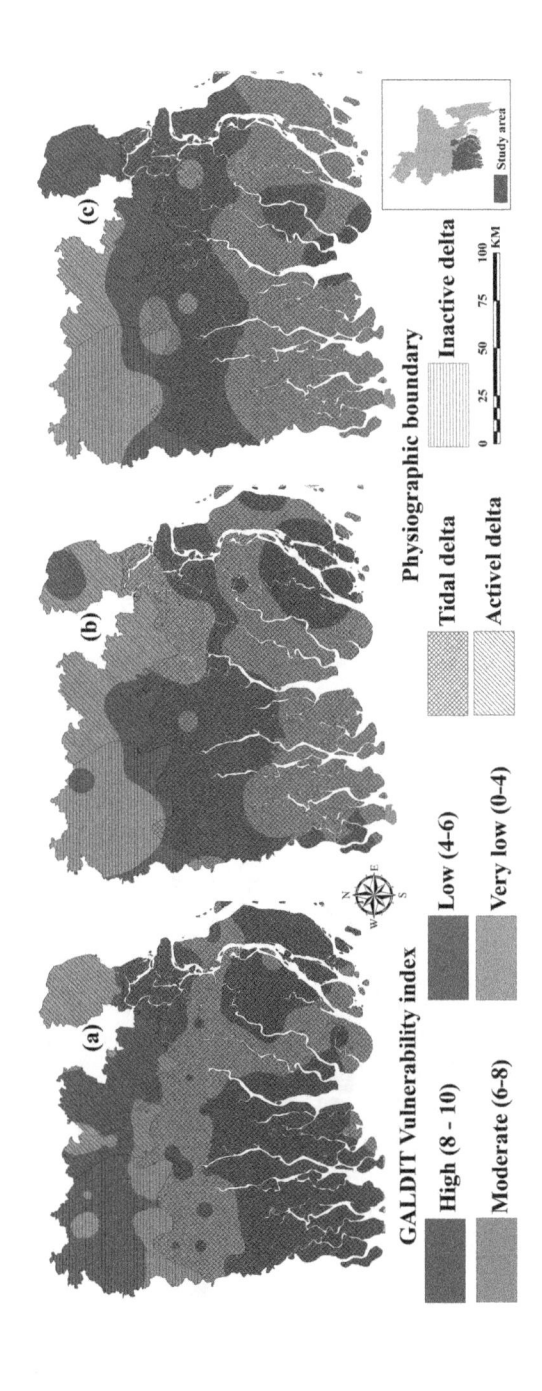

FIGURE 14.4 Vulnerability map of the aquifer systems in the study area in the wet season: (a) first aquifer; (b) second aquifer; and (c) third aquifer.

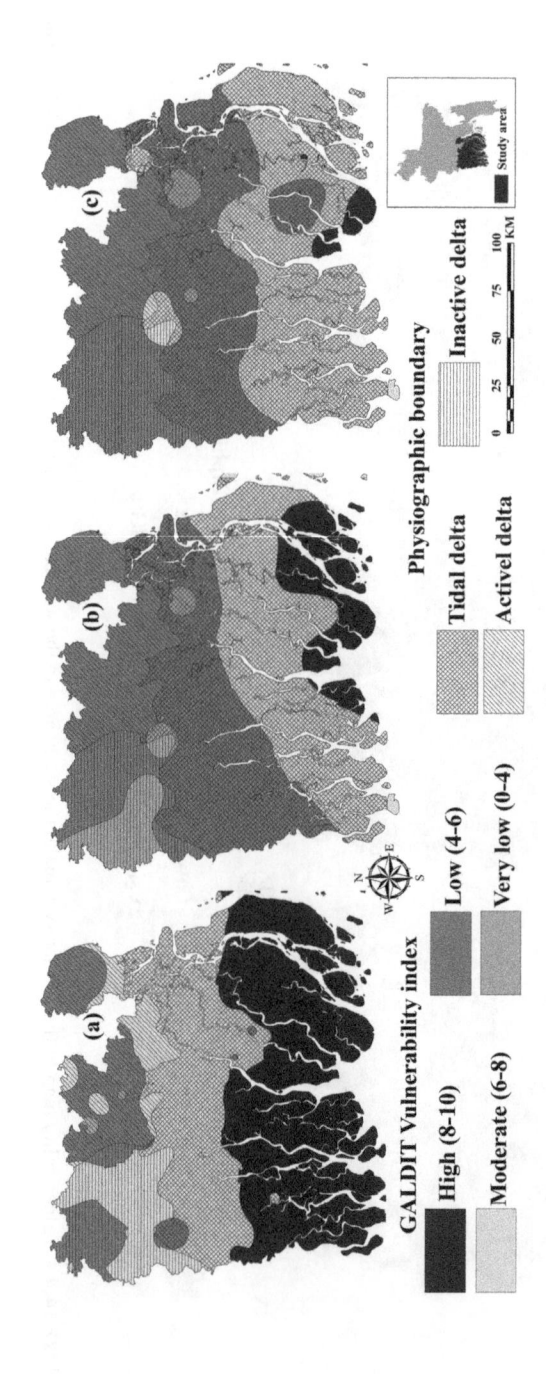

FIGURE 14.5 Vulnerability map of the aquifer systems in the study area in the dry season: (a) first aquifer; (b) second aquifer; and (c) third aquifer.

the northern periphery of the study area. The very low vulnerability class was 28%, 7,308 km², in the wet season but only 7%, 1,878 km², in the dry season.

The high-vulnerability class was merely 2%, 564 km², of the total area of the third aquifer during the wet season, and 4%, 471 km², during the dry season, in the mid-southern tip of the coast and extending around 15 km inland (Figures 14.4 and 14.5). The moderate-vulnerability class was 38%, 10,105 km², in the wet season and 36%, 9,688 km², during the dry season, in the southern part of the study area.

The low-vulnerability class covered 44%, 11,517 km², of the total area in the wet season and 59%, 15,792 km², in the dry season, covering the middle, and to some extent, the northern part of the study area. The very low vulnerability class was only 16%, 4,138 km², of the total area in the wet season and 2%, 371 km², in the dry season, in the extreme northern part of the study area.

The high-vulnerability class covered 54% of the total area of the first aquifer during the wet season, increasing to 59% during the dry season. The major part of the second aquifer was occupied by the moderate- and low-vulnerability classes, 41% of the total area in both the wet and dry seasons, with the remaining area dominated by the low-vulnerability class, 38% and 43% in the wet and dry seasons, respectively. The major part of the third aquifer was occupied by the moderate-vulnerability class; 53% of the total area in the wet season and 52% in the dry season. However, here the low-vulnerability class had the second-largest coverage, 42% and 44% in the wet and dry seasons, respectively.

In the *active deltaic regions*, the first aquifer was characterized by low vulnerability, embracing 57% of its total area during the wet season and 67% in the dry season. The rest of the area fell into either the moderate-vulnerability or very low vulnerability class. The high-vulnerability class was noteworthy by its covering only a very small area. The second aquifer was dominated by the low-vulnerability class, 28% of the total area in both the wet and dry season. The third aquifer was also dominated by the low vulnerability class, covering 66% of its total area in the wet season and 98% in the dry season.

In the *inactive deltaic regions*, the first aquifer was dominated by the moderate- and low-vulnerability classes. In the wet season, 27% of the area was characterized as moderately vulnerable; this increased to 65% during the dry season. The low-vulnerability class covered 68% of the area during the wet season but decreased to 35% during the dry season. The second aquifer was dominated by the very low vulnerability class, 71% of the area during the wet season, but decreased to 42% during the dry season. Both high- and moderate-vulnerability classes were barely represented, with the rest of the area in the low-vulnerability class, covering 28% of the total area in the wet season and 58% in the dry season. The third aquifer was also dominated by the low-vulnerability and very low vulnerability classes like the second aquifer.

The statistics for each of the 13 districts in the study area were compiled (Table 14.4) to gain insight into their detailed vulnerability. Most of the districts within the tidal deltaic region had a highly vulnerable first aquifer. For example, 68% of the area in the Bhola district was found to be highly vulnerable during the wet season, followed by 66% in the Pathuakhali District and 65% in the Bagerhat District; the corresponding values for the dry season were 75%, 97%, and 61%. As far as the first aquifer is concerned, these three districts are closest to the coast and have been affected by saltwater for a long time.

Similar statistics were found for the second aquifer, but with relatively fewer highly vulnerable areas than the first aquifer. Two districts, Patuakhali and Bhola, made up 54% and 34%, respectively, of the highly vulnerable areas during the wet season, 53% and 28% during the dry season. The second and third aquifers in most of the other districts in the tidal deltaic region were found to be moderately vulnerable in both seasons (Figures 14.6–14.8).

The first aquifer in districts in *active deltaic regions*, Gopalganj, Shariatpur, Narail, and part of Barisal, mostly had low vulnerability in the wet and dry seasons. The relative areas occupied by the low-vulnerability class in these districts were 92%, 4%, 77%, and 94%, respectively, in the wet season and 72%, 79%, 81%, and 6%, respectively, in the dry season. Almost 90% of the area of the second and third aquifers in these districts had low vulnerability, with little or no presence of the other three vulnerability classes.

The only district in an *inactive deltaic setting*, Jessore, had no highly vulnerable areas. In the first aquifer, 11% of the area was moderately vulnerable during the wet season, 58% during the dry season; 82% had low vulnerability during the wet season and 42% during the dry season. In the second aquifer, 86% of this district had low vulnerability during the wet season and 52% during the dry season. In the third aquifer, 10% had low vulnerability and 90% very low vulnerability in the wet season, with 100% low vulnerability during the dry season.

14.8 CONCLUSION

This study was the first attempt to measure the intrinsic vulnerability to saltwater intrusion into multi-layer coastal aquifers in the Ganges Delta. The the popular GALDIT model was used to Determine the seasonal vulnerability of the heterogeneous aquifer system comprising three aquifers up to 350 m in depth. This particular model considers the representative and context specific hydrogeological parameters which mainly control saltwater intrusion in the study area. Despite large groundwater storage with a moderate-to-good groundwater potential, together with plenty of rainfall (around 2,000 mm year^{-1}), the Ganges aquifers are stressed by salinity exacerbated by sea-level rise. From the analysis, it is evident that the Ganges Delta aquifer system is highly vulnerable to saltwater intrusion during the dry season (pre-monsoon), with lower vulnerability during the wet season (post-monsoon). An area of about 26,300 km², covering three major aquifers on the Ganges deltaic coast of Bangladesh was taken as the study area. The results from the GALDIT model indicate that, in both the wet and dry seasons, the first (upper) aquifer is predominantly a high-vulnerability zone, the second (middle) aquifer a moderate-to-low vulnerability zone and the third aquifer a low to very low vulnerability zone.

The tidal deltaic physiographic region is predominately highly to moderately vulnerable, whereas the active deltaic regions are predominately of low vulnerability and the inactive deltaic region low to very low vulnerability. Of the districts in the study area, Khulna, Satkhira, and Bagerhat Districts are highly vulnerable to saltwater intrusion, Jalokhati, Barisal, and Bhola districts moderately vulnerable

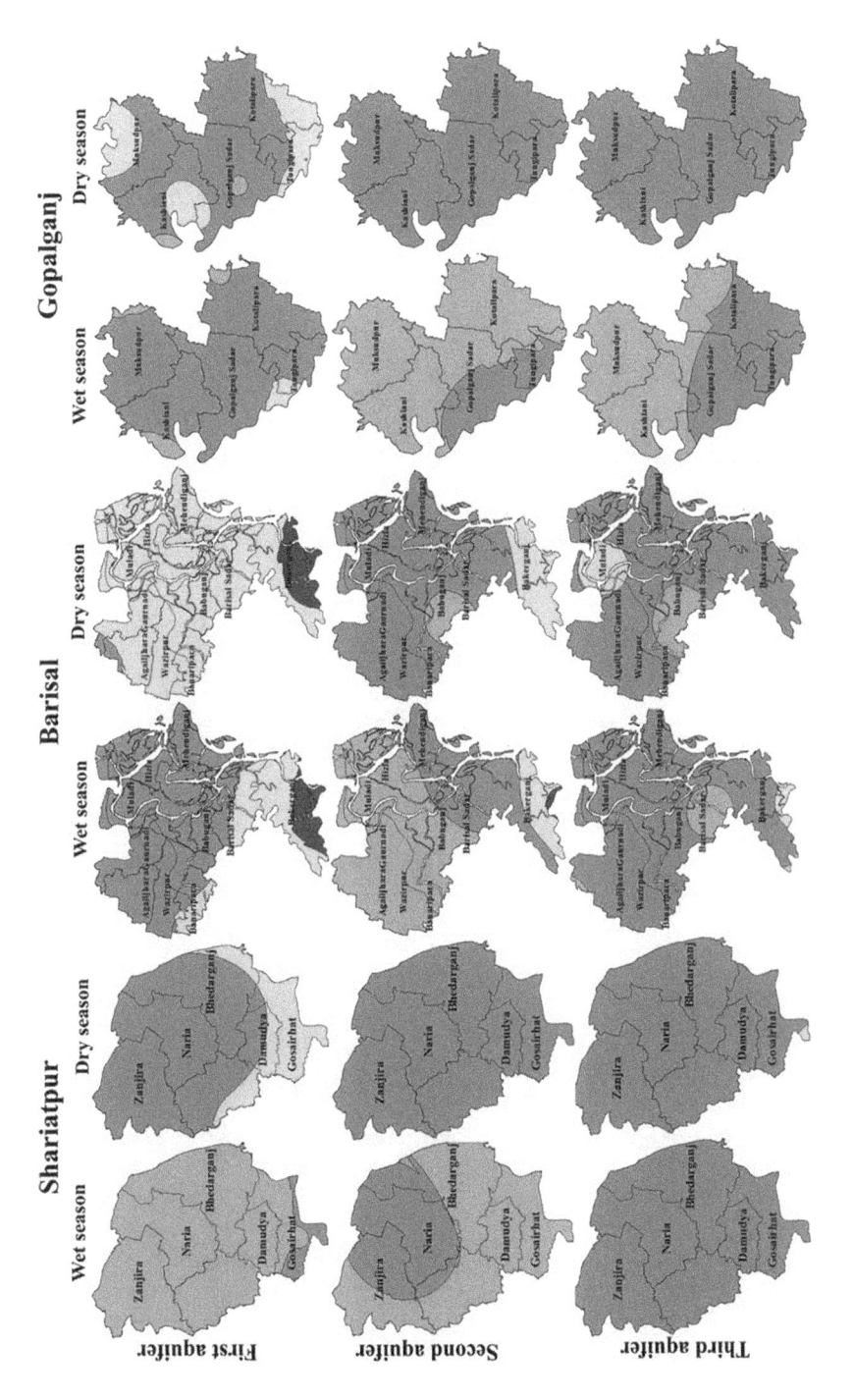

FIGURE 14.6 District GALDIT Vulnerability Map 1 for the three aquifer systems.

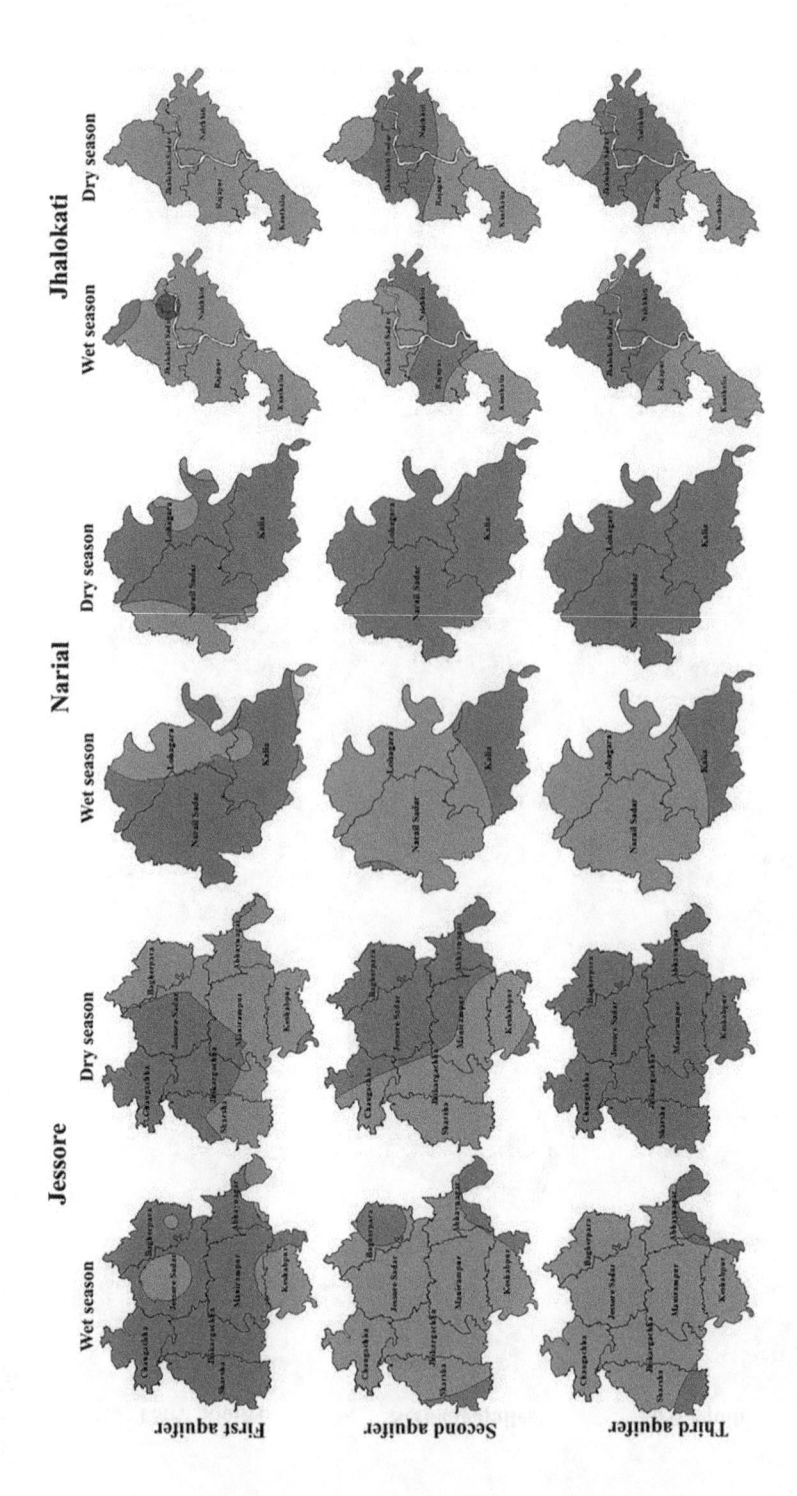

FIGURE 14.7 District GALDIT Vulnerability Map 2 for the three aquifer systems.

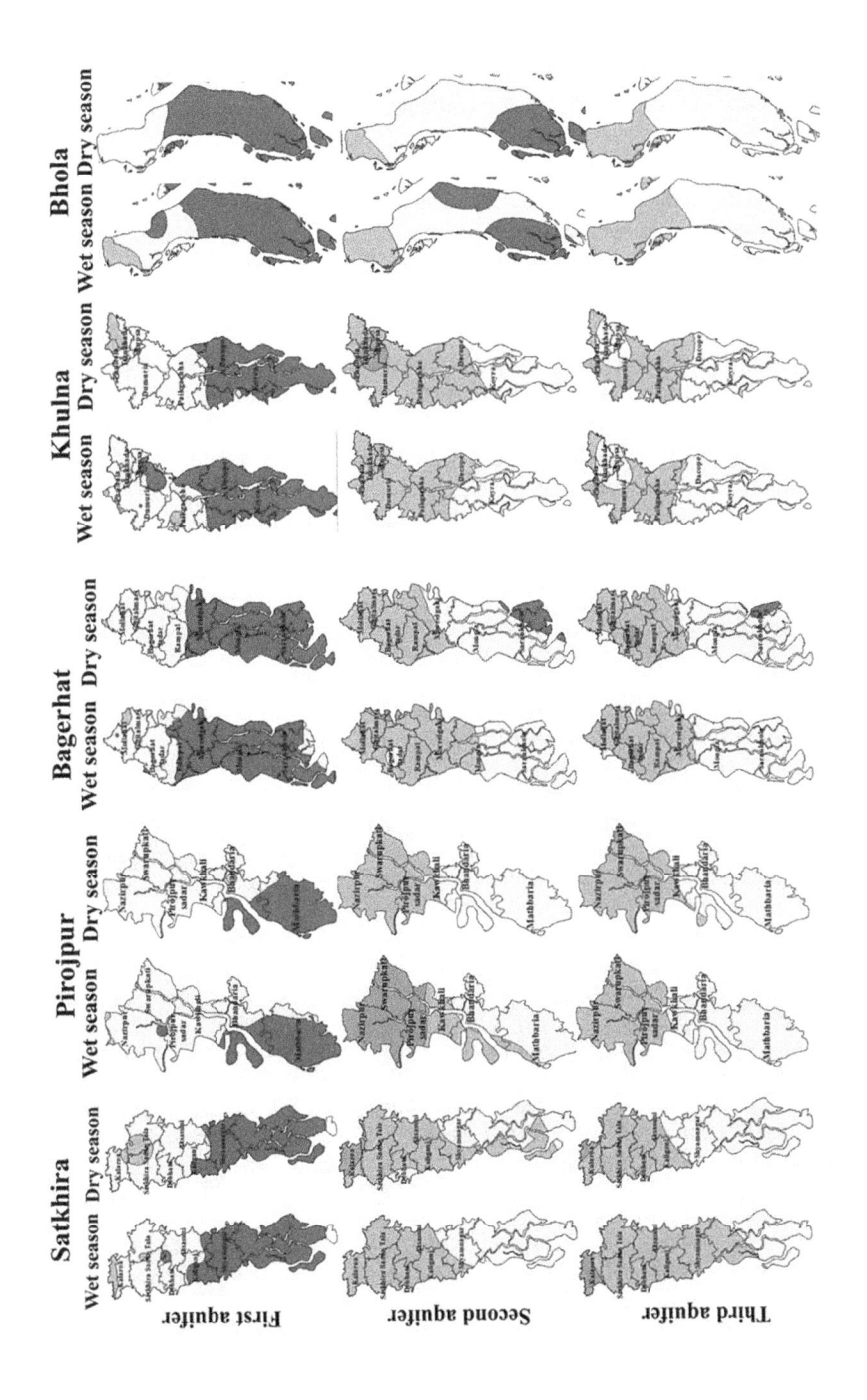

FIGURE 14.8 District GALDIT Vulnerability Map 3 for the three aquifer systems.

whereas the remainder, Gopalganj, Sariatpur, Jessore, and Narail Districts, have low vulnerability.

The GALDIT model predicts the vulnerability to saltwater intrusion of the aquifers in the study area but this needs to be confirmed by field checking and simulation models. As past chemical data were not available, the detection of changes in vulnerability was not possible. The approach here was to use the model to give the seasonal variations in aquifer vulnerability. The study area has already experienced the effect of climate change in the form of saltwater intrusion due to sea-level rise, and seasonal variations in salinity concentration were also evident. The present research demonstrated this variability using the GALDIT approach.

Assessment of aquifer vulnerability to saltwater intrusion is of strategic importance for an area that is highly dependent on groundwater resources for both drinking water and irrigation. As the coastal region of the Ganges Delta consumes substantial quantities of groundwater, the vulnerability map proposed in this research provides a fundamental framework for the management of the groundwater resources. The outcomes could also contribute to conservation and management policies, and the identification of sensitive recharge zones, which should eventually be protected for water supply. Moreover, this study could be used as an aid in regional groundwater management planning to achieve sustainable use of groundwater resources.

REFERENCES

Abedin, M.A., Habiba, U. and Shaw, R., 2014. Community perception and adaptation to safe drinking water scarcity: Salinity, arsenic, and drought risks in coastal Bangladesh. *International Journal of Disaster Risk Science*, 5(2), pp. 110–124.

Aggarwal, P.K., Basu, A.R. and Poreda, R.J., 2000. Isotope hydrology of groundwater in Bangladesh: Implications for characterization and mitigation of arsenic in groundwater. *Preliminary Report of Investigations. IAEA-TC project (BGD/8/016). IAEA, Vienna.*

Aller, L., Benette, T., Lehr, J. H. and Petty, R. J., 1987. DRASTIC: A standardized system for evaluating groundwater pollution potential using hydrogeologic settings. *U.S. Environmental Protection Agency Rep. 600/ 2–85/018.*

Ammar, S.B., Taupin, J.D., Zouari, K. and Khouatmia, M., 2016. Identifying recharge and salinization sources of groundwater in the Oussja Ghar el Melah plain (northeast Tunisia) using geochemical tools and environmental isotopes. *Environmental Earth Sciences*, 75(7), p. 606.

Barlow, P.M., 2003. Groundwater in freshwater-saltwater environments of the Atlantic Coast. *US Department of the Interior. US Geological Survey, Reston, Virginia.*

BBS, 2011. Population and housing census 2011, *Bangladesh Bureau of Statistics (BBS), Ministry of Planning, Government of the People's Republic of Bangladesh, Dhaka. National Series (4)*, p.363.

Brammer, H., 2014. Bangladesh's dynamic coastal regions and sea-level rise. *Climate Risk Management, 1*, pp. 51–62.

BWDB, 2013. Establishment of Monitoring Network and Mathematical Model Study to Assess Salinity Intrusion in Groundwater in the Coastal Area of Bangladesh due to Climate Change. *Main Report, Package, 3*(1), p. 213.

BWDB-UNDP, 1982. Ground-Water Survey – The Hydrologic Conditions of Bangladesh, New York, United Nations, United Nations Development Program.

Chachadi, A.G. and Lobo Ferreira, J.P., 2001. Seawater intrusion vulnerability mapping of aquifers using the GALDIT method. *Coastin, 4*, pp. 7–9.

Chachadi, A.G., 2005. Seawater intrusion mapping using modified GALDIT indicator model-case study in Goa. *Jalvigyan Sameeksha, 20*, pp. 29–45.

Custodio, E., 1997. Seawater Intrusion in Coastal Aquifers: Guidelines for Study, Monitoring and Control, *Water Report 11. Food and Agriculture Organization of the United Nation. Rome, Italy.* P.161.

DPHE-BGS, 1999.Groundwater studies for arsenic contamination in Bangladesh. Final report, Rapid Investigation Phase, Department of Public Health, Government of Bangladesh, Mott MacDonald and British Geological Survey.

DPHE-BGS, 2001. Arsenic Contamination of Groundwater in Bangladesh. D.G. Kinniburgh, P.L. Smedley (Eds.), Final report. BGS Tech. Rep. WC/00/19 Keyworth, British Geological Survey, 267 pp.

Essink, G.H.O., 2001. Improving fresh groundwater supply—problems and solutions. *Ocean & Coastal Management, 44*(5–6), pp. 429–449.

Fetter, C.W., 1994. Applied Hydrogeology. 3rd Edition, Macmillan College Publishing Company, New York.

Illangasekare, T., Tyler, S.W., Clement, T.P., Villholth, K.G., Perera, A.P.G.R.L., Obeysekera, J., Gunatilaka, A., Panabokke, C.R., Hyndman, D.W., Cunningham, K.J. and Kaluarachchi, J.J., 2006. Impacts of the 2004 tsunami on groundwater resources in Sri Lanka. *Water Resources Research, 42*(5), pp. 1–9.

Khan, A.E., Ireson, A., Kovats, S., Mojumder, S.K., Khusru, A., Rahman, A. and Vineis, P., 2011. Drinking water salinity and maternal health in coastal Bangladesh: Implications of climate change. *Environmental Health Perspectives, 119*(9), pp. 1328–1332.

Khanom, T., 2016. Effect of salinity on food security in the context of interior coast of Bangladesh. *Ocean & Coastal Management, 130*, pp. 205–212.

Minderhoud, P.S.J., Erkens, G., Pham, V.H., Bui, V.T., Erban, L., Kooi, H. and Stouthamer, E., 2017. Impacts of 25 years of groundwater extraction on subsidence in the Mekong delta, Vietnam. *Environmental Research Letters, 12*(6), p. 064006.

Mulligan, A.E., Evans, R.L. and Lizarralde, D., 2007. The role of paleochannels in groundwater/seawater exchange. *Journal of Hydrology, 335*(3–4), pp. 313–329.

Pustry, P. and Farook, S., 2020. Salt water intrusion in coastal aquifers of India- A review. *HydroResearch, 3*, pp. 61–74

Rahman, A.K.M.M., Ahmed, K.M., Butler, A.P. and Hoque, M.A., 2018. Influence of surface geology and micro-scale land use on the shallow subsurface salinity in deltaic coastal areas: a case from southwest Bangladesh. *Environmental Earth Sciences, 77*(12), p. 423.

Rahman, M.M., Hassan, M.Q., Islam, M.S. and Shamsad, S.Z.K.M., 2000. Environmental impact assessment on water quality deterioration caused by the decreased Ganges outflow and saline water intrusion in south-western Bangladesh. *Environmental Geology, 40*(1–2), pp. 31–40.

Rakib, M.A., Sasaki, J., Matsuda, H., Quraishi, S.B., Mahmud, M.J., Bodrud-Doza, M., Ullah, A.A., Fatema, K.J., Newaz, M.A. and Bhuiyan, M.A., 2020. Groundwater salinization and associated co-contamination risk increase severe drinking water vulnerabilities in the southwestern coast of Bangladesh. *Chemosphere, 246*, p. 125646.

Rapti-Caputo, D., 2010. Influence of climatic changes and human activities on the salinization process of coastal aquifer systems. *Italian Journal of Agronomy, 3*, pp. 67–80.

Sarker, M.M.R., Van Camp, M., Islam, M., Ahmed, N. and Walraevens, K., 2018. Hydrochemistry in coastal aquifer of southwest Bangladesh: Origin of salinity. *Environmental Earth Sciences, 77*(2), p. 39.

Shammi, M., Rahman, M.M., Islam, M.A., Bodrud-Doza, M., Zahid, A., Akter, Y., Quaiyum, S. and Kurasaki, M., 2017. Spatio-temporal assessment and trend analysis of surface water salinity in the coastal region of Bangladesh. *Environmental Science and Pollution Research, 24*(16), pp. 14273–14290.

Shamsudduha, M., Chandler, R.E., Taylor, R.G. and Ahmed, K.M., 2009. Recent trends in groundwater levels in a highly seasonal hydrological system: the Ganges-Brahmaputra-Meghna Delta. *Hydrology and Earth System Sciences*, *13*(12), pp. 2373–2385.

Sharma, K.L., Rao, C.S., Chandrika, D.S., Nandini, N., Munnalal, Reddy, K.S., Indoria, A.K. and Kumar, T.S., 2016. Assessment of GMean biological soil quality indices under conservation agriculture practices in rainfed Alfisol soils. *Current Science*, *111(8)*, pp. 1383–1387.

Sultana, S., Ahmed, K.M., Mahtab-Ul-Alam, S.M., Hasan, M., Tuinhof, A., Ghosh, S.K., Rahman, M.S., Ravenscroft, P. and Zheng, Y., 2015. Low-cost aquifer storage and recovery: implications for improving drinking water access for rural communities in coastal Bangladesh. *Journal of Hydrologic Engineering*, *20*(3), p. B5014007.

Uddin, M.N. and Abdullah, S.K.M., 2003. Quaternary geology and aquifer systems in the Ganges-Brahmaputra-Meghna delta complex, Bangladesh. *Proceedings of GEOSAS-IV, Geological Survey of India*, pp. 400–416.

Zahid, A., Hassan, M.Q., Imes, J.L. and Clark, D.W., 2009. Hydraulic characterization of aquifer (s) and pump test data analysis of deep aquifer in the arsenic affected Meghna River floodplain of Bangladesh. *Journal of Environmental Research*, *3*(2/3), pp. 325–355.

15 Salinity Dynamics in the Hooghly-Matla Estuarine System and Its Impact on the Mangrove Plants of Indian Sundarbans

Subrata Mitra, Abhra Chanda, Sourav Das, Tuhin Ghosh, and Sugata Hazra
Jadavpur University

CONTENTS

15.1 INTRODUCTION

Mangrove plants typically inhabit the intertidal zones of tropics and subtropics and can tolerate highly saline seawater (Liang et al., 2008). Mangroves can withstand a wide range of salinity (Hutchings and Saenger, 1987; Lugo and Snedaker, 1974).

However, the ability of salt tolerance varies from species to species (Barik et al., 2018). However, mangrove plants can withstand salinity better than any other plant species (Ball, 1988); the growth and productivity of mangrove plants are often found to be regulated by salinity (Clough and Sim, 1989; Ball, 2002). Usually, the adjoining water column of the estuaries and coasts through regular tidal activities (Lin et al., 2003) regulates the salinity of mangrove sediments. Hence, a change in the salinity level in the aquatic column can substantially change the salinity of the adjacent mangrove sediments. An increase in salinity beyond a certain threshold is often found to be physiologically challenging for many mangrove species, as it makes water uptake extremely difficult and leads to ion toxicity due to over-accumulation of salts within the mangrove plant parts (Reef and Lovelock, 2015). The CO_2 assimilation potential of several mangrove species is also compromised under elevated salinity, as these trees are forced to restrict photosynthetic activities (Ball, 1988; Bjorkman et al., 1988; Nandy and Ghose, 2001). An increase in salinity is also held accountable for the loss of some mangrove species (Dasgupta et al., 2017). Mangroves are considered to be a highly productive ecosystem and are known for their high 'blue carbon' stock in the aboveground and belowground compartments (Pendleton et al., 2012). Recent researches showed that changes in salinity regime within a mangrove forest can potentially change the floral species composition, which in turn can significantly alter the carbon stock of the forest (Chanda et al., 2016).

The Indian Peninsula covering a long coastline of ~7,500 km shelters many mangrove forests and are found to be under potential threat from an increase in salinity (Sandilyan et al., 2010). With regard to the above-mentioned background, the present review tried to characterize the long-term salinity dynamics in the Hooghly-Matla estuarine complex (from the existing literature on measured salinity at different times throughout the last seven decades) which feeds the world's largest mangrove forest – the Sundarbans and collated the physiological observations on the mangrove plants of this region from the perspective of salt stress and salinity-induced changes in the species assemblage of this forest. Quite a few studies have indicated that Sundarbans is suffering lately from the increase in salinity, in both the Indian (Banerjee et al., 2017; Chowdhury et al., 2019; Raha et al., 2012; Trivedi et al., 2016) as well as Bangladesh counterpart (Ahmed et al., 2011; Islam and Gnauck, 2011; Nasrin et al., 2019). However, a holistic synthesis of preexisting salinity data in the different zones of the Indian Sundarbans was lacking. This prompted us to take up the present endeavour, wherein the main stress was given to collate all the reported salinity data from the entire Hooghly-Matla estuarine complex from the year 1955–2017 and accordingly synthesize the temporal trend in salinity in the different zones within this estuarine complex.

15.2   THE HOOGHLY-MATLA ESTUARINE COMPLEX

The River Ganges, originating in the western Himalayas, flows for a distance of ~2,600 km and empties in the Bay of Bengal. The estuarine part of the Ganges shelters the world's largest continuous mangrove forest of Sundarbans. The Sundarbans mangroves is a cross-country forest shared by India and Bangladesh. In the Indian part, the Hooghly River Estuary, a distributary and the first deltaic offshoot of the River Ganges, acts as the principal source of freshwater to the mangroves of Sundarbans

(Biswas et al., 2007). Besides Hooghly, there are a number of other estuaries like the Saptamukhi, Thakuran, Matla, Bidya, Gosaba, Gomdi, Raimangal, Harinbhanga, etc., that intersect the Indian Sundarbans and this criss-cross intricate network of estuaries is collectively referred to as Hooghly-Matla estuarine system. The Raimangal Estuary on the eastern end serves as the international boundary between the Indian and Bangladesh part of Sundarban and still carries substantial freshwater discharge from the Ganges into the Bay of Bengal (Aziz, 2017; Chatterjee et al., 2013). However, all the other estuaries (except Hooghly and Raimangal) have lost their connection with the freshwater flow in the upstream due to multiple factors like (i) change in the course of lower Ganges towards the east, influenced by the natural east-ward tilt of Ganga-Brahmaputra-Meghna Delta; (ii) anthropogenic interferences, especially in and around the city of Kolkata (situated in the bank of Hooghly River); and (iii) widespread sedimentation in the upper reaches of these rivers (Darby et al., 2020; Morgan and McIntire, 1959; Raha et al., 2012; Sanyal and Bal, 1986). These estuaries at present act as arms of the Bay of Bengal and their estuarine character is maintained by the semi-diurnal tides and the freshwater from monsoonal runoffs and depression and cyclone-induced rainfall that takes place during the non-monsoon dry seasons (Attri and Tyagi, 2010; Chatterjee et al., 2013; Cole and Vaidyaraman, 1966).

15.3 MANGROVES OF INDIAN SUNDARBANS

The mangrove ecosystem of Sundarban is a highly productive system. The large spatial and temporal variability in the hydrology, topography and texture of the substratum coupled with varying salinity, and their interactions, result in very high habitat heterogeneity within this mangrove ecosystem, which in turn led to the formation of highly biodiverse floral communities in the Sundarbans (Chanda, 2015). Several workers have quantified the floral species composition of Indian Sundarbans based on extensive ground surveys. Blasco (1975) and Blasco et al. (1996) reported that the Indian mangrove plants comprise 58 species. Rao (1986) reported 60 mangrove species under 41 genera and 29 families. Naskar (1988) reported 35 true mangroves in addition to 28 mangrove associates and 7 obligate mangroves under 29 families and 49 genera from the Indian Sundarbans. Furthermore, Naskar and Guha Bakshi (1987) have reported 96 species of mangrove associates or back mangals or weed flora or species. Recently Sanyal et al. (2008) reported 25 species of true mangroves, 9 semi-mangroves (grasses), 37 mangrove associates and 22 mangrove commensals from the Indian part of Sundarbans. Naskar (1988) and many other workers indicated that the mangroves of Sundarban are richer than any other tropical mangrove formation of the world. *Avicennia* sp., *Bruguierra* sp., *Sonneratia* sp., *Rhizophora* sp., *Ceriops* sp. *Excoecaria* sp., *Xylocarpus* sp. and *Phoneix* sp. are some of the dominant mangroves found in the Indian Sundarbans.

15.4 METHODOLOGY

15.4.1 DEMARCATION OF ZONES WITHIN THE AREA OF INTEREST

A large number of studies were carried out in the past decades which measured and reported salinity in the Hooghly estuary and the several creeks and estuaries within

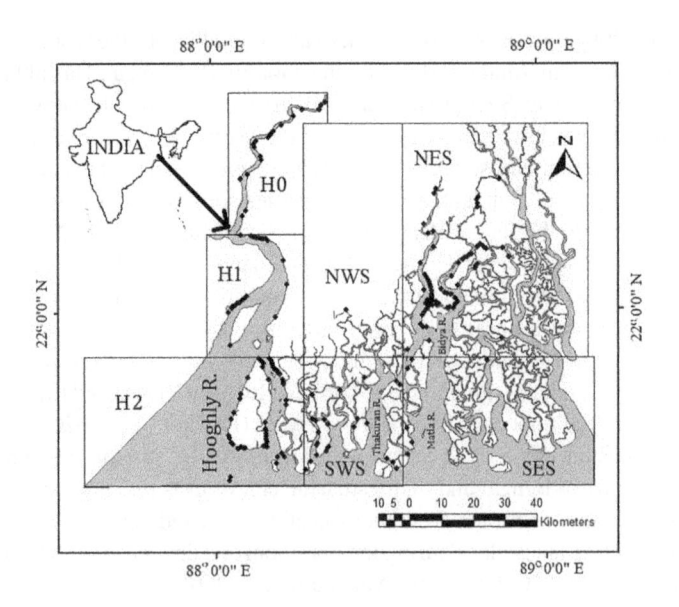

FIGURE 15.1   The study area map showing the seven zones of the Hooghly-Matla estuarine complex out of which Hooghly Estuary covers three zones (H0, H1 and H2 zone), and the estuarine network within the Indian Sundarbans covers four zones, namely, the north-eastern Sundarbans (NES zone), the north-western Sundarbans (NWS zone), the south-eastern Sundarbans (SES zone) and the south-western Sundarbans (SWS zone). The black dots are showing the sampling locations that we retrieved from the literature considered in the present study.

the Indian Sundarbans. The sampling locations in these studies varied in accordance with the objective or hypothesis of the respective studies. Thus synthesizing the temporal trend of salinity at a particular pin-pointed location was absolutely impossible from the existing literature available. Hence we divided the area of interest in this study into seven broad zones. For the sake of better understanding, the longitudinal stretch of Hooghly Estuary was divided into three zones (H0, H1 and, H2 zone), and the estuarine network within the Indian Sundarbans was divided into four quadrants, namely, the north-eastern Sundarbans (NES zone), the north-western Sundarbans (NWS zone), the south-eastern Sundarbans (SES zone) and the south-western Sundarbans (SWS zone) (Figure 15.1).

15.4.2   DATA MINING AND SYNTHESIS OF TEMPORAL TRENDS

Literatures available only on the internet have been downloaded and assembled for the present study. A wide range of keywords has been used to search for the scientific literature like 'salinity', 'biogeochemistry', 'water quality', 'physicochemical parameters', 'Hooghly estuary', 'Indian estuaries', 'Sundarban estuary', 'Mooriganga estuary', 'Saptamukhi estuary', 'Bidya estuary', 'Raimangal estuary', 'Matla estuary', 'Haribhanga estuary', 'Thakuran estuary', 'creeks of Sundarban', 'rivers of Sundarban', 'waterways of Sundarban', 'creeks of Sundarban', etc. Search engines like Google Scholar, Scopus and ScienceDirect were used to search and download the papers. A total of 141 papers were scanned for the present study, out of which salinity measurements from more than 60 papers were used to construct the salinity trends. Several studies despite having reported salinities within the present area of interest could not be taken into account because (i) the date and time of sampling were not specifically mentioned, (ii) the study was carried out during only high or low tide conditions, and (iii) the location of the sampling was not clearly mentioned.

In this study, salinity data were used only from those papers that reported either annual mean salinity or seasonal mean salinity. Studies which reported short term salinity measurements (like one or two discrete days or few hours) were not taken into account. In some cases, the seasonal mean salinities were averaged to compute the annual mean salinities. The salinity measurements reported by various studies within each of the seven zones were separately clustered and the temporal trend within these zones was separately synthesized and discussed in this study. In cases where we found more than one study which reported salinity from the same zone for the same year, we carried out an arithmetic mean of the reported salinities by the different studies and considered the averaged salinity for that particular year, while constructing the temporal trend. Depending upon the available data, the time span of temporal salinity trends varied in the different zones. Only linear trend lines were constructed to examine the salinity dynamics over the years.

15.5 TEMPORAL CHANGE IN SALINITY

15.5.1 SALINITY DYNAMICS IN H0 ZONE

Year old data have always been hard to retrieve due to lack of documentation and maintenance of proper and systematic records. The earliest record of salinity that we could retrieve from available literature in the H0 zone dated back to the year 1955. Prior to the commissioning of Farakka Barrage, the salinity in this zone was significantly high which indicated an acute shortage of freshwater during those days. Bose (1956) observed the annual mean salinity in this region to be 9.75 in the year 1955. In the year 1961, Shetty et al. (1961) recorded mean salinity values as high as 20.2. Basu et al. (1970) observed a mean salinity of ~7. However, a detailed report on the seasonal variation was not mentioned in any of these studies. After the inauguration of Farakka Barrage, the salinity values were found to decrease substantially which signifies that Farakka Barrage once started operating provided ample freshwater to the Hooghly River Estuary. Farakka Barrage was commissioned and brought into operation in the year 1975. Shortly after the barrage started operating, Nandy et al. (1983) observed mean salinity values of ~10.2 in the year 1976, which was found to reduce to 3.8 in the year 1987, reported by Chakraborty and Chattopadhyay (1989). Since the onset of the present century, the number and frequency of studies that reported annual mean salinity increased to a great extent in his region. Sarkar et al. (2007) and Roy and Nandi (2012) observed annual mean salinity to be 0.5 and 0.4 in the years 2002 and 2005, respectively, signifying an ongoing increase in freshwater inflow. Analyzing the temporal data, the lowest annual mean salinity was observed by Manna et al. (2013) in this zone (0.05 in the year 2010–2011). However, three studies in the following years of 2012 [Samanta et al. (2015); salinity ~ 0.18], 2013

[Samanta et al. (2015); salinity ~ 0.21] and 2014 [Mitra et al. (2018); salinity ~ 0.92] observed a very gradual increase in annual mean salinity. This short-term increasing trend in salinity could be due to natural inter-annual variability in parameters like decreased freshwater discharge from Farakka or lesser rainfall and catchment runoff; however, due to the lack of studies reporting salinity during the years 2015–2020, we could not confirm whether this increasing trend is still continuing or not. Taking into account the linear trend from the year 1955–2014, a steady decrease in salinity is found to take place (0.25 per year; $R^2 = 0.76$, $p < 0.01$) (Figure 15.2a) (Please note that the magnitude of slope in all the regression equations were rounded to two significant digits after decimal point). However, upon scrutinizing separately the span from 2010 to 2014, a mild increase in salinity was observed (0.06 per year; $R^2 = 0.76$, $p < 0.01$) (Figure 15.2b). This short-term increase in salinity was observed in all three seasons, with the highest increase during the pre-monsoon season, followed by post-monsoon season, and the least during monsoon season (Figure 15.2c).

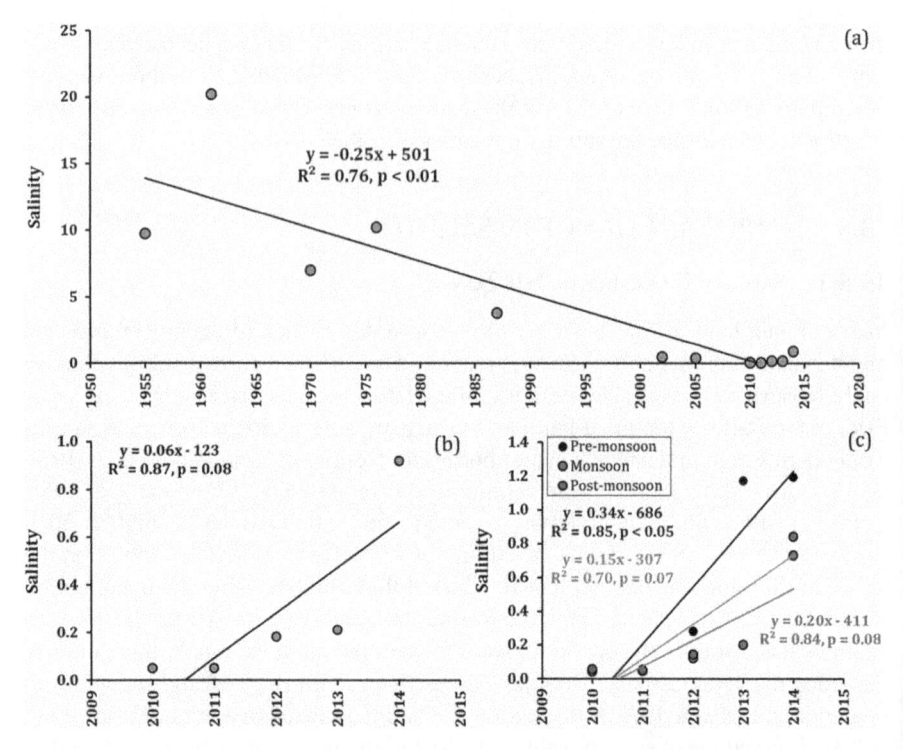

FIGURE 15.2 The plot of salinity data extracted from Bose (1956) (for the year 1955); Shetty et al. (1961) (for the year 1961); Basu et al. (1970) (for the year 1970); Nandy et al. (1983) (for the year 1976); Chakraborty and Chattopadhyay (1989) (for the year 1987); Sarkar et al. (2007) (for the year 2002); Roy and Nandi (2012) (for the year 2005); Manna et al. (2013) (for the years 2010 and 2011); Samanta et al. (2015) (for the years 2012 and 2013); Mitra et al. (2018) (for the year 2014) showing (a) the linear temporal trend of annual mean salinity between the years 1955 and 2014, (b) the linear temporal trend of annual mean salinity between the years 2010 and 2014 and (c) the linear temporal trend of seasonal mean salinity between the years 2010 and 2014 in the H0 zone.

15.5.2 Salinity Dynamics in H1 Zone

Similar to the H0 zone, there was also a stark difference in annual mean salinity in the H1 zone between the pre-Farakka Barrage and post-Farakka Barrage time span. This zone (H1) being comparatively closer to the Bay of Bengal experiences higher dominance of marine water; hence the salinity values are expected to be higher in this zone compared to H0 zone. Bose (1956) in the year 1955 observed a mean salinity of ~23 in the Diamond Harbour station located in the middle of H1 zone. The very same study of Bose (1956) observed a mean salinity of ~9.7 in the H0 zone, which shows that marine water influence was more than double in the H1 zone compared to that in the H0 zone. This observation further signifies that the H1 zone was significantly devoid of freshwater input from the upstream before the construction of Farakka Barrage. Few other studies like Rao (1967) carried out salinity measurements during the year 1963–1964; however, nothing specific was mentioned about the tidal timing of the sampling nor was the annual mean reported, which prevented us to use the data in delineating the trend line. The very next study in due course of time which measured the annual mean salinity in H1 zone was that of Lal (1990) carried out in 1986 and mean salinity of 0.9 was reported from this region. This signifies that post-Farakka Barrage operation the freshwater content increased to a great extent in H1 zone also. In between the years 1990 and 2010, Chakraborty et al. (1995), Mukhopadhyay et al. (2006), Sarkar et al. (2007) and Manna et al. (2014) observed mean salinity values of ~1.9, ~1.2, ~1.1 and ~0.6 in the years 1994, 2000, 2002 and 2010, respectively. However, during the same span two studies, namely, Roy et al. (2002) and Roy and Nandi (2012) observed comparatively higher mean salinity values of ~4.6 and ~3.8 in the years 1998 and 2005, respectively, in H1 zone. Analyzing the temporal trend in salinity from year 1955 to 2014, a significant decrease in salinity was observed. The rate of decrease in salinity was marginally higher (0.32 per year; $R^2 = 0.72$, $p < 0.01$) (Figure 15.3a) compared to the rate of salinity decrease observed in H0 zone during the same period.

Similar to H0 zone, a short-term increase in annual mean salinity was also observed in the H1 zone from 2010 to 2014; however, the rate of increase was substantially higher than that observed in H0 zone (0.25 per year; $R^2 = 0.76$, $p < 0.01$) (Figure 15.3b). Compared to H0 zone, studies that have recorded a seasonal variation of salinity are much more in the H1 zone. From the year 1991 till the year 2014, 15 studies have reported seasonal variation. The temporal trend in salinity variation in the respective seasons during this 24-year span (1991–2014) showed a significant decrease in salinity during the pre-monsoon season (0.22 per year; $R^2 = 0.37$, $p < 0.05$) and monsoon season (0.25 per year; $R^2 = 0.42$, $p < 0.05$); however, during post-monsoon season no clear trend was observed ($R^2 = 0.02$, $p > 0.05$) (Figure 15.3c).

15.5.3 Salinity Dynamics in H2 Zone

This zone of Hooghly Estuary is very wide (width ranging from 13 to 22 km). The marine influence of the Bay of Bengal is very strong in this region. During the pre-Farakka Barrage time, three studies, namely, Bose (1956), Shetty et al. (1961) and Rao (1967) reported annual salinity range (maximum and minimum) in this region in the years 1955, 1961 and 1963, respectively. The maximum salinity ranged between

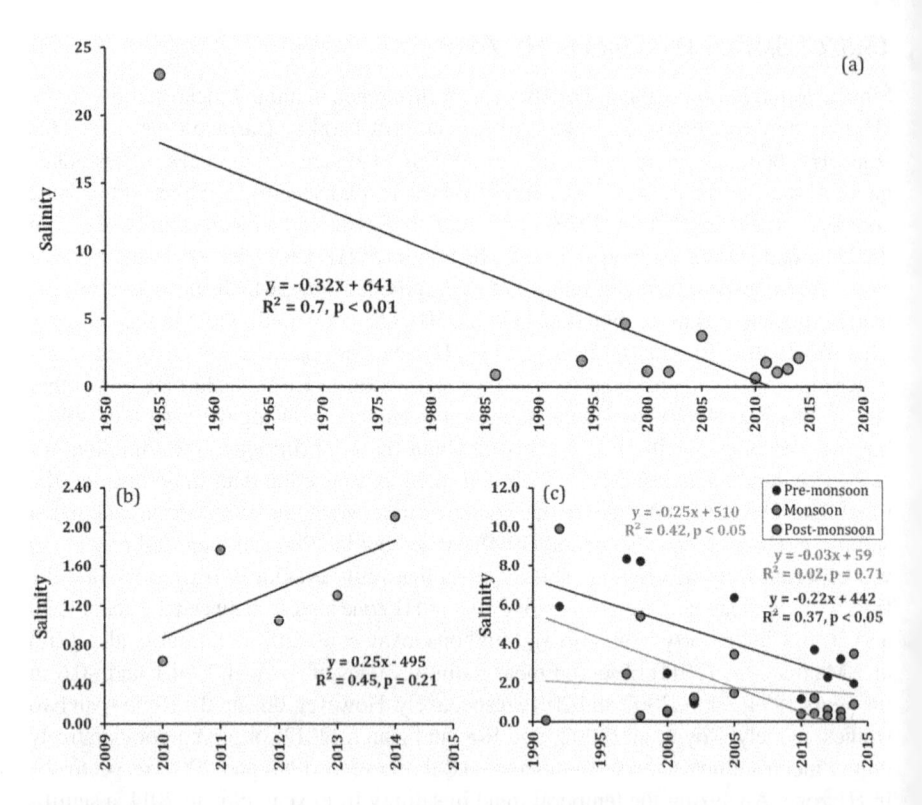

FIGURE 15.3   The plot of salinity data extracted from Bose (1956) (for the year 1955); Lal (1990) (for the year 1986); Somayajulu et al. (2002) (for the year 1991); Sinha et al. (1996, 1999) (for the year 1992); Chakraborty et al. (1995) (for the year 1994); Sadhuram et al. (2005) for the year 1997); Roy et al. (2002) (for the year 1998); Mukhopadhyay et al. (2006) (for the year 2000); Sarkar et al. (2007) (for the year 2002); Roy and Nandi (2012) (for the year 2005); Manna et al. (2014) (for the year 2010); Manna et al. (2013, 2014) (for the year 2011); Samanta et al. (2015) and Rakshit et al. (2014) (for the year 2012); Samanta et al. (2015) (for the year 2013); Ray et al. (2015, 2018) and Mitra et al. (2018) (for the year 2014); showing (a) the linear temporal trend of annual mean salinity between the years 1955 and 2014, (b) the linear temporal trend of annual mean salinity between the years 2010 and 2014 and (c) the linear temporal trend of seasonal mean salinity between the years 1991 and 2014 in the H1 zone.

30 and 33 in these studies. After the commissioning of Farakka Barrage, salinity in this region was reported by Bhunia and Choudhury (1981), Chakraborty and Chattopadhyay (1989), and Nath and De (1998) in the years 1976, 1986 and 1997, respectively. The maximum salinity in these post-Farakka Barrage studies ranged between 18 and 28, which was substantially less than the pre-Farakka Barrage maximums (Figure 15.4b). However, none of these earlier studies reported annual mean salinity, which barred us from using these data to draw a temporal trend during the years 1955–1997. The year 1998 onward, several studies were conducted which measured and reported annual as well as seasonal mean salinities. Between the years 1998 and 2017, annual mean salinity ranged between 11.2 and 19.2 and exhibited a significant decreasing trend (0.26 per year; $R^2 = 0.41$, $p < 0.05$) (Figure 15.4a).

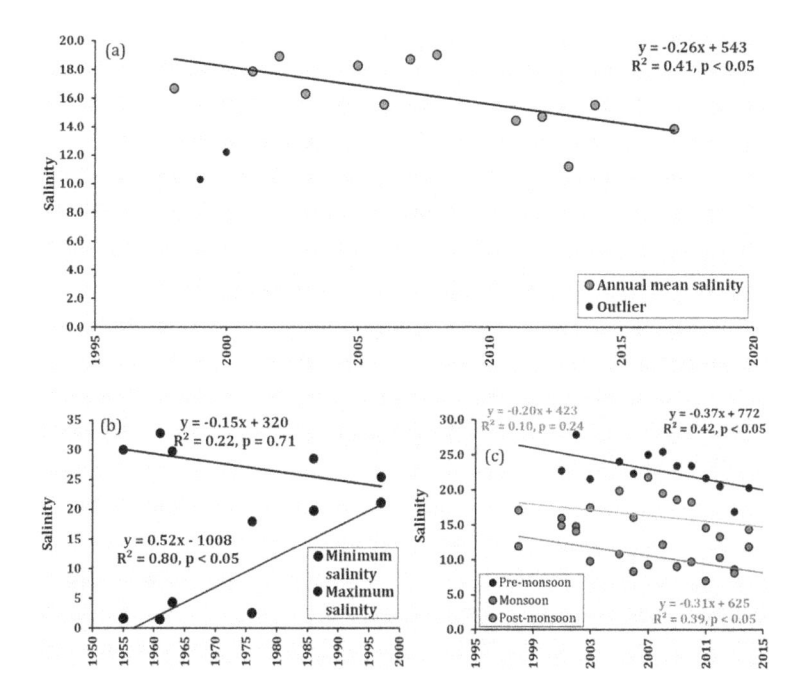

FIGURE 15.4 The plot of salinity data extracted from Bose (1956) (for the year 1955); Shetty et al. (1961) (for the year 1961); Rao (1967) (for the year 1963); Biswas et al. (2010) (for the year 1970 and 1980); Bhunia and Choudhury (1981) and Chakraborty and Chattopadhyay (1989) (for the year 1976); Chakraborty and Chattopadhyay (1989) and Lal (1990) (for the year 1986); Nath (1998) and Nath and De (1998) (for the year 1997); Roy et al. (2002), Niyogi et al. (2001) and Sarkar and Bhattacharya (2003) (for the year 1998); Sarkar and Bhattacharya (2003) (for the year 1999); Mukhopadhyay et al. (2006) and Ghosh et al. (2013) (for the year 2000); Biswas et al. (2004) (for the year 2001); Saha et al. (2006) and Sarkar et al. (2007) (for the year 2002); Dey et al. (2005) (for the year 2003); Mitra et al. (2010), Roy and Nandi (2012), Bhattacharjee et al. (2013a), Sarkar et al. (2013), Banerjee and Mitra (2015) and Banerjee et al. (2017) (for the year 2005); Mitra et al. (2010), Mandal et al. (2012a,b), Bhattacharjee et al. (2013a), Banerjee and Mitra (2015) and Banerjee et al. (2017) (for the year 2006); Mitra et al. (2010), Bhattacharjee et al. (2013a), Banerjee and Mitra (2015) and Banerjee et al. (2017) (for the year 2007); Mitra et al. (2010), De et al. (2011), Banerjee et al. (2013), Bhattacharjee et al. (2013a), Banerjee and Mitra (2015), Padhy et al. (2016) and Banerjee et al. (2017) (for the year 2008); Mitra et al. (2010), Roy (2010), Massolo et al. (2012), Banerjee et al. (2013), Bhattacharjee et al. (2013b), Banerjee and Mitra (2015) and Banerjee et al. (2017) (for the year 2009); Mitra et al. (2010), Das (2012), Banerjee et al. (2013), Bhattacharjee et al. (2013c), Manna et al. (2013), Banerjee and Mitra (2015) and Choudhury et al. (2015) (2015) (for the year 2010); Akhand et al. (2013), Mitra et al. (2012), Manna et al. (2013), Manna et al. (2014) and Banerjee et al. (2017) (for the year 2011); Ghosh and Banerjee (2013), Rakshit et al. (2014), Banerjee and Mitra (2015), Bhattacharya et al. (2014), Samanta et al. (2015) and Banerjee et al. (2017) (for the year 2012); Choudhury et al. (2015), (2015), Choudhury and Bhadury (2015), Samanta et al. (2015), Akhand et al. (2016) and Banerjee et al. (2017) (for the year 2013); Ray et al. (2015), Banerjee et al. (2017), Mitra et. al. (2018) and Ray et al. (2018) (for the year 2014); Chakraborty et al. (2016) (for the year 2016) and Agarwal and Mitra (2018) (for the year 2017) showing (a) the linear temporal trend of annual mean salinity between the years 1998 and 2017, (b) the linear temporal trend of maximum and minimum salinity between the years 1955 and 1997 and (c) the linear temporal trend of seasonal mean salinity between the years 1998 and 2017 in the H2 zone.

The annual mean salinity computed from the seasonal mean reported by Sarkar and Bhattacharya (2003) for the year 1999, and Mukhopadhyay et al. (2006) and Ghosh et al. (2013) for the year 2000 was 10.3 and 12.2 only. These two very low values were omitted while delineating the trend in annual mean salinity. The computed annual mean for the year 2013 from the data reported by Akhand et al. (2016), Banerjee et al. (2017), Choudhury et al. (2015), Choudhury and Bhadury (2015) and Samanta et al. (2015) was also very low (11.2). Similar to H1 zone, the temporal trend in salinity variation showed significant decreasing trend during the pre-monsoon season (0.37 per year; $R^2 = 0.42$, $p < 0.05$) and monsoon season (0.31 per year; $R^2 = 0.39$, $p < 0.05$); however, during the post-monsoon season the trend was not significant ($R^2 = 0.10$, $p > 0.05$) (Figure 15.4c). Taking into account, the seasonal average salinity of all the years between 1998 and 2017, a prominent difference in salinity was observed between the pre-monsoon (~22), monsoon (~10) and post-monsoon (~16) seasons.

15.5.4 SALINITY DYNAMICS IN SWS ZONE

Compared to the Hooghly River Estuary, much fewer studies have reported annual and seasonal salinity within the estuaries of the Sundarbans mangroves. The SWS zone of this study mainly includes the Thakuran and Saptamukhi Estuaries. Unlike the Hooghly Estuary, there is no proper record of pre-Farakka Barrage salinity in any of the zones within the Sundarbans mangroves. The oldest annual mean salinity for this region could be retrieved for the year 1991 (Bose et al., 2012). Bose et al. (2012) reported the data acquired by Central Pollution Control Board, India, in the Saptamukhi Estuary during the years 1991–1998; however, the exact locations were not mentioned in their paper. Bose et al. (2012)'s data exhibited a significant increasing trend in salinity (0.83 per year; $R^2 = 0.53$, $p < 0.05$) (Figure 15.5a). The rate of increase in salinity in this duration was very high. Plotting annual mean salinity retrieved from other studies during the span 2001–2017 also exhibited a significant increasing trend (0.12 per year; $R^2 = 0.59$, $p < 0.05$) (Figure 15.5b); however, the rate of increase in salinity was much less than that observed by Bose et al. (2012) during 1991–1998. Critically analyzing the seasonal mean salinity during the same span (2001–2017), it was observed that the pre-monsoon salinity (0.36 per year; $R^2 = 0.62$, $p < 0.05$) and post-monsoon salinity (0.25 per year; $R^2 = 0.65$, $p < 0.05$) exhibited a steep increasing trend; however, there was no statistically significant trend during the monsoon ($R^2 = 0.08$, $p > 0.05$) (Figure 15.5c). This showed that during the dry seasons (i.e. pre- and post-monsoon seasons), the salinity in this region is steadily increasing due to the increasing reduction of freshwater; however, during the monsoon season, this region receives freshwater from rain as well as surface runoff from upstream, which has more or less remained constant. The deficit of freshwater in this zone is especially observed during the pre-monsoon season (which includes the dry summer months of April and May), as the composite mean seasonal salinity (during 2001–2017) was highest for the pre-monsoon season (~27), followed by post-monsoon season (~18) and the monsoon season (~14).

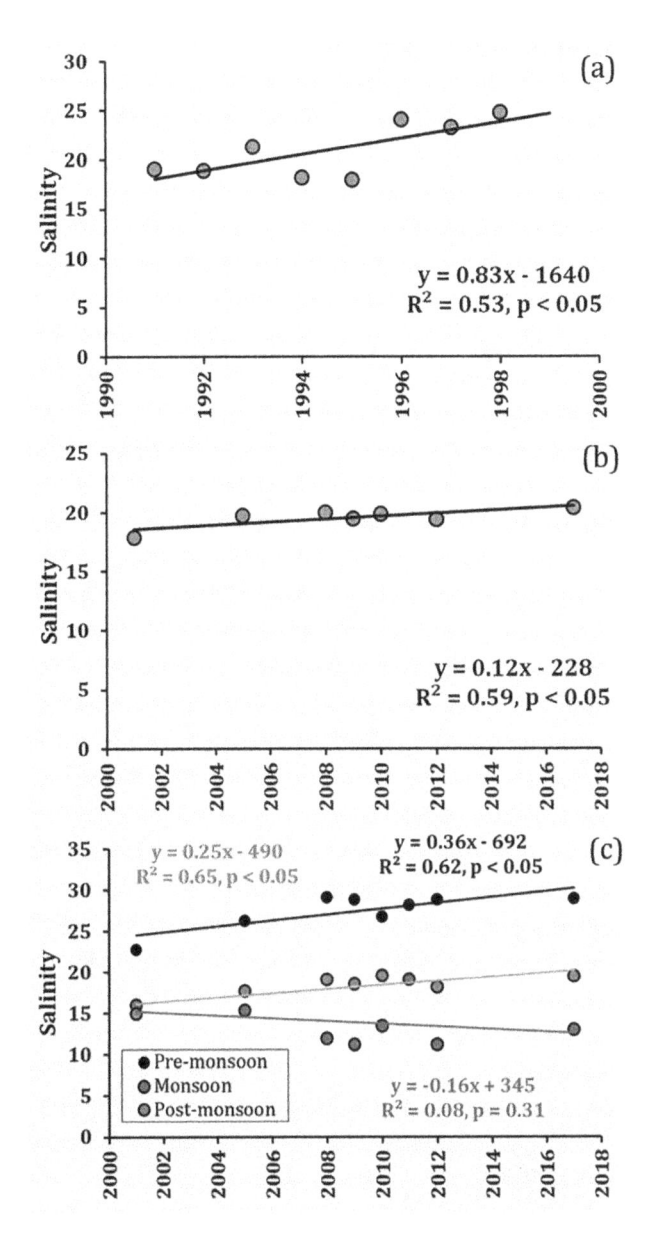

FIGURE 15.5   The plot of salinity data extracted from Bose et al. (2012) (for the years 1991–1999); Biswas et al. (2004) (for the year 2001); Biswas et al. (2007) (for the year 2003); Sarkar et al. (2013) (for the year 2005); Banerjee et al. (2013) (for the years 2008 and 2009); Das (2012), Dutta et al. (2013) and Banerjee et al. (2013) (for the year 2010); Akhand et al. (2013), Mitra et al. (2012) and Dutta et al. (2015) (for the year 2011); Banerjee and Mitra (2015) (for the year 2012) and Agarwal and Mitra (2018) (for the year 2017) showing (a) the linear temporal trend of annual mean salinity between the years 1991 and 1999, (b) the linear temporal trend of annual mean salinity between the years 2001 and 2017 and (c) the linear temporal trend of seasonal mean salinity between the years 2001 and 2017 in the SWS zone.

15.5.5 Salinity Dynamics in NES Zone

The majority of the studies carried out in the estuarine waters of Sundarbans have concentrated in this zone. This zone includes the Matla Estuary and the upper part of Bidya Estuary. The oldest salinity data from this zone date back to the year 1963, where the minimum and maximum salinity in the Matla Estuary was reported to be 10.8 and 30.4, respectively (Rao, 1967). In the following years, Sengupta et al. (2013) reported an increase in the central sector of Sundarbans (which matches with this zone to a large extent) from the year 1989 to 2012 (at a rate of 0.12 per year); however, neither the data source nor the exact sampling locations were mentioned in detail. Measured seasonal and annual salinity from several studies was plotted during the time span 2004–2011, and a significant increasing trend in annual mean salinity was observed (0.9 per year; $R^2 = 0.62$, $p < 0.05$) (Figure 15.6a). During the same span (2004–2011), many studies reported seasonal mean salinity. Analyzing

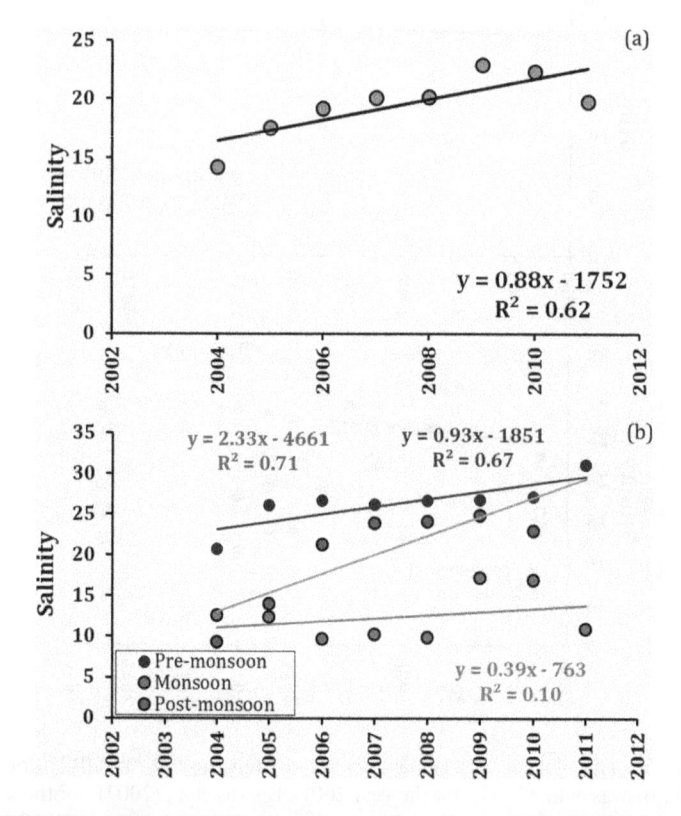

FIGURE 15.6 The plot of salinity data extracted from Roy and Nandi (2012) and Banerjee et al. (2017) (for the year 2004); Sarkar et al. (2013) (for the year 2005); Banerjee and Mitra (2015) (for the years 2006 and 2007); Bhattacharjee et al. (2013a) (for the year 2008); Chaudhuri et al. (2013) (for the year 2009); Das (2012) (for the year 2010) and Chaudhuri et al. (2013) (for the year 2013) showing (a) the linear temporal trend of annual mean salinity between the years 2004 and 2011 ($p < 0.05$), and (b) the linear temporal trend of seasonal mean salinity between the years 2004 and 2011 in the NES zone ($p < 0.05$).

the seasonal mean salinity, it was observed that post-monsoon seasonal mean salinity exhibited a very steep increasing trend (2.3 per year; $R^2 = 0.71$, $p < 0.05$), followed by pre-monsoon seasonal mean salinity (0.9 per year; $R^2 = 0.67$, $p < 0.05$). However, like the SWS zone, no significant trend in salinity variation was observed during the monsoon season ($R^2 = 0.10$, $p > 0.05$) (Figure 15.6b). Unlike SWS, the highest rate of increase in salinity was observed during the post-monsoon season, instead of the pre-monsoon season.

However, very much like SWS, the freshwater deficit was observed to be highest during the pre-monsoon season, as reflected from the composite seasonal mean salinity during the pre-monsoon season (~27), followed by the post-monsoon season (~20) and the monsoon season (~13). During the years, 2012 to date, many studies have been carried out in the Sundarbans; however, the salinity data could not be retrieved properly in accordance with the zones described in this study.

15.5.6 SALINITY DYNAMICS IN NWS AND SES

These two zones of Indian Sundarbans are least studied from the perspective of aquatic biogeochemistry. Due to the lack of a significant number of works in the different time periods, the temporal trends could not be estimated for the NWS and SES zones. The NWS zone encompasses the dead ends of the Thakuran and Saptamukhi Estuaries. The only work of Manna et al. (2014) reported a mean salinity of 28.8 in the Raidighi sector of NWS; however, the season or time of sampling was not clearly mentioned. Similarly, very few studies reported salinity from the SES zone of the Indian Sundarbans. The SES zone encompasses the Sundarbans National Park and access to these zones is extremely restricted. Though entry to this core forest zone is allowed for research work, it requires prior approval from the West Bengal State Forest Department, and under very rare circumstances, these waters are allowed to venture. This could be one of the possible reasons behind such a lower number of works reported from this zone. Scrutinizing the existing literature, we found that Sarkar et al. (2013) during the years 2004–2006 measured salinity in all the seasons in this zone and reported a seasonal mean salinity of ~28, ~14 and ~20 during pre-monsoon, monsoon and post-monsoon seasons, respectively. Later in the year 2013, Akhand et al. (2016) while working throughout the Indian Sundarbans observed similar seasonal variations in salinity in the SES zone (~28, ~13 and ~21 during pre-monsoon, monsoon and post-monsoon seasons, respectively). The spatial variability of salinity in the entire study area is portrayed for the pre-monsoon (Figure 15.7), monsoon (Figure 15.8) and the post monsoon (Figure 15.9) season.

15.6 EFFECT OF SALT STRESS ON THE MANGROVES PLANTS OF SUNDARBANS

The synthesized salinity trends in the seven zones covering the entire Hooghly-Matla estuarine complex clearly portrayed that perennial freshwater supply in the Hooghly estuary has increased steadily since the commissioning of the Farakka barrage; however, the estuaries and creek networks that are interspersed throughout the

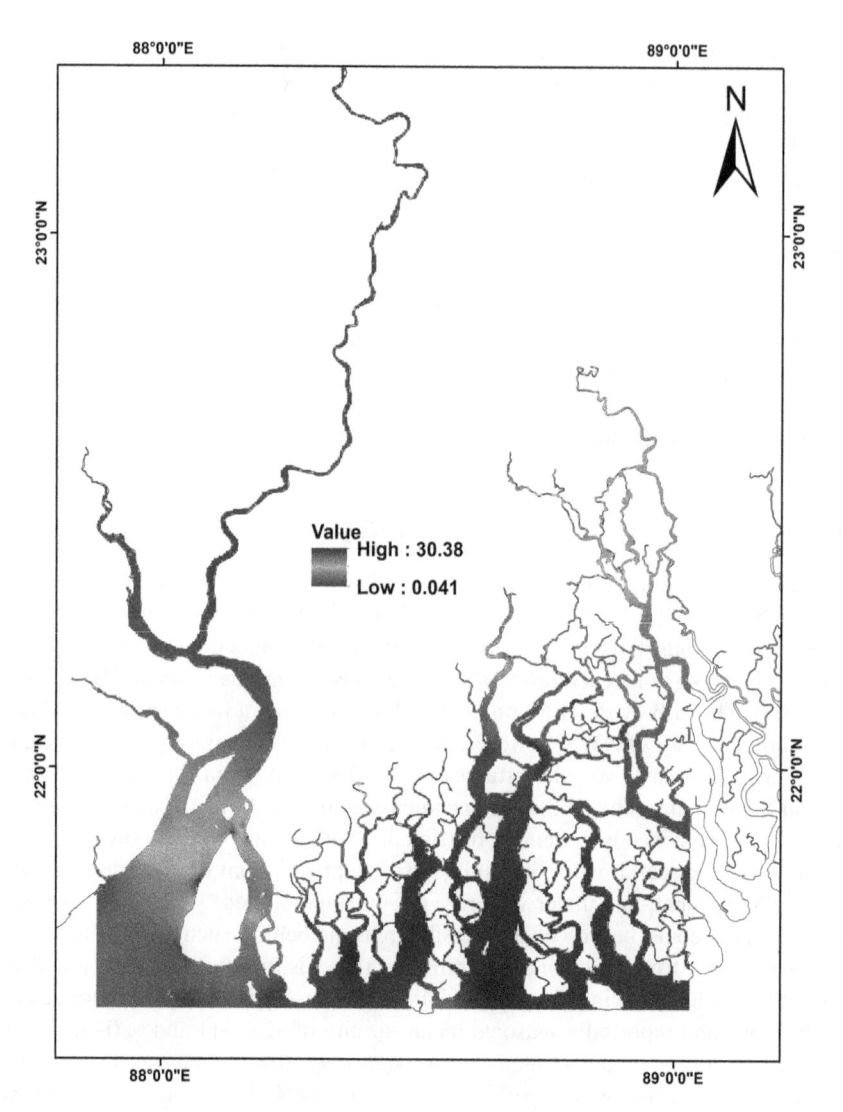

FIGURE 15.7 The map showing a tentative spatial variability of salinity in the study area during the pre-monsoon season

mangroves of Sundarbans are facing acute scarcity of freshwater. Chowdhury et al. (2019), in this regard, observed recently that the enhanced aquatic salinization of the estuaries of Sundarbans is enhancing the salinity of the mangrove sediments, which in turn are making these sediments highly anoxic and impoverished in nutrient content. They further reported that the microbial communities which are responsible for the release of nutrients are also unable to cope up with the increase in salinity, leading to high sulphide build-up and nutrient depletion in the soils. Chowdhury et al. (2019) estimated that due to the ongoing increase in salinity the structure and health of the forests degraded substantially (the extent of degradation varied between 11%

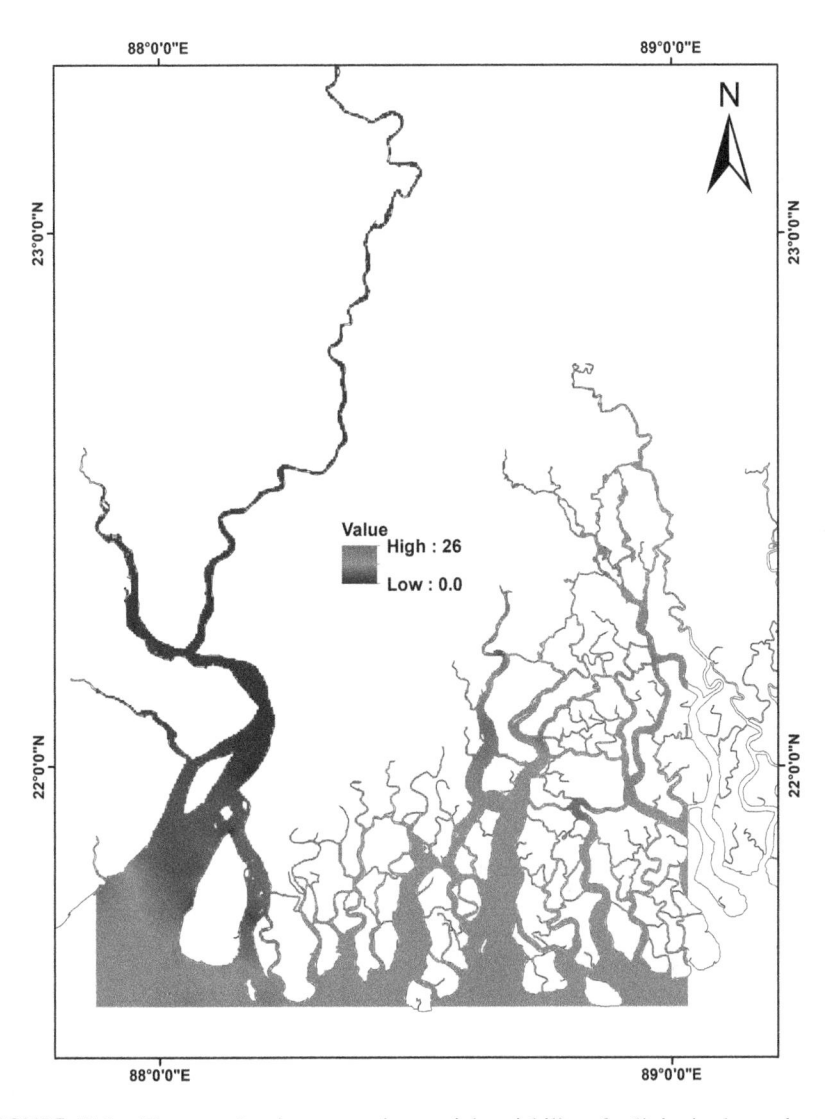

FIGURE 15.8 The map showing a tentative spatial variability of salinity in the study area during the monsoon season.

and 98% over the spatial coverage of the forest). Dasgupta et al. (2012) observed that with the increase in ambient salinity, the leaf protein depleted substantially in mangrove species like *Heritiera fomes*, *Aegialitis rotundifolia*, *Xylocarpus mekongensis* and *Xylocarpus granatum* coupled with the increase in hydrolyzing and antioxidative enzymes like esterase, acid phosphates, superoxide dismutase and peroxidase to defend the reactive oxygen species build-up within the trees, which can damage several macromolecules and the overall cellular structure of these trees (Goudarzi and Pakniyal, 2009). Nandy et al. (2009) observed that many of the mangrove species of Indian Sundarbans exhibited improved physiological functioning like enhanced

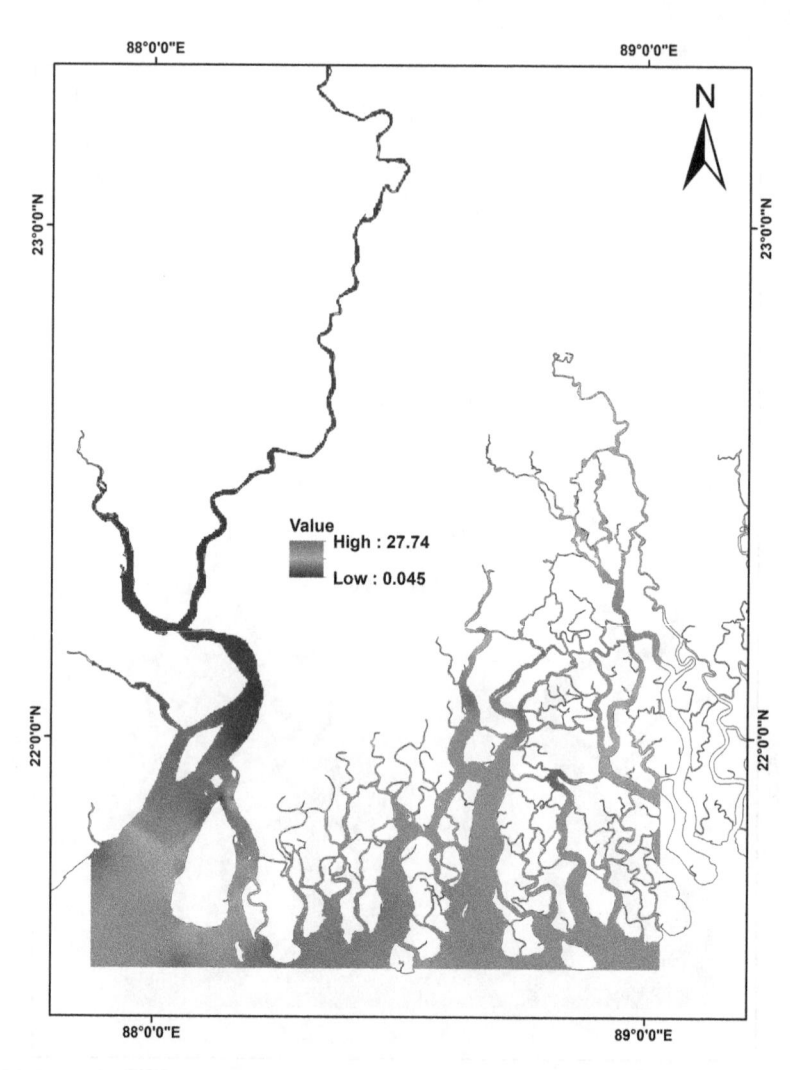

FIGURE 15.9 The map showing a tentative spatial variability of salinity in the study area during the post-monsoon season.

stomatal conductance, elevated carbon assimilation rate, increased chlorophyll content and higher specific leaf area under low saline conditions, which implied that increase in salinity can lead to severe stress on these species and would bar these species from carrying out optimum photosynthesis and negatively impact the overall well-being of these species. Nandy et al. (2009) further concluded that many of the mangrove species in Sundarbans can tolerate high salt levels to some extent, but very few of these species are actually salt-loving. Nandy and Ghose (2003) highlighted that the mangrove plants of Sundarbans usually maintain a high negative leaf osmotic potential in order to maintain optimum water uptake rate from the saline substrate. In order to facilitate such high negative leaf osmotic potential, the presence of the amino acid proline plays a very crucial role as the principal osmoticum. Nandy and Ghose (2003) inferred that under salt-stressed conditions, the free proline in the mangrove leaves is substantially substituted by free alanine (another strong osmoticum); however, the reduction in amino acids like cysteine and proline in the mangrove leaves under extreme NaCl shock compels these trees to curb many photosynthetic functioning to survive. In addition to this, Nandy et al. (2007) observed that under increased salinity the mangrove plants of Sundarbans exhibited a tendency to accumulate trace metals at a higher rate which can have toxic effects on these plants, and the photosynthetic nitrogen use efficiency is significantly compromised leading to much lower photosynthetic peaks compared to that observed in low saline conditions. Thus it can be observed that a number of studies have indicated that the ongoing rise in salinity proves to be severely detrimental to the entire mangrove community of the Sundarbans.

15.7 FUTURE OF THE MANGROVE PLANTS: AN EXPECTED SHIFT IN SPECIES COMPOSITION

The species composition and assemblage of mangroves are largely dependent on the salinity of the sediments and the adjoining water column (Mukhopadhyay et al., 2015). Depending upon the range of the sediment and water salinity, mangrove species are usually classified as oligohaline (salinity: 0.5–5), mesohaline (salinity: 5–18 ppt) and polyhaline (salinity: >18 ppt) (Barik et al., 2018). Previous studies have shown that with the change in the salinity regime, an overall shift in the species composition takes place within a mangrove forest (Mukhopadhyay et al., 2015; Chanda et al., 2016). As many of the mangrove species can serve as an indicator species of a particular range of salinity, monitoring the changes in the abundance and spatial distribution of such species over time helps us to infer the role of salinity in regulating the species composition pattern of mangrove forest (Barik et al., 2018). Chowdhury et al. (2019) reported that salt-sensitive species like *Phoenix paludosa*, *Xylocarpus granatum* and *Xylocarpus mekongensis* and especially freshwater loving species like *Heritiera fomes* could not cope with the salinity levels already reached and hence are almost absent through the natural mangrove stands of Indian Sundarbans forest. On the contrary, salt-loving species like *Avicennia* spp. and *Excoecaria agallocha* are rapidly filling up the voids created from the absence of freshwater loving species. Banerjee et al. (2017) also reported that *Heritiera fomes* is the most hard-hit mangrove species due to this ongoing rise in salinity in the Sundarbans. Many people argue that the name Sundarbans was coined due to the massive abundance of the Sundari tree (common name of *Heritiera fomes*) in this region; however, at present this species is almost absent in the central Indian Sundarban, and the main root cause behind its absence has been the increasing salinity (Banerjee et al., 2017). An increase in salinity is reported to hamper the proper functioning of the pneumatophores of *Heritiera fomes* and these trees are consequently found to be infected by top-dying disease (the top branches of the tree shed their leaves abnormally making the tree absolutely devoid of any canopy), which in turn leads to stunted growth and eventually, the trees perish (Rahman, 1994; Hoque et al., 2006). Similar degradation of *Heritiera fomes* is also feared to take place in the Bangladesh Sundarbans due to an increase in salinity, and this is expected to have a negative impact on the livelihood of many people who are directly dependent on this high timber value of this species (Dasgupta et al., 2017). Besides *Heritiera fomes*, other freshwater loving species like *Nypa fruticans* are also found to be steadily diminishing in number in the Indian Sundarbans (Banerjee et al., 2017).

This change in species composition is also supposed to take a toll on the total blue carbon stock of this forest, as previous studies like Nandy and Ghose (2001) observed that oligohaline haline species *Aegialitis rotundifolia* and *Heritiera fomes*, which are reducing in numbers, happen to have very high photosynthetic carbon assimilation potential. Though the stable species like *Avicennia* spp. and *Excoecaria agallocha* which are found to adapt well with the increasing salinity have moderately high photosynthetic carbon assimilation potential, their photosynthetic efficiency might get hampered in the future with a further rise in salinity. Thus it can be summed up from the observations made till date that the freshwater loving or low-saline mangrove flora like *Heritiera fomes, Nypa fruticans, Xylocarpus granatum* and *Xylocarpus mekongensis* have diminished significantly at present and are very much on the verge of total extinction from Indian Sundarbans, unless strict policy intervention and proper conservation measures are being taken.

15.8 SUMMARY AND CONCLUSION

By extracting and analyzing the salinity data from various literature spread across different time periods, it can be inferred that the freshwater supply in the main channel of Hooghly estuary has increased since the commissioning of the Farakka Barrage; however, the same has decreased substantially in the estuaries and tidal creeks situated in the north-eastern part and south-western part of the Indian Sundarbans over the last two to three decades. Availability of salinity data in the north-western and south-eastern parts of the Sundarbans (which encompass the core area of the Sundarbans National Park) was very scarce to draw any temporal trend line, but the few reported magnitudes of salinity indicated that these sectors are also experiencing an acute scarcity of freshwater. The literature reviewed clearly indicates that the mangrove plants of Sundarbans are at present thriving under potential salt-stress, which is manifested by stunted growth, poor health, improper physiological functioning and degradation of forest cover in several places within this ecosystem. The freshwater loving mangrove species are diminishing in large numbers and these are getting replaced by more salt-tolerant species, resulting in a net loss of species biodiversity. This study emphasized on the fact that the present situation warrants immediate policy intervention and conservation measures to rejuvenate the mangroves with freshwater, or else, we are nearing towards permanently jeopardizing one of the most productive eco-regions of the world.

CONFLICT OF INTEREST

The authors declare that they have no conflict of interest.

ACKNOWLEDGMENTS

The first author is grateful to the Department of Higher Education, Government of West Bengal, India, for providing the Swami Vivekananda Merit-cum-Means fellowship. All the authors also take this opportunity to thank Jadavpur University, for facilitating the access to various subscribed browsers to download several papers required for the present analysis.

REFERENCES

Agarwal, S.K. and Mitra, A., 2018. Salinity: A Primary Growth Driver of Mangrove Flora. *Current Trends in Forest Research: CTFR-114*. doi: 10.29011/ CTFR-114.1000014.

Ahmed, A., Aziz, A., Khan, A.N.A., Islam, M.N., Iqubal, K.F., Nazma, M. and Islam, M.S., 2011. Tree diversity as affected by salinity in the Sundarban mangrove forests, Bangladesh. *Bangladesh Journal of Botany*, 40(2), pp. 197–202.

Akhand, A., Chanda, A., Dutta, S., Manna, S., Sanyal, P., Hazra, S., Rao, K.H. and Dadhwal, V.K., 2013. Dual character of Sundarban estuary as a source and sink of CO_2 during summer: an investigation of spatial dynamics. *Environmental Monitoring and Assessment*, 185(8), pp. 6505–6515.

Akhand, A., Chanda, A., Manna, S., Das, S., Hazra, S., Roy, R., Choudhury, S.B., Rao, K.H., Dadhwal, V.K., Chakraborty, K. and Mostofa, K.M.G., 2016. A comparison of CO_2 dynamics and air-water fluxes in a river-dominated estuary and a mangrove-dominated marine estuary. *Geophysical Research Letters*, 43(22), pp. 11–726.

Attri, S.D. and Tyagi, A., 2010. Climate Profile of India; Met. Monograph No. Environment Meteorology–01/2010. India Meteorological Department, Ministry of Earth Sciences, Government of India, p. 129.

Aziz, M.A., 2017. *Population status, threats, and evolutionary conservation genetics of Bengal tigers in the Sundarbans of Bangladesh* (Doctoral dissertation, University of Kent).

Ball, M.C., 1988. Ecophysiology of mangroves. *Trees*, 2(3), pp. 129–142.

Ball, M.C., 2002. Interactive effects of salinity and irradiance on growth: implications for mangrove forest structure along salinity gradients. *Trees*, 16(2–3), pp. 126–139.

Banerjee, K., Gatti, R.C. and Mitra, A., 2017. Climate change-induced salinity variation impacts on a stenoecious mangrove species in the Indian Sundarbans. *Ambio*, 46(4), pp. 492–499.

Banerjee, K. and Mitra, A., 2015. Mangrove biomass and salinity: Is the match made in heaven? *International Journal of Biological Sciences and Engineering*, 6, pp. 45–54.

Banerjee, K., Sengupta, K., Raha, A. and Mitra, A., 2013. Salinity based allometric equations for biomass estimation of Sundarban mangroves. *Biomass and Bioenergy*, 56, pp. 382–391.

Barik, J., Mukhopadhyay, A., Ghosh, T., Mukhopadhyay, S.K., Chowdhury, S.M. and Hazra, S., 2018. Mangrove species distribution and water salinity: An indicator species approach to Sundarban. *Journal of Coastal Conservation*, 22(2), pp. 361–368.

Basu, A.K., Ghosh, B.B. and Pal, R.N., 1970. Comparison of the polluted Hooghly estuary with the unpolluted Matlah estuary, India. *Journal (Water Pollution Control Federation)*, 42, pp. 1771–1781.

Bhattacharjee, A.K., Zaman, S., Raha, A.K., Gadi, S.D. and Mitra, A., 2013a. Impact of salinity on above ground biomass and stored carbon in a common mangrove Excoecaria agallocha of Indian Sundarbans. *The American Journal of Bio-Pharmacology Biochemistry and Life Sciences*, 2, pp. 1–11.

Bhattacharjee, D., Samanta, B., Danda, A. and Bhadury, P., 2013b. Impact of climate change in the sundarban aquatic ecosystems: Phytoplankton as proxies. In *Climate Change and Island and Coastal Vulnerability* (pp. 126–140). Springer, Dordrecht.

Bhattacharjee, D., Samanta, B., Danda, A.A. and Bhadury, P., 2013c. Temporal succession of phytoplankton assemblages in a tidal creek system of the Sundarbans mangroves: An integrated approach. *International Journal of Biodiversity*, 2013, p. 824543.

Bhattacharya, S., Dash, J.R., Patra, P.H., Dubey, S.K., Das, A.K., Mandal, T.K. and Bandyopadhyay, S.K., 2014. Spatio-temporal variation of mercury in Bidyadhari River of Sundarban delta, India. *Exploratory Animal and Medical Research*, 4(1), pp. 19–32.

Bhunia, A.B. and Choudhury, A., 1981. Observations on the hydrology and the quantitative studies on benthic macrofauna in a tidal creek of Sagar Island, Sunderbans, West Bengal, India. In *Proceedings of Indian National Science Academy B* (Vol. 47, pp. 398–407).

Biswas, H., Dey, M., Ganguly, D., De, T.K., Ghosh, S. and Jana, T.K., 2010. Comparative analysis of phytoplankton composition and abundance over a two-decade period at the land–ocean boundary of a tropical mangrove ecosystem. *Estuaries and Coasts*, *33*(2), pp. 384–394.

Biswas, H., Mukhopadhyay, S.K., De, T.K., Sen, S. and Jana, T.K., 2004. Biogenic controls on the air—water carbon dioxide exchange in the Sundarban mangrove environment, northeast coast of Bay of Bengal, India. *Limnology and Oceanography*, *49*(1), pp. 95–101.

Biswas, H., Mukhopadhyay, S.K., Sen, S. and Jana, T.K., 2007. Spatial and temporal patterns of methane dynamics in the tropical mangrove dominated estuary, NE coast of Bay of Bengal, India. *Journal of Marine Systems*, *68*(1–2), pp. 55–64.

Bjorkman, O., Demmig, B. and Andrews, T.J., 1988. Mangrove photosynthesis: response to high-irradiance stress. *Functional Plant Biology*, *15*(2), pp. 43–61.

Blasco, F., 1975. Mangroves of India. French Institute of Pondicherry. Travaux de la section scientifique et technique, *14*, p. 180.

Blasco, F., Saenger, P. and Janodet, E., 1996. Mangroves as indicators of coastal change. *Catena*, *27*(3–4), pp. 167–178.

Bose, B.B., 1956. Observations on the hydrology of the Hooghly estuary. *Indian Journal of Fisheries*, *3*(1), pp. 101–118.

Bose, R., De, A., Sen, G. and Mukherjee, A.D., 2012. Comparative study of the physico-chemical parameters of the coastal waters in rivers Matla and Saptamukhi: Impacts of coastal water coastal pollution. *Journal of Water Chemistry and Technology*, *34*(5), pp. 246–251.

Chakraborty, P.K. and Chattopadhyay, G.N., 1989. Impact of Farakka barrage on the estuarine ecology of the Hooghly-Matlah system. *Conservation and Management of Inland Capture Fisheries Resources of India. Barrackpore, Inland Fisheries Sopciety of India*, pp. 189–196.

Chakraborty, N., Chakrabarti, P.K., Vinci, G.K. and Sugunan, V.V., 1995. Spatio-temporal distribution pattern of certain plankton of river Hooghly. *Journal of Inland Fisheries Society of India*, *27*(1), pp. 6–12.

Chakraborty, S., Biswas, S., Banerjee, K. and Mitra, A., 2016. Concentrations of Zn, Cu and Pb in the Muscle of Two Edible Finfish Species in and around Gangetic Delta Region. *International Journal of Life Science & Pharma Research*, *6*(3), pp. 14–22.

Chanda, A., 2015. Assessment of Terrestrial Carbon Dioxide Fluxes Above and Below the Canopy of the Indian Sundarban (Doctoral dissertation, Jadavpur University).

Chanda, A., Mukhopadhyay, A., Ghosh, T., Akhand, A., Mondal, P., Ghosh, S., Mukherjee, S., Wolf, J., Lázár, A.N., Rahman, M.M. and Salehin, M., 2016. Blue carbon stock of the Bangladesh Sundarban mangroves: what could be the scenario after a century? *Wetlands*, *36*(6), pp. 1033–1045.

Chatterjee, M., Shankar, D., Sen, G.K., Sanyal, P., Sundar, D., Michael, G.S., Chatterjee, A., Amol, P., Mukherjee, D., Suprit, K. and Mukherjee, A., 2013. Tidal variations in the Sundarbans estuarine system, India. *Journal of Earth System Science*, *122*(4), pp. 899–933.

Chaudhuri, A., Mukherjee, S. and Homechaudhuri, S., 2013. Seasonal dynamics of fish assemblages in an intertidal mudflat of Indian Sundarbans. *Scientia Marina*, *77*(2), pp. 301–311.

Choudhury, A.K. and Bhadury, P., 2015. Relationship between N: P: Si ratio and phytoplankton community composition in a tropical estuarine mangrove ecosystem. *Biogeosciences Discussions*, *12*(3), pp. 2307–2355.

Choudhury, A.K., Das, M., Philip, P. and Bhadury, P., 2015. An assessment of the implications of seasonal precipitation and anthropogenic influences on a mangrove ecosystem using phytoplankton as proxies. *Estuaries and Coasts*, *38*(3), pp. 854–872.

Chowdhury, R., Sutradhar, T., Begam, M.M., Mukherjee, C., Chatterjee, K., Basak, S.K. and Ray, K., 2019. Effects of nutrient limitation, salinity increase, and associated stressors on mangrove forest cover, structure, and zonation across Indian Sundarbans. *Hydrobiologia*, *842*(1), pp. 191–217.

Clough, B. and Sim, R.G., 1989. Changes in gas exchange characteristics and water use efficiency of mangroves in response to salinity and vapour pressure deficit. *Oecologia*, *79*(1), pp. 38–44.

Cole, C.V. and Vaidyaraman, P.P., 1966. Salinity distribution and effect of freshwater flows in the Hooghly River. In *Proceedings Tenth Conference on Coastal Engineering, Tokyo (American Society of Civil Engineers, New York)* (pp. 1312–1434).

Darby, S.E., Addo, K.A., Hazra, S., Rahman, M.M. and Nicholls, R.J., 2020. Fluvial sediment supply and relative sea-level rise. In *Deltas in the Anthropocene* (pp. 103–126). Palgrave Macmillan, Cham.

Das, G.K., 2012. Impact of water quality on the changing environmental scenario of Sunderbans. *Reason. Kalyani Govt. Engineering College XI*, pp. 57–66.

Dasgupta, N., Nandy, P., Sengupta, C. and Das, S., 2012. Protein and enzymes regulations towards salt tolerance of some Indian mangroves in relation to adaptation. *Trees*, *26*(2), pp. 377–391.

Dasgupta, S., Sobhan, I. and Wheeler, D., 2017. The impact of climate change and aquatic salinization on mangrove species in the Bangladesh Sundarbans. *Ambio*, *46*(6), pp. 680–694.

De, T.K., De, M., Das, S., Chowdhury, C., Ray, R. and Jana, T.K., 2011. Phytoplankton abundance in relation to cultural eutrophication at the land-ocean boundary of Sunderbans, NE Coast of Bay of Bengal, India. *Journal of Environmental Studies and Sciences*, *1*(3), p. 169.

Dey, M., Jamadar, Y.A. and Mitra, A.B.H.I.J.I.T., 2005. Distribution of intertidal malacofauna at Sagar Island. *Records of the Zoological Survey of India*, *105*(1&2), pp. 25–35.

Dutta, M.K., Chowdhury, C., Jana, T.K. and Mukhopadhyay, S.K., 2013. Dynamics and exchange fluxes of methane in the estuarine mangrove environment of the Sundarbans, NE coast of India. *Atmospheric Environment*, *77*, pp. 631–639.

Dutta, M.K., Mukherjee, R., Jana, T.K. and Mukhopadhyay, S.K., 2015. Biogeochemical dynamics of exogenous methane in an estuary associated to a mangrove biosphere; the Sundarbans, NE coast of India. *Marine Chemistry*, *170*, pp. 1–10.

Ghosh, P., Chakrabarti, R. and Bhattacharya, S.K., 2013. Short-and long-term temporal variations in salinity and the oxygen, carbon and hydrogen isotopic compositions of the Hooghly Estuary water, India. *Chemical Geology*, *335*, pp. 118–127.

Ghosh, R. and Banerjee, K., 2013. Inter-relationship between Physico-chemical Variables and Litter Production in Mangroves of Indian Sundarbans. *Journal of Marine Science. Research & Development*, *S11*, p. 1. doi: 10.4172/2155-9910.S11-001.

Goudarzi, M. and Pakniyat, H., 2009. Peroxidase activity in wheat cultivars. *Journal of Applied Sciences*, *9*(2), pp. 348–353.

Hoque, M.A., Sarkar, M.S.K.A., Khan, S.A.K.U., Moral, M.A.H. and Khurram, A.K.M., 2006. Present status of salinity rise in Sundarbans area and its effect on Sundari (Heritiera fomes) species. *Research Journal of Agriculture and Biological Sciences*, *2*(3), pp. 115–121.

Hutchings, P. and Saenger, P., 1987. Ecology of mangroves. *Ecology of mangroves*. St Lucia, Australia, University of Queensland Press.

Islam, S.N. and Gnauck, A., 2011. Water salinity investigation in the Sundarbans rivers in Bangladesh. *International Journal of Water*, *6*(1–2), pp. 74–91.

Lal, B.A.B.U., 1990. Impact of Farakka barrage on the hydrological changes and productivity potential of Hooghly Estuary. *Journal of Inland Fisheries Society of India (India)*, 22, pp. 38–42.

Liang, S., Zhou, R., Dong, S. and Shi, S., 2008. Adaptation to salinity in mangroves: Implication on the evolution of salt-tolerance. *Chinese Science Bulletin*, 53(11), p. 1708.

Lin, H.J., Shao, K.T., Chiou, W.L., Maa, C.J.W., Hsieh, H.L., Wu, W.L., Severinghaus, L.L. and Wang, Y.T., 2003. Biotic communities of freshwater marshes and mangroves in relation to saltwater incursions: implications for wetland regulation. *Biodiversity & Conservation*, 12(4), pp. 647–665.

Lugo, A.E. and Snedaker, S.C., 1974. The ecology of mangroves. *Annual Review of Ecology and Systematics*, 5(1), pp. 39–64.

Mandal, S., Debnath, M., Ray, S., Ghosh, P.B., Roy, M. and Ray, S., 2012a. Dynamic modelling of dissolved oxygen in the creeks of Sagar Island, Hooghly–Matla estuarine system, West Bengal, India. *Applied Mathematical Modelling*, 36(12), pp. 5952–5963.

Mandal, S., Ray, S. and Ghosh, P.B., 2012b. Modeling nutrient (dissolved inorganic nitrogen) and plankton dynamics at Sagar island of Hooghly–Matla estuarine system, West Bengal, India. *Natural Resource Modeling*, 25(4), pp. 629–652.

Manna, R.K., Roshith, C.M., Das, S.K., Suresh, V.R. and Sharma, A.P., 2014. Salinity Regime and Fish Species Distribution in the Hooghly. Matlah Estuary. IUCN.

Manna, R.K., Satpathy, B.B., Roshith, C.M., Naskar, M., Bhaumik, U. and Sharma, A.P., 2013. Spatio-temporal changes of hydro-chemical parameters in the estuarine part of the river Ganges under altered hydrological regime and its impact on biotic communities. *Aquatic Ecosystem Health & Management*, 16(4), pp. 433–444.

Massolo, S., Bignasca, A., Sarkar, S.K., Chatterjee, M., Bhattacharya, B.D. and Alam, A., 2012. Geochemical fractionation of trace elements in sediments of Hugli River (Ganges) and Sundarban wetland (West Bengal, India). *Environmental Monitoring and Assessment*, 184(12), pp. 7561–7577.

Mitra, A., Chowdhury, R., Sengupta, K. and Banerjee, K., 2010. Impact of salinity on mangroves. *Journal of Coastal Environment*, 1(1), pp. 71–82.

Mitra, A., Zaman, S., Ray, S.K., Sinha, S. and Banerjee, K., 2012. Inter-relationship between phytoplankton cell volume and aquatic salinity in Indian sundarbans. *National Academy Science Letters*, 35(6), pp. 485–491.

Mitra, S., Ghosh, S., Satpathy, K.K., Bhattacharya, B.D., Sarkar, S.K., Mishra, P. and Raja, P., 2018. Water quality assessment of the ecologically stressed Hooghly River Estuary, India: A multivariate approach. *Marine Pollution Bulletin*, 126, pp. 592–599.

Morgan, J.P. and McIntire, W.G., 1959. Quaternary geology of the Bengal basin, East Pakistan and India. *Geological Society of America Bulletin*, 70(3), pp. 319–342.

Mukhopadhyay, A., Mondal, P., Barik, J., Chowdhury, S.M., Ghosh, T. and Hazra, S., 2015. Changes in mangrove species assemblages and future prediction of the Bangladesh Sundarbans using Markov chain model and cellular automata. *Environmental Science: Processes & Impacts*, 17(6), pp. 1111–1117.

Mukhopadhyay, S.K., Biswas, H.D.T.K., De, T.K. and Jana, T.K., 2006. Fluxes of nutrients from the tropical River Hooghly at the land–ocean boundary of Sundarbans, NE Coast of Bay of Bengal, India. *Journal of Marine Systems*, 62(1–2), pp. 9–21.

Nandy, P. and Ghose, M., 2001. Photosynthesis and water-use efficiency of some mangroves from Sundarbans, India. *Journal of Plant Biology*, 44(4), pp. 213–219.

Nandy, A.C., Bagchi, M.M. and Majumder, S.K., 1983. Ecological changes in the Hooghly estuary due to water release from Farakka Barrage. *Mahasagar*, 16(2), pp. 209–220.

Nandy, D.P. and Ghose, M., 2003. Estimation of osmotic potential and free amino acids in some mangroves of the Sundarbans, India. *Acta Botanica Croatica*, 62(1), pp. 37–45.

Nandy, P., Das, S., Ghose, M. and Spooner-Hart, R., 2007. Effects of salinity on photosynthesis, leaf anatomy, ion accumulation and photosynthetic nitrogen use efficiency in five Indian mangroves. *Wetlands Ecology and Management, 15*(4), pp. 347–357.

Nandy, P., Dasgupta, N. and Das, S., 2009. Differential expression of physiological and biochemical characters of some Indian mangroves towards salt tolerance. *Physiology and Molecular Biology of Plants, 15*(2), pp. 151–160.

Naskar, K.R., 1988. Economic potentialities of the tidal mangrove forests of Sundarbans in India. *Journal of the Indian Society of Coastal Research, 6*(2), pp. 149–158.

Naskar, K.R. and Guha Bakshi, D.N., 1987. *Mangrove swamps of the Sundarbans.* Naya Prokash.

Nasrin, S., Hossain, M. and Rahman, M.M., 2019. Adaptive responses to salinity: nutrient resorption efficiency of *Sonneratia apetala* (Buch.-Ham.) along the salinity gradient in the Sundarbans of Bangladesh. *Wetlands Ecology and Management, 27*(2–3), pp. 343–351.

Nath, D., 1998. Zonal distribution of nutrients and their bearing on primary production in Hooghly estuary. *Journal of Inland Fisheries Society of India, 30*(2), pp. 64–74.

Nath, D. and De, D.K., 1998. Preliminary studies on the changes in the physicochemical characteristics of Hooghly estuary in relation to tides. *Journal of Inland Fisheries Society of India, 30*(2), pp. 29–36.

Niyogi, S., Biswas, S., Sarker, S. and Datta, A.G., 2001. Antioxidant enzymes in brackishwater oyster, Saccostrea cucullata as potential biomarkers of polyaromatic hydrocarbon pollution in Hooghly Estuary (India): Seasonality and its consequences. *Science of the Total Environment, 281*(1–3), pp. 237–246.

Padhy, P.C., Nayak, R.K., Dadhwal, V.K., Salim, M., Mitra, D., Chaudhury, S.B., Rao, P.R., Rao, K.H. and Dutt, C.B.S., 2016. Estimation of partial pressure of carbon dioxide and air-sea fluxes in Hooghly estuary based on in situ and satellite observations. *Journal of the Indian Society of Remote Sensing, 44*(1), pp. 135–143.

Pendleton, L., Donato, D.C., Murray, B.C., Crooks, S., Jenkins, W.A., Sifleet, S., Craft, C., Fourqurean, J.W., Kauffman, J.B., Marbà, N. and Megonigal, P., 2012. Estimating global "blue carbon" emissions from conversion and degradation of vegetated coastal ecosystems. *PloS One, 7*(9), p. e43542.

Raha, A., Das, S., Banerjee, K. and Mitra, A., 2012. Climate change impacts on Indian Sunderbans: A time series analysis (1924–2008). *Biodiversity and Conservation, 21*(5), pp. 1289–1307.

Rahman, M.A., 1994. Final report on mangrove plant pathology of the Sundarbans reserved forest in Bangladesh.

Rakshit, D., Biswas, S.N., Sarkar, S.K., Bhattacharya, B.D., Godhantaraman, N. and Satpathy, K.K., 2014. Seasonal variations in species composition, abundance, biomass and production rate of tintinnids (Ciliata: Protozoa) along the Hooghly (Ganges) River Estuary, India: A multivariate approach. *Environmental Monitoring and Assessment, 186*(5), pp. 3063–3078.

Rao, A.N., 1986. *Mangrove ecosystems of Asia and the Pacific.* National Resources Management Center and National Mangrove Committe, Ministry of Natural Resources.

Rao, R.M. 1967. Studies of the prawn fisheries of the hooghly estuarine system. *Central Inland Fisheries Research Institute, 35*, pp. 1–26.

Ray, R., Baum, A., Rixen, T., Gleixner, G. and Jana, T.K., 2018. Exportation of dissolved (inorganic and organic) and particulate carbon from mangroves and its implication to the carbon budget in the Indian Sundarbans. *Science of the Total Environment, 621*, pp. 535–547.

Ray, R., Rixen, T., Baum, A., Malik, A., Gleixner, G. and Jana, T.K., 2015. Distribution, sources and biogeochemistry of organic matter in a mangrove dominated estuarine system (Indian Sundarbans) during the pre-monsoon. *Estuarine, Coastal and Shelf Science, 167*, pp. 404–413.

Reef, R. and Lovelock, C.E., 2015. Regulation of water balance in mangroves. *Annals of Botany*, *115*(3), pp. 385–395.

Roy, A., 2010. Vulnerability of the Sundarbans ecosystem. *Journal of Coastal Environment*, 1, pp. 169–180

Roy, M. and Nandi, N.C., 2012. Distribution pattern of macrozoobenthos in relation to salinity of Hugli-Matla estuaries in India. *Wetlands*, *32*(6), pp. 1001–1009.

Roy, S., Hens, D., Biswas, D., Biswas, D. and Kumar, R., 2002. Survey of petroleum-degrading bacteria in coastal waters of Sunderban Biosphere Reserve. *World Journal of Microbiology and Biotechnology*, *18*(6), pp. 575–581.

Sadhuram, Y., Sarma, V.V., Murthy, T.R. and Rao, B.P., 2005. Seasonal variability of physico-chemical characteristics of the Haldia channel of Hooghly estuary, India. *Journal of Earth System Science*, *114*(1), pp. 37–49.

Saha, M., Sarkar, S.K. and Bhattacharya, B., 2006. Interspecific variation in heavy metal body concentrations in biota of Sunderban mangrove wetland, northeast India. *Environment International*, *32*(2), pp. 203–207.

Samanta, S., Dalai, T.K., Pattanaik, J.K., Rai, S.K. and Mazumdar, A., 2015. Dissolved inorganic carbon (DIC) and its $\delta13C$ in the Ganga (Hooghly) River estuary, India: Evidence of DIC generation via organic carbon degradation and carbonate dissolution. *Geochimica et Cosmochimica Acta*, *165*, pp. 226–248.

Sandilyan, S., Thiyagesan, K., Nagarajan, R. and Vencatesan, J., 2010. Salinity rise in Indian mangroves–a looming danger for coastal biodiversity. *Current Science*, *98*(6), pp. 754–756.

Sanyal, P. and Bal, A., 1986. Some observations on abnormal adaptations of mangrove in Indian Sundarbans. *Journal of Indian Society of Coastal Agricultural Research*, 4, pp. 9–15.

Sanyal, P., Mukhopadhyay, A., Das, I., 2008. Sundarban – the greatest Mangal diversity of the planet. *Journal of Indian Society of Coastal Agricultural Research*, 26, pp. 132–134.

Sarkar, S., Ghosh, P.B., Das, T.M., Som Mazumdar, S. and Saha, T., 2013. Environmental assessment in terms of salinity distribution in the Tropical Mangrove forest of Sundarban, North East Coast of Bay of Bengal, India. *Archives of Applied Science Research*, *5*(6), pp. 109–118.

Sarkar, S.K. and Bhattacharya, A.K., 2003. Conservation of biodiversity of the coastal resources of Sundarbans, Northeast India: An integrated approach through environ-mental education. *Marine Pollution Bulletin*, *47*(1–6), pp. 260–264.

Sarkar, S.K., Saha, M., Takada, H., Bhattacharya, A., Mishra, P. and Bhattacharya, B., 2007. Water quality management in the lower stretch of the river Ganges, east coast of India: An approach through environmental education. *Journal of Cleaner Production*, *15*(16), pp. 1559–1567.

Sengupta, K., Roy Chowdhury, M., Bhattacharya, S.B., Raha, A., Zaman, S. and Mitra, A., 2013. Spatial variation of stored carbon in *Avicennia alba* of Indian Sundarbans. *Discovery Nature*, *3*(8), pp. 19–24.

Shetty, H.P.C., Saha, S.B. and Ghosh, B.B., 1961. Observations on the distribution and fluctua-tions of plankton in the Hooghly-Matlah estuarine system, with notes on their relation to commercial fish landings. *Indian Journal of Fisheries*, *8*(2), pp. 326–363.

Sinha, P.C., Rao, Y.R., Dube, S.K., Rao, A.D. and Chatterjee, A.K., 1996. Modeling of circu-lation and salinity in Hooghly estuary. *Marine Geodesy*, *19*(2), pp. 197–213.

Sinha, P.C., Rao, Y.R., Dube, S.K., Murthy, C.R. and Chatterjee, A.K., 1999. Application of two turbulence closure schemes in the modelling of tidal currents and salinity in the Hooghly Estuary. *Estuarine, Coastal and Shelf Science*, *48*(6), pp. 649–663.

Somayajulu, B.L.K., Rengarajan, R. and Jani, R.A., 2002. Geochemical cycling in the Hooghly estuary, India. *Marine Chemistry*, *79*(3–4), pp. 171–183.

Trivedi, S., Zaman, S., Chaudhuri, T.R., Pramanick, P., Fazli, P., Amin, G. and Mitra, A., 2016. Inter-annual variation of salinity in Indian Sundarbans. *Indian Journal of Geo-Marine Sciences*, 45, pp. 410–415.

Index

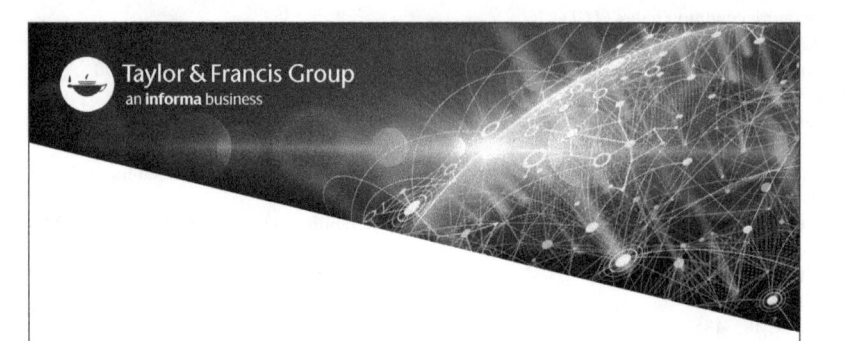